Lecture Notes in Computer Science 11059

Advanced Research in Computing and Software Science

Subline of Lecture Notes in Computer Science

More information about this series at http://www.springer.com/series/7409

Xiaotie Deng (Ed.)

Algorithmic Game Theory

11th International Symposium, SAGT 2018
Beijing, China, September 11–14, 2018
Proceedings

Editor
Xiaotie Deng
Peking University
Beijing
China

ISSN 0302-9743 ISSN 1611-3349 (electronic)
Lecture Notes in Computer Science
ISBN 978-3-319-99659-2 ISBN 978-3-319-99660-8 (eBook)
https://doi.org/10.1007/978-3-319-99660-8

Library of Congress Control Number: 2018952342

LNCS Sublibrary: SL3 – Information Systems and Applications, incl. Internet/Web, and HCI

This Springer imprint is published by the registered company Springer Nature Switzerland AG
The registered company address is: Gewerbestrasse 11, 6330 Cham, Switzerland

Preface

This volume of LNCS 11059 contains the collection of the papers presented at the 11th International Symposium on Algorithmic Game Theory (SAGT 2018), held during September 11–14, 2018, in Beijing, China. This year, we received a number of 54 submissions, among which 19 regular papers and six short papers were accepted by the Program Committee. Each submission was evaluated by at least three Program Committee members.

In addition to the submitted presentations, the program also included five plenary talks by distinguished researchers in algorithmic game theory: Edith Elkind (Oxford University), Ron Lavi (Technion), Pinyan Lu (Shanghai University of Finance and Economics), Paul Spirakis (University of Liverpool), and Andrew Yao (Tsinghua University).

With the generous support of Springer, the Program Committee selected the best paper award for this year: "The Complexity of Cake-Cutting with Unequal Shares," by Agnes Cseh of the Institute of Economics, Hungarian Academy of Sciences, and Tamas Fleiner, of Department of Computer Science and Information Theory, Budapest University of Technology and Economics.

Works accepted covered major aspects of algorithmic game theory, including market equilibrium, auctions and applications, two-sided markets, cake-cutting, cooperative games, voting games, multi-agent scheduling, price of stability, various mechanism design problems, online dynamics, and multi-stages as well as revenue maximization, resource allocation and applications such as hide-and-seek.

I would like to thank the SAGT Steering Committee for all their support and direction during the organization of SAGT 2018. The SAGT 2017 organizers also provided generous help with their experiences in handling the previous conference.

I would also like to thank all authors who submitted their work for evaluation, all the Program Committee members, and all the external reviewers – together you made this symposium possible. Even though we could not accept the submissions, we hope the feedback from the reviewers can be of some help in further improvement of the work.

Finally, we thank Peking University for the conference venue on their beautiful campus and the National Natural Science Foundation of China for their financial support. We are also indebted to Springer, especially Anna Kramer and Alfred Hofmann, for helping with the proceedings, as well as the EasyChair system for the review process management.

July 2018 Xiaotie Deng

Organization

Program Committee

Xiaohui Bei	Nanyang Technological University, Singapore
Kaigui Bian	Virginia Tech, USA
Felix Brandt	Technical University of Munich, Germany
Simina Branzei	Purdue University, USA
Yang Cai	McGill University, USA
Jing Chen	Stony Brook University, USA
Xujin Chen	Institute of Applied Mathematics, Chinese Academy of Sciences, China
Yiwei Chen	Singapore University of Design and Technology
Xiaotie Deng	Peking University, China
Kousha Etessami	The University of Edinburgh, UK
Michele Flammini	University of L'Aquila, Italy
Vasilis Gkatzelis	Drexel University, USA
Paul Goldberg	University of Oxford, UK
Yannai A. Gonczarowski	The Hebrew University of Jerusalem and Microsoft Research, Israel
Rica Gonen	Yahoo!, *now* The Open University of Israel
Mingyu Guo	The University of Adelaide, Australia
Kristoffer Arnsfelt Hansen	Aarhus University, Denmark
Martin Hoefer	Goethe University Frankfurt/Main, Germany
Ron Lavi	Technion – Israel Institute of Technology, Israel
Pinyan Lu	Shanghai University of Finance and Economics, China
David Manlove	University of Glasgow, UK
Marios Mavronicolas	University of Cyprus, Cyprus
Ruta Mehta	University of Illinois at Urbana-Champaign, USA
Herve Moulin	University of Glasgow, UK
Giuseppe Persiano	Università degli Studi di Salerno, Italy
Georgios Piliouras	Singapore University of Technology and Design
Qi	The Hong Kong University of Science and Technology, SAR China
Arunava Sen	Indian Statistical Institute, India
Paul Spirakis	Research Academic Computer Technology Institute and University of Patras, Greece
Warut Suksompong	Stanford University, USA
Adrian Vetta	McGill University, Canada
Lirong Xia	Rensselaer Polytechnic Institute (RPI)

Cheng Yukun	Zhejiang University of Finance and Economics, China
Jie Zhang	University of Southampton, UK
Weinan Zhang	Shanghai Jiao Tong University, China
Yair Zick	National University of Singapore

Additional Reviewers

Ben-Porat, Omer
Brandl, Florian
Cao, Zhigang
Chen, Zhou
Deligkas, Argyrios
Eden, Alon
Fearnley, John
Feigenbaum, Itai
Ferraioli, Diodato
Gerstgrasser, Matthias
Hofbauer, Johannes
Hollender, Alexandros
Jin, Yaonan
Kahng, Anson
Kulkarni, Rucha
Lazos, Philip
Lev, Omer
Li, Bo
Li, Weian
Livanos, Vasilis
Marmolejo Cossio, Francisco Javier
Mattei, Nicholas
Mauro, Manuel
Mei, Lili
Melissourgos, Themistoklis
Moses, William
Peters, Dominik
Piliouras, Georgios
Plaut, Benjamin
Saile, Christian
Shen, Jian
Shi, Yangguang
Sikdar, Sujoy
Ventre, Carmine
Vinci, Cosimo
Wang, Jun
Wang, Wenwei
Wong, Prudence W. H.
Xiao, Tao
Yu, Wei
Zhang, Yuanxing
Zhao, Mingfei

Plenary Talks

On Revenue Monotonicity in Combinatorial Auctions

Andrew Chi-chih Yao

Tsinghua University, Beijing, China
andrewcyao@tsinghua.edu.cn

Abstract. Along with substantial progress made recently in designing near-optimal mechanisms for multi-item auctions, interesting structural questions have also been raised and studied. In particular, is it true that the seller can always extract more revenue from a market where the buyers value the items higher than another market? In this paper we obtain such a revenue monotonicity result in a general setting. Precisely, consider the revenue-maximizing combinatorial auction for m items and n buyers in the Bayesian setting, specified by a valuation function v and a set F of nm independent item-type distributions. Let $REV(v, F)$ denote the maximum revenue achievable under F by any incentive compatible mechanism. Intuitively, one would expect that $REV(v, G) \geq REV(v, F)$ if distribution G stochastically dominates F. Surprisingly, Hart and Reny (2012) showed that this is not always true even for the simple case when v is additive. A natural question arises: Are these deviations contained within bounds? To what extent may the monotonicity intuition still be valid? We present an approximate monotonicity theorem for the class of fractionally subadditive (XOS) valuation functions v, showing that $REV(v, G) \geq cREV(v, F)$ if G stochastically dominates F under v where $c > 0$ is a universal constant. Previously, approximate monotonicity was known only for the case $n = 1$: Babaioff et al. (2014) for the class of additive valuations, and Rubinstein and Weinberg (2015) for all subaddtive valuation functions.

Keywords: Mechanism design • Subadditive valuation • Maximum revenue

An Update on the Price of Stability (Invited Talk)

Paul G. Spirakis[1,2]

[1] Department of Computer Science, University of Liverpool, UK
P.Spirakis@liverpool.ac.uk
[2] Computer Engineering and Informatics Department, University of Patras, Greece

Abstract. Here we review very recent results about the Price of Stability of weighted congestion games, derived by joint work of the author and G. Christodoulou, M. Gairing and Y. Giannakopoulos [4]. We also discuss older works (some joint with the author and also other works on the topic). This abstract corresponds to an invited talk (of the author) in SAGT 2018.

A main line of research in Algorithmic Game Theory has focused on how to quantify the inefficiency of equilibria (compared to optimal solutions). Two standard measures of this inefficiency have been adopted by the community: (a) The Price of Anarchy, introduced by the seminal work of Koutsoupias and Papadimitriou [10], which compares the worst-case equilibrium with the optimum for the system where the game is played. (b) And also the Price of Stability (PoS) [1, 18] which uses the best equilibrium for this comparison. The estimation of PoS is the main subject of this abstract.

The main issues that were raised about the Price of Anarchy have been resolved. The most well studied models, with respect to this, are the (atomic and non-atomic) variants of congestion games (see [12] Chapter 18 for a detailed discussion). Congestion games capture many scenarios in which users compete for resources (such as routing games). The seminal work of Roughgarden and Tardos [16, 17] provided the answer for the Price of Anarchy in non-atomic congestion games. Awerbuch et al. [2] and Christodoulou and Koutsoupias [5] resolved the Price of Anarchy for atomic congestion games.

Allowing the players to have different loads lead to the definition of weighted congestion games. This class was defined quite early by Rosenthal [14, 15] and can model numerous applications in selfish scheduling and routing. However, although unweighted congestion games always have Pure Equilibria, weighted congestion games may not have pure equilibria (see for example [8]) and the problem of finding whether a weighted congestion game has a Pure Equilibrium, is strongly NP-hard [7]. Moreover in such games there does exist, in general, a potential function [11]. Exact pure Nash Equilibria exist for linear and exponential latency functions (see [9, 13]).

As a result of this dichotomy about the existence of potentials, we have a sharp contrast with respect to understanding the Price of Anarchy and the Price of Stability. The asymptotic behaviour of weighted and unweighted congestion games with respect to the Price of Anarchy is the same: it is $\Theta((d/\log d)^d)$ for both game classes when the delays in resources are polynomials of degree d.

[1] Supported by the ERC Project ALGAME.

The picture for the Price of Stability is very different. On one hand, for unweighted congestion games, the PoS upper and lower bounds are quite tight and much lower than the Price of Anarchy values. For example $\Theta(d)$ for polynomials [3]. For weighted congestion games however there was a huge gap for many years: The lower bound was known to be $\Theta(d)$ and the upper bound $\Theta((d/\log d)^d)$.

Since weighted congestion games are not potential games, we cannot use any more the idea of the global minimizer of Rosenthals potential as an equilibrium refinement. A completely fresh approach is hence required. We discuss here briefly this new approach which is presented in detail in [4]. Basically we are able to construct a congestion game, which is weighted and has a unique Nash Equilibrium. That game allows us to lower bound the PoS for weighted congestion games and to show that the lower bound is $\Omega(\Phi_d^{d+1})$, where Φ_d is the unique positive solution of the equation $(x+1)^d = x^{d+1}$. This bound closes the previous huge gap between linear lower bounds and exponential upper bounds for the PoS of weighted congestion games and asymptotically matches the upper bound on the Price of Anarchy! The PoS (as we show) remains exponential even for singleton games. More generally we provide a lower bound of $\Omega((1+1/\alpha)^d/d)$ on the PoS of α-approximate Nash Equilibria (even for singleton games). All our results extend to network congestion games and hold also for mixed and correlated equilibria.

Our lower bound constructions of the "bad" game use players weights that form basically a geometric increasing sequence. In particular in our constructions the ratio R of the largest over the smallest weight grows very large as the number of players grows. But for equal weights we know that the PoS is at most $d+1$. It is hence natural to ask how the performance of equilibria captured by the notion of PoS varies with the ratio R. We derive a general upper bound for α-approximate equilibria which is sensitive to both R and α. For example, even for equal weights, we get a PoS upper bound for α-approximate equilibria of $(d+1)/\alpha$. This was also not known before to the best of our knowledge.

In order to derive our positive results (the upper bounds for small ranges of weight ratios) we define a new approximate Potential Function (see also [6]). This function generalizes that of Rosenthal's in a smooth way. The technique for defining this approximate potential function is purely analytical and may be of independent interest.

References

1. Anshelevich, E., Dasgupta, A., Kleinberg, J.M., Tardos, É., Wexler, T., Roughgarden, T.: The price of stability for network design with fair cost allocation. SIAM J. Comput. **38**(4), 1602–1623 (2008)
2. Awerbuch, B., Azar, Y., Epstein, A.: The price of routing unsplittable flow. SIAM J. Comput. **42**(1), 160–177 (2013)
3. Christodoulou, G., Gairing, M.: Price of stability in polynomial congestion games. ACM Trans. Econ. Comput. (TEAC), **4**(2), 10 (2016)
4. Christodoulou, G., Gairing, M., Giannakopoulos, Y., Spirakis, P.: The price of stability of weighted congestion games. In: 45th International Colloquium on Automata, Languages, and Programming, ICALP 2018, Prague, Czech Republic, 9–13 July 2018

5. Christodoulou, G., Koutsoupias, E.: The price of anarchy of finite congestion games. In: Proceedings of the 37th Annual ACM Symposium on Theory of Computing, pp. 67–73, Baltimore, MD, USA, 22–24 May (2005)
6. Christodoulou, G., Koutsoupias, E., Spirakis, P.G.: On the performance of approximate equilibria in congestion games. Algorithmica, **61**(1), 116–140 (2011)
7. Fabrikant, A., Papadimitriou, C.H., Talwar, K.: The complexity of pure nash equilibria. In: Proceedings of the 36th Annual ACM Symposium on Theory of Computing, pp. 604–612, Chicago, IL, USA, 13–16 June (2004)
8. Fotakis, D., Kontogiannis, S., Spirakis, P.: Selfish unsplittable flows. Theor. Comput. Sci. **348**(2–3), 226–239 (2005)
9. Harks, T., Klimm, M.: On the existence of pure nash equilibria in weighted congestion games. Math. Oper. Res. **37**(3), 419–436 (2012)
10. Koutsoupias, E., Papadimitriou, C.H.: Worst-case equilibria. In: Proceedings of 16th Annual Symposium on Theoretical Aspects of Computer Science. STACS 99, pp. 404–413, Trier, Germany, 4–6 March (1999)
11. Monderer, D., Shapley, L.S.: Potential games. Games Econ. Behav. **14**(1), 124–143 (1996)
12. Nisan, N., Roughgarden, T., Tardos, E., Vazirani, V.V.: Algorithmic game theory. Cambridge University Press (2007)
13. Panagopoulou, P.N., Spirakis, P.G.: Algorithms for pure nash equilibria in weighted congestion games. J. Exp. Algorithmics (JEA), **11**, 2–7 (2007)
14. Rosenthal, R.W.: A class of games possessing pure-strategy nash equilibria. Int. J. Game Theor. **2**(1), 65–67 (1973)
15. Rosenthal, R.W.: The network equilibrium problem in integers. Netw. **3**(1), 53–59 (1973)
16. Roughgarden, T., Tardos, É.: How bad is selfish routing? J. ACM (JACM), **49**(2), 236–259 (2002)
17. Roughgarden, T., Tardos, É.: Bounding the inefficiency of equilibria in nonatomic congestion games. Games Econ. Behav. **47**(2), 389–403 (2004)
18. Schulz, A.S., Stier Moses, N.E.: On the performance of user equilibria in traffic networks. In: Proceedings of the Fourteenth Annual ACM-SIAM Symposium on Discrete Algorithms, pp. 86–87, Baltimore, Maryland, USA, 12–14 January (2003)

Correlation-Robust Mechanism Design

Pinyan Lu

ITCS, Shanghai University of Finance and Economics, Shanghai, China
lu.pinyan@mail.shufe.edu.cn

In this talk we will discuss the correlation-robust framework proposed by Carroll [Econometrica 2017] and our recent developments. In this framework, the seller only knows marginal distributions for each separate item or values of an individual bidder but has no information about correlation across different items or different bidders in the joint distribution. Any mechanism is then evaluated according to its expected profit in the worst-case, over all possible joint distributions with given marginal distributions. We illustrate the correlation-robust framework in the context of two well studied revenue maximization settings: (i) single-item action with correlated bidders and (ii) multi-item linear monopoly problem with a single buyer. In the first setting, a single item is sold via sealed-bid auction to n bidders with potentially interdependent (correlated) values for the good. In the second setting, a monopolist seller has n heterogeneous items to sell to a single buyer with additive value for the items.

Carroll's main result states that in multi-item monopoly problem with additive buyer, i.e., buyer's value for any set of items is the sum of values of individual item in the set, the optimal correlation-robust mechanism should sell items separately. We extend the separation result to the case where buyer has a budget constraint on her total payment. Namely, we show that the optimal robust mechanism splits the total budget in a fixed way across different items independent of the bids, and then sells each item separately with a respective per item budget constraint.

For single-item auctions in this correlation-robust framework, we focus on one specific class of single-item auctions, the sequential posted-price mechanism (SPM). We show that (1) the best SPM is a constant approximation to the optimal revenue in this correlation-robust framework; (2) when buyers have the same marginal distribution, SPM has almost better worst-correlation revenue than the best second price auction with common reserve price; (3) SPM also becomes the optimal correlation-robust auction when the number of buyers is large enough.

In this talk we highlight our approach via a dual Linear Programming formulation for the optimal correlation-robust mechanism design problem. This LP avoids explicit construction of the worst-case distribution, which in general may have exponential in the number of items and/or the number of bidders size. We show how our approach can be adopted in a broad range of Bayesian settings with uncertainty about correlated prior distribution.

Job Security, Stability and Production Efficiency, with Applications to Auctions (Invited Plenary Talk)

Ron Lavi[1]

Technion – Israel Institute of Technology
ronlavi@ie.technion.ac.il
https://ronlavi.net.technion.ac.il/

Abstract. This talk describes research results joint with Hu Fu, Robert Kleinberg, and Rann Smorodinsky [1–3]. We study a two-sided matching market with a set of heterogeneous firms and workers in an environment where jobs are secured by regulation. Without job security Kelso and Crawford have shown that stable outcomes and efficiency prevail when all workers are gross substitutes to each firm. It turns out that by introducing job security, stability and efficiency may still prevail, and even for a significantly broader class of production functions. Connections to stability and equilibrium notions in combinatorial auctions and in simultaneous single item auctions will be discussed as well.

References

1. Fu, H., Kleinberg, R., Lavi, R.: Conditional equilibrium outcomes via ascending price processes with applications to combinatorial auctions with item bidding. In: The 13th ACM Conference on Electronic Commerce (EC'12). pp. 586 (2012)
2. Fu, H., Kleinberg, R., Lavi, R., Smorodinsky, R.: Stability and auctions in labor markets with job security. Econ. Lett. **154**, 55–58 (2017)
3. Fu, H., Kleinberg, R.D., Lavi, R., Smorodinsky, R.: Job security, stability, and production efficiency. Theor. Econ. **12**(1), 1–24 (2017)

Restricted Preference Domains in Social Choice: Two Perspectives

Edith Elkind

University of Oxford, UK

Abstract. Preference aggregation is a challenging task: Arrow's famous impossibility theorem [1] tells us that there is no perfect voting rule. One of the best-known ways to circumvent this difficulty is to assume that voters' preferences satisfy a structural constraint, such as, e.g, being single-peaked. Indeed, under this assumption many impossibility results in social choice disappear. Restricted preference domains also play an important role in computational social choice: for instance, there are voting rules that are NP-hard to compute in general, but admit efficient winner determination algorithms when voters' preferences belong to a restricted domain. However, restricted domains that have nice social choice-theoretic properties are not necessarily attractive from an algorithmic perspective, and vice versa. In this note, we will discuss some domain restrictions that have proved to be useful from a computational perspective, and compare the use of restricted domains in computational and classic social choice theory.

Contents

On Revenue Monotonicity in Combinatorial Auctions 1
Andrew Chi-chih Yao

Restricted Preference Domains in Social Choice: Two Perspectives 12
Edith Elkind

The Complexity of Cake Cutting with Unequal Shares 19
Ágnes Cseh and Tamás Fleiner

A Near Optimal Mechanism for Energy Aware Scheduling 31
Antonios Antoniadis and Andrés Cristi

Information Elicitation for Bayesian Auctions. 43
Jing Chen, Bo Li, and Yingkai Li

Coreness of Cooperative Games with Truncated Submodular
Profit Functions 56
Wei Chen, Xiaohan Shan, Xiaoming Sun, and Jialin Zhang

Simple Games Versus Weighted Voting Games 69
Frits Hof, Walter Kern, Sascha Kurz, and Daniël Paulusma

Hide and Seek Game with Multiple Resources 82
Marcin Dziubiński and Jaideep Roy

An Improved Envy-Free Cake Cutting Protocol for Four Agents. 87
Georgios Amanatidis, George Christodoulou, John Fearnley,
Evangelos Markakis, Christos-Alexandros Psomas,
and Eftychia Vakaliou

A Truthful Mechanism for Interval Scheduling 100
Jugal Garg and Peter McGlaughlin

On Revenue-Maximizing Mechanisms Assuming Convex Costs 113
Amy Greenwald, Takehiro Oyakawa, and Vasilis Syrgkanis

On the Price of Stability of Social Distance Games. 125
Christos Kaklamanis, Panagiotis Kanellopoulos,
and Dimitris Patouchas

Schelling Segregation with Strategic Agents. 137
Ankit Chauhan, Pascal Lenzner, and Louise Molitor

Efficient Rational Proofs with Strong Utility-Gap Guarantees 150
Jing Chen, Samuel McCauley, and Shikha Singh

Removal and Threshold Pricing: Truthful Two-Sided Markets with Multi-dimensional Participants 163
Moran Feldman and Rica Gonen

A Two-Stage Mechanism for Ordinal Peer Assessment 176
Zhize Li, Le Zhang, Zhixuan Fang, and Jian Li

The Equilibrium Existence of a Robust Routing Game Under Interval Uncertainty 189
Xujin Chen, Xiaodong Hu, and Chenhao Wang

Online Trading as a Secretary Problem 201
Elias Koutsoupias and Philip Lazos

Constrained Swap Dynamics over a Social Network in Distributed Resource Reallocation 213
Abdallah Saffidine and Anaëlle Wilczynski

A Hashing Power Allocation Game in Cryptocurrencies. 226
Yukun Cheng, Donglei Du, and Qiaoming Han

Resource Based Cooperative Games: Optimization, Fairness and Stability . . . 239
Ta Duy Nguyen and Yair Zick

Short Paper: Strategic Contention Resolution in Multiple Channels with Limited Feedback 245
George Christodoulou, Themistoklis Melissourgos, and Paul G. Spirakis

The Communication Burden of Single Transferable Vote, in Practice 251
Manel Ayadi, Nahla Ben Amor, and Jérôme Lang

Mechanism Design for Two-Opposite-Facility Location Games with Penalties on Distance 256
Xujin Chen, Xiaodong Hu, Xiaohua Jia, Minming Li, Zhongzheng Tang, and Chenhao Wang

An Optimal Strategy for Static Black-Peg Mastermind with Three Pegs. 261
Gerold Jäger and Frank Drewes

Tight Bounds on the Relative Performances of Pricing Mechanisms in Storable Good Markets 267
Gerardo Berbeglia, Shant Boodaghians, and Adrian Vetta

What is the Optimal Deferral Number in Waitlist Mechanism. 273
Zhou Chen, Qi Qi, Changjun Wang, and Wenwei Wang

Author Index 275

On Revenue Monotonicity in Combinatorial Auctions

Andrew Chi-chih Yao(✉)

Tsinghua University, Beijing, China
andrewcyao@tsinghua.edu.cn

Abstract. Along with substantial progress made recently in designing near-optimal mechanisms for multi-item auctions, interesting structural questions have also been raised and studied. In particular, is it true that the seller can always extract more revenue from a market where the buyers value the items higher than another market? In this paper we obtain such a revenue monotonicity result in a general setting. Precisely, consider the revenue-maximizing combinatorial auction for m items and n buyers in the Bayesian setting, specified by a valuation function v and a set F of nm independent item-type distributions. Let $REV(v, F)$ denote the maximum revenue achievable under F by any incentive compatible mechanism. Intuitively, one would expect that $REV(v, G) \geq REV(v, F)$ if distribution G stochastically dominates F. Surprisingly, Hart and Reny (2012) showed that this is not always true even for the simple case when v is additive. A natural question arises: Are these deviations contained within bounds? To what extent may the monotonicity intuition still be valid? We present an approximate monotonicity theorem for the class of fractionally subadditive (XOS) valuation functions v, showing that $REV(v, G) \geq c\,REV(v, F)$ if G stochastically dominates F under v where $c > 0$ is a universal constant. Previously, approximate monotonicity was known only for the case $n = 1$: Babaioff et al. (2014) for the class of additive valuations, and Rubinstein and Weinberg (2015) for all subaddtive valuation functions.

Keywords: Mechanism design · Subadditive valuation
Maximum revenue

1 Introduction

Along with substantial progress made recently in designing near-optimal mechanisms for multi-item auctions, interesting structural questions have also been raised and studied. In particular, is it true that the seller can always extract more revenue from a market where the buyers value the items higher than another market? In this paper we obtain such a revenue monotonicity result in a general setting, leveraging on recent progress made in the mechanism design literature.

In the simplest case of Myerson's 1-item auction [13], let $REV(\mathcal{F})$ denote the optimal revenue for independent valuation distributions $\mathcal{F} = F_1 \times \cdots \times F_n$.

X. Deng (Ed.): SAGT 2018, LNCS 11059, pp. 1–11, 2018.
https://doi.org/10.1007/978-3-319-99660-8_1

Is it true that $REV(\mathcal{G}) \geq REV(\mathcal{F})$ when $\mathcal{G} = G_1 \times \cdots \times G_n$ stochastically dominates $\mathcal{F}$ (i.e. G_i stochastically dominates F_i for each buyer i)? Intuitively, if each buyer i is prepared to pay more for the item, it seems reasonable that the seller should be able to extract more revenue. This is indeed true for the 1-item auctions, as remarked in Rubinstein and Weinberg [14], as a consequence of Myerson's characterization.

The revenue monotonicity question becomes much subtler when there are $m > 1$ items in the auction. Hart and Reny [10] showed that revenue monotonicity is not universally true even with just one buyer ($n = 1$) and two items ($m = 2$). They gave examples with distributions $\mathcal{G}$ stochastically dominating $\mathcal{F}$, yet $REV(\mathcal{G}) < REV(\mathcal{F})$. Thus, when there are $m > 1$ items, the target can only be *approximate revenue monotonicity*, e.g. $REV(\mathcal{G}) \geq cREV(\mathcal{F})$ for some absolute positive constant c.

The following monotonicity results have been shown for $n = 1$ buyer and any number of items: if $\mathcal{G}$ stochastically dominates $\mathcal{F}$, then $REV(\mathcal{G}) \geq \frac{1}{6}REV(\mathcal{F})$ when the valuation function is additive (Babaioff et al. [1]), and $REV(\mathcal{G}) \geq \frac{1}{338}REV(\mathcal{F})$ for combinatorial auctions with any subadditive valuations (Rubinstein and Weinberg [14]). These results were obtained as immediate corollaries of their respective near-optimal mechanisms which are revenue monotone in distributions. Recently, near-optimal mechanism were found for any n, m in Yao [15] and Cai, Devanur and Weinberg [2] for additive valuations, and in Cai and Zhao [3] (also Feldman, Gravin and Lucier [7] for welfare maximization) for XOS valuation functions. However, these mechanisms are not obviously revenue monotone in distributions; as such, no general monotonicity results are known for $n > 1$.

Our main result is to resolve this question in the affirmative for the class of XOS valuation functions. For any m, n, and combinatorial auctions with XOS valuations v, we show that $REV(\mathcal{G}) \geq cREV(\mathcal{F})$ for $c = \frac{1}{1448}$ if $\mathcal{G}$ stochastically dominates $\mathcal{F}$ with respect to v. We also prove two auxiliary theorems which are needed in proving our main result, and also useful in their own right. Firstly, for any single-parameter environment auction $A \subseteq [0,1]^n$, we show that the optimal revenue satisfies $REV_A(\mathcal{G}) \geq REV_A(\mathcal{F})$ if $\mathcal{G}$ stochastically dominates $\mathcal{F}$. This implies that revenue monotonicity is true not just for Myerson's 1-item optimal auctions, but also for general 1-dimensional auction problems in which the allocation vectors are restricted to an arbitrary allowable set of patterns. Secondly, as a consequence of the single-parameter monotonicity above, we infer that $REV(\mathcal{G}) \geq \frac{1}{24}REV(\mathcal{F})$ in the unit-demand multi-item auctions.

Contributions of This Work: (1) We have given answers of great generality to the revenue monotonicity question, applicable to all XOS subadditive valuation functions, while previously little was known even for the additive valuations. (2) Our key innovation in proof technique is a conceptual one (see Sect. 4.2), requiring no complicated calculations or analysis. A main difficulty in proving XOS monotonicity is the fact that the near-optimal mechanism given in [3] is not monotone. To overcome this obstacle, we first embed our auction into a more relaxed context (that of *digital goods*). In this larger space, we can then establish revenue monotonicity via a connecting path (in the new space) between the two embedded distributions. (3) At a more philosophical level, we agree with the sentiment (such as expressed in Hart and Nisan [9]) that the goal of designing mechanisms is not only to produce an algorithm that works, but also to reveal mathematical structures that allow interesting questions such as revenue monotonicity to be answered. Our present work serves as another validation of the fruitfulness of this approach.

2 Main Results

We present results in three standard auction models in the *independent setting*, in which all the mn item-types are drawn from independent distributions. Some familiar terminologies are reviewed below.

For any two random variables X, Y, we write $X \simeq Y$ if X and Y are *equal in distribution*, i.e., $Pr\{X \in S\} = Pr\{Y \in S\}$ for any measurable set S. Let F, G be distributions over $[0, \infty)$. We write $F \preceq G$ if F is *(stochastically) dominated* by G (i.e. $Pr\{F > t\} \leq Pr\{G > t\}$ for all $t \in [0, \infty)$). Equivalently, we may write $G \succeq F$ if G dominates F. Let $\mathcal{F} = F_1 \times \cdots \times F_n$ and $\mathcal{G} = G_1 \times \cdots \times G_n$ be product distributions over $[0, \infty)^n$. We say $\mathcal{F} \preceq \mathcal{G}$ (or equivalently $\mathcal{G} \succeq \mathcal{F}$) if $F_i \preceq G_i$ for each i.

Let $M = (x, p)$ be an n-buyer DIC-IR mechanism, and let $\mathcal{F}$ be an input valuation distribution over $[0, \infty)^n$. Let $X_{M,\mathcal{F}}$ and $P_{M,\mathcal{F}}$ stand for the random variables $X_{M,\mathcal{F}} = x(t)$, $P_{M,\mathcal{F}} = p(t)$ in the probability space $\{t | t \sim \mathcal{F}\}$.

A *randomized* DIC-IR mechanism is a family of mechanisms $\{M^r | r \sim \mathcal{H}\}$, where each $M^r = (x^r, p^r)$ is a DIC-IR mechanism, and r is randomly chosen according to some distribution $\mathcal{H}$. Let $x^r = (x_1^r, \ldots, x_n^r)$, $p^r = (p_1^r, \ldots, p_n^r)$. The *revenue* yielded by M^R under input distribution $\mathcal{F}$ is defined as $M^R(\mathcal{F}) = \sum_{i=1}^n E_{r\sim\mathcal{H}, t\sim\mathcal{F}}(p_i^r(t))$. Let $X_{M^R,\mathcal{F}}$ and $P_{M^R,\mathcal{F}}$ stand for the random variables $X_{M^R,\mathcal{F}} = x^r(t)$, $P_{M^R,\mathcal{F}} = p^r(t)$ in the probability space $\{(r,t) | r \sim \mathcal{H}, t \sim \mathcal{F}\}$.

A *single-parameter environment* (see e.g. Gonczarowski and Nisan [8]) is specified by a set of *possible outcomes* $A \subseteq [0, 1]^n$. For a valuation distribution $\mathcal{F}$ over $[0, \infty)^n$, $REV_A(\mathcal{F})$ is defined as the maximum revenue yielded by any DIC-IR mechanism, where the allocation $x(t)$ for any type profile t is restricted to be in the set A.

Theorem 1 (Single-Parameter Environment). *Let $M = (x, p)$ with allocation function x and payment function p be an n-player DIC-IR mechanism with valuation distribution $\mathcal{F}$ over $[0, \infty)^n$. Let $X_{M,\mathcal{F}}, P_{M,\mathcal{F}}$ denote the random variables corresponding to x, p. Then for any valuation distribution $\mathcal{G} \succeq \mathcal{F}$, there exists a randomized DIC-IR mechanism M^R such that*
(A) $(X_{M^R,\mathcal{G}}, P_{M^R,\mathcal{G}}) \simeq \{(x(t), p_G(t)) | t \sim \mathcal{F}\}$ where $p_G : [0, \infty)^n \to [0, \infty)^n$ and $p_G \geq p$.
(B) $X_{M^R,\mathcal{G}} \simeq X_{M,\mathcal{F}}$, and $M^R(\mathcal{G}) \geq M(\mathcal{F})$.

Corollary 1. *For any single-parameter environment A, we have $REV_A(\mathcal{G}) \geq REV_A(\mathcal{F})$ if $\mathcal{G} \succeq \mathcal{F}$.*

Theorem 1 can be used to prove a monotonicity theorem for unit-demand multi-item auctions. Let $DREV^{UD}(\mathcal{F})$ denote the optimal revenue achievable by any deterministic DIC-IR mechanism for distribution $\mathcal{F}$ in the unit-demand model, and let $REV^{UD}(\mathcal{F})$ be the optimal revenue achievable by any incentive-compatible mechanism allowing randomized lotteries. It is known (Chawla, Malec and Sivan [5]) that allowing lotteries sometimes can generate more revenues.

Theorem 2 (Unit-Demand Multi-item Auction). *If $\mathcal{G} \succeq \mathcal{F}$, then*

$$DREV^{UD}(\mathcal{G}) \geq \max\{\frac{1}{6}DREV^{UD}(\mathcal{F}), \frac{1}{24}REV^{UD}(\mathcal{F})\}.$$

Theorem 2 is useful in the proof of the next theorem, which is the main result of this paper. Let $v = (v_1, \cdots, v_n)$ be a valuation function, and each v_i has $\alpha_v(\geq 1)$-*supporting prices* for all t_i (see Sect. 5 for definitions). It is known that $\alpha_v = 1$ if v is fractionally subadditive (XOS), and more generally $\alpha_v = O(\log m)$ if v is subadditive. Let D and D' be distributions over some type space. We say that D *v-stochastically dominates* D' if a coupled random pair (t, t') can be generated such that the marginal distributions satisfy $t \sim D$, $t' \sim D'$ and $v(t, S) \geq v(t', S)$ for all $S \subseteq [m]$.

In the next theorem, $REV_{DIC}(v, \mathcal{G})$ refers to the maximum revenue achievable by any deterministic DIC-IR mechanism under valuation v and distribution $\mathcal{G}$; $REV_{BIC}(v, \mathcal{F})$ refers to the maximum revenue achievable by any randomized BIC-BIR mechanism under valuation v and distribution $\mathcal{F}$.

Theorem 3 (Subadditive Combinatorial Auction). *Let v be a valuation satisfying monotonicity, subadditivity and with no externalities. If $\mathcal{G} \succeq_v \mathcal{F}$, then for any $0 < b < 1$ we have*

$$REV_{DIC}(v, \mathcal{G}) \geq \frac{1}{\lambda} REV_{BIC}(v, \mathcal{F})$$

where $\lambda = 32\alpha_v + 6\left(12 + \frac{8}{1-b} + \alpha_v(\frac{16}{b(1-b)} + \frac{96}{1-b})\right)$.

Corollary 2. *If v is XOS, then $\alpha_v = 1$ and choosing $b = \frac{1}{4}$ gives*

$$REV_{DIC}(v, \mathcal{G}) \geq \frac{1}{1448} REV_{BIC}(v, \mathcal{F}).$$

Theorem 1 is easy to prove for the special case when all the distributions F_i, G_i over $[0, \infty)$ are continuous and strictly increasing, using Myerson's theory. The general case needs greater care, due to discontinuities and plateaus of the distributions. We omit the proof here.

We prove Theorems 2 and 3 in Sects. 3 and 4, respectively. More background information and discussions will be presented along the way in proving these results.

3 Unit-Demand Auctions

Consider revenue maximization in an auction with m heterogenous items to sell, and n buyers who are *unit-demand*, i.e., each buyer is allocated either 0 or 1 item. Buyer i has for item j valuation distribution F_{ij} over $[0, \infty)$, and all F_{ij} are independent. Thus a deterministic mechanism $M = (x, p)$ satisfies

$$x_{ij}(t) \in \{0, 1\} \text{ and } \sum_{\ell \in [m]} x_{i\ell}(t) \leq 1, \sum_{\ell \in [n]} x_{\ell j}(t) \leq 1 \text{ for all } i, j. \tag{1}$$

Let $DREV^{(UD)}(\mathcal{F})$ be the maximum revenue achievable under $\mathcal{F}$ by any deterministic DIC-IR mechanism. Let $REV^{(UD)}(\mathcal{F})$ denote the maximum revenue achievable under $\mathcal{F}$ by any randomized BIC-BIR mechanism (equivalent to incentive-compatible lottery-based mechanisms [5]).

An interesting connection was established in Chawla et al. [4,5] between unit-demand auctions and single-parameter environment auctions. A unit-demand auction with valuation distribution $\mathcal{F}$ induces a single-parameter auction $OPT^{COPIES-UD}$ for nm buyers B_{ij} with independent valuation distributions F_{ij}, satisfying the same allocation constraints as Eq. 1. Let $OPT^{COPIES-UD}(\mathcal{F})$ denote the maximum revenue achievable under $\mathcal{F}$.

Theorem A (Chawla et al. [4], Kleinberg and Weinberg [12])

$$DREV^{UD}(\mathcal{F}) \leq OPT^{COPIES-UD}(\mathcal{F}) \leq 6 \cdot DREV^{UD}(\mathcal{F}).$$

The upper bound $6 \cdot DREV^{UD}(\mathcal{F})$ in Theorem A is accomplished by certain sequential posted-price mechanism.

Theorem B (Chawla, Malec and Sivan [5], Cai, Devanur and Weinberg [2])

$$REV^{UD}(\mathcal{F}) \leq 4 \cdot OPT^{COPIES-UD}(\mathcal{F}).$$

Proof of Theorem 2. From Theorem A and B, we have

$$DREV^{UD}(\mathcal{G}) \geq \frac{1}{6} OPT^{COPIES-UD}(\mathcal{G}),$$
$$OPT^{COPIES-UD}(\mathcal{F}) \geq \max\{DREV^{UD}(\mathcal{F}), \frac{1}{4} REV^{UD}(\mathcal{F})\}.$$

Now, $OPT^{COPIES-UD}$ is a single-parameter environment auction, and thus by Theorem 1,

$$OPT^{COPIES-UD}(\mathcal{G}) \geq OPT^{COPIES-UD}(\mathcal{F}).$$

Theorem 2 follows from the above three inequalities. □

4 Subadditive Combinatorial Auctions

4.1 Background

We consider revenue maximization in the combinatorial auction with n independent buyers and m heterogeneous items. The auction is specified by a *type profile distribution* $\mathcal{F}$ and *valuation* function v as elaborated below. We follow the convention used in [3,14].

For each $i \in [n]$, buyer i receives a *type* $t_i = (t_{i1}, \cdots, t_{im})$, where t_i is drawn from a product distribution $F_i = F_{i1} \times \cdots \times F_{im}$ over some *type space* $T_i = T_{i1} \times \cdots \times T_{im}$. Let $\mathcal{F} = F_1 \times \cdots \times F_n$ and $t = (t_1, \cdots, t_m)$. Let $T = T_1 \times \cdots \times T_n$. We also regard the type profile $t \in T$ as an $n \times m$ matrix $t = (t_{ij})$, and $\mathcal{F} = (F_{ij})$ as an $n \times m$ matrix of independent distributions. For each $i \in [n]$, buyer i has a *valuation* function $v_i : T_i \times 2^{[m]} \to [0, \infty)$. The quantity $v_i(t_i, S)$ expresses the value buyer i attaches to the collection of items $S \subseteq [m]$ if t_i is buyer i's received type. Let $v = (v_1, \cdots, v_n)$.

A *deterministic* mechanism $M = (x, p)$ specifies the *allocation* $x = (x_1, \cdots, x_n)$ and *payment* $p = (p_1, \cdots, p_n)$, where $x_i : T \to 2^{[m]}$ and $p_i : T \to [0, \infty)$, satisfying the condition that no item can be allocated to more than 1 buyer. A *randomized* mechanism is specified by a distribution over the set of deterministic mechanisms. Let $REV_{DIC}(v, \mathcal{F})$ and $REV_{BIC}(v, \mathcal{F})$ be the maximum achievable revenue of any deterministic DIC-IR mechanism, and any randomized BIC-BIR mechanism, respectively.

We are interested in valuation function v_i that have the following properties:
(1) *No Extenalities:* For each $t_i \in T_i$ and $S \subseteq [m]$, $v_i(t_i, S)$ depends only on $(t_{ij} | j \in S)$.
(2) *Monotone:* For each $t_i \in T_i$ and $U \subseteq V \subseteq [m]$, $v_i(t_i, U) \leq v_i(t_i, V)$.
(3) *Subadditive:* For each $t_i \in T_i$ and $U \subseteq V \subseteq [m]$, $v_i(t_i, U \cup V) \leq v_i(t_i, U) + v_i(t_i, V)$.

We are particularly interested in valuations v_i that are *XOS* (or called *fractionally subadditive*). Namely, $v_i(t_i, S) = \max_{k \in [K]} v_i^{(k)}(t_i, S)$ where K is finite for each k and $v_i^{(k)}(t_i, \cdot)$ is an *additive* function, i.e., $v_i^{(k)}(t_i, S) = \sum_{j \in S} v_i^{(k)}(t_i, \{j\})$ for all $S \subseteq [m]$.

In Rubinstein and Weinberg [14], it was shown that, in the case of 1 buyer ($n = 1$), for any subadditive valuation and $\mathcal{F}$, there is a simple mechanism achieving a constant approximation of optimal revenue. It was also shown in Cai and Zhao [3] that for any n, $\mathcal{F}$, and XOS valuation v, there exist certain sequential posted price mechanisms that achieve a constant approximation of optimal revenue. We review below some fact from [3] that are essential for our proof of Theorem 3.

Definition 1. Rational Sequential Posted Price Mechanism (RSPM)
Let $\xi = (\xi_{ij})$ where $\xi_{ij} \geq 0$ is a posted price for buyer i and item j. Let $p = (p_{ij})$ where $p_{ij} \geq 0$ is payment of buyer i if item j is chosen. For $i = 1$ to n, buyer i picks at most 1 item j from the set S of available items, maximizing buyer i's utility $V_{ij} - \xi_{ij}$ (where $V_{ij} = v_i(t_i, \{j\})$) is a function of t_{ij}); buyer i pays p_{ij}

if j is picked, and pays 0 if no item is picked. We use $RSPM_\xi$ to denote this mechanism.

Definition 2. Anonymous Sequential Posted Price with Entry Fee Mechanism (ASPE)
This mechanism is specified by $Q = (Q_1, \cdots, Q_m)$, $\delta = (\delta_1, \cdots, \delta_n)$, where for each $j \in [m]$, Q_j is posted price *for item j, and for each $i \in [n]$, $\delta_i(t_i, S)$ is the* entry fee *paid by buyer i with reported type t_i when the set of available items is $S \subseteq [m]$. The mechanism proceeds as follows. For $i = 1$ to n, buyer i either picks no item (and pays 0), or picks a subset I from the set $S \subseteq [m]$ of still available items and pays $\delta_i(t_i, S) + \sum_{j\in I} Q_j$; we update S. The chosen subset I is determined by maximizing buyer i's utility $v(t_i, I) - (\delta_i(t_i, S) + \sum_{j\in I} Q_j)$ among all $I \subseteq S$; ties are broken arbitrarily. We use $ASPE_{Q,\delta}$ to denote this mechanism.*

Definition 3 Supporting Price (Dobzinski, Nisan and Schapira [6]). *For any $\alpha \geq 1$, a type t_i and a subset $S \subseteq [m]$,* prices *$(p_j | j \in S)$ are α-*supporting prices *for $v_i(t_i, S)$ if*
(1) $v_i(t_i, S') \geq \sum_{j\in S'} p_j$ for all $S' \subseteq S$, and
(2) $\sum_{j\in S} p_j \geq \frac{1}{\alpha} v_i(t_i, S)$.

Definition 4. *Given any valuation v, let α_v be such that for all i, t_i and S, there exist α_v-supporting prices for $v_i(t_i, S)$. Clearly, for XOS valuations v_i, we can take $\alpha_v = 1$. It is also known that for any subadditive valuations v, we can take $\alpha_v = O(\log m)$.*

Theorem C (Cai and Zhao [3]). *For any $v, \mathcal{F}$ and constant $b \in (0, 1)$, there exist ξ, Q, δ such that*

$$REV_{BIC}(v, \mathcal{F}) \leq (12 + \frac{8}{1-b}) RSPM_\xi(v, \mathcal{F}) + 8\alpha_v \sum_{j\in[m]} Q_j,$$

$$ASPE_{Q,\delta}(v, \mathcal{F}) \geq \frac{1}{4} \sum_{j\in[m]} Q_j - C \cdot RSPM_\xi(v, \mathcal{F}).$$

where $C = \frac{5}{2(1-b)} + \frac{b+1}{2b(1-b)}$.

4.2 Proof of Theorem 3

To prove Theorem 3, our plan is to prove the following two lemmas for the $\{\xi, Q, \delta\}$ satisfying Theorem C above. As $REV_{DIC}(v, \mathcal{G}) \geq ASPE_{Q,\delta}(v, \mathcal{G})$, these lemmas together with Theorem C immediately imply Theorem 3.

Lemma 1. *If $\mathcal{G} \succeq_v \mathcal{F}$, then $RSPM_\xi(v, \mathcal{F}) \leq 6REV_{DIC}(v, \mathcal{G})$.*

Lemma 2. *If $\mathcal{G} \succeq_v \mathcal{F}$, then $ASPE_{Q,\delta}(\mathcal{G}) \geq \frac{1}{4} \sum_{j\in[m]} Q_j - C \cdot RSPM_\xi(v, \mathcal{F})$.*

Proof of Lemma 1. First we observe that $RSPM_\xi$ (under $v, \mathcal{F}$) can be regarded as a mechanism for a standard unit-demand auction where buyer i has valuation $y_{ij} \in [0, \infty)$ with distribution $Y_{ij} = \{y_{ij} = v_i(t_i, \{j\}) \,|\, t_i \sim F_i\}$. Let $\mathcal{F}_v$ denote the product distribution of all Y_{ij}. It is easy to see that $RSPM_\xi(v, \mathcal{F}) \leq DREV^{(UD)}(\mathcal{F}_v)$. Now, as $\mathcal{G} \succeq_v \mathcal{F}$, we have $\mathcal{G}_v \succeq \mathcal{F}_v$. By Theorem 2, we have

$$DREV^{(UD)}(\mathcal{F}_v) \leq 6DREV^{(UD)}(\mathcal{G}_v) \leq 6REV_{DIC}(v, \mathcal{G}).$$

Lemma 1 follows immediately. □

The rest of Sect. 4 is denoted to the proof of Lemma 2. We first define some notations related to the operations of $ASPE_{Q,\delta}$. For any $i \in [n]$ and type profile t, let $t_{<i} = (t_1, \ldots, t_{i-1})$ and denote by $S_i(t_{<i})$ the *available set* of items (i.e., not purchased by buyers $1, \ldots, i-1$) for buyer i to choose from. For any t_i and $I \subseteq [m]$, let $u_i(t_i, I) = \max_{I' \subseteq I}(v(t_i, I') - \sum_{j \in I'} Q_j)$. By definition of $ASPE_{Q,\delta}$, if $u_i(t_i, I) < \delta_i(I)$ where $I = S_i(t_{<i})$, then buyer i receives no item and pays 0; if $u_i(t_i, I) \geq \delta_i(I)$, buyer i receives a bundle $I' \subseteq I$ maximizing $v(t_i, I') - \sum_{j \in I'} Q_j$ and pays $\delta_i(I) + \sum_{j \in I'} Q_j$ (regarded as *entry fee* $\delta_i(I)$ plus *item price* $\sum_{j \in I'} Q_j$). Let $SOLD(t)$ denote the set of items not picked by any buyer after $ASPE_{Q,\delta}$ is finished.

For any distribution $\mathcal{G}$, the revenue $ASPE_{Q,\delta}(\mathcal{G})$ consists of two parts: $EntryFee(\mathcal{G})$ and $ItemPrice(\mathcal{G})$ which are the expected value of entry fees and item prices, respectively, paid by all buyers.

We are now ready to prove Lemma 2. We show that

$$EntryFee(\mathcal{G}) \geq \frac{1}{4} \sum_{j \in [m]} Pr_{t' \sim \mathcal{G}}\{j \notin SOLD(t')\} \cdot Q_j - C \cdot RSPM_\xi(v, \mathcal{F}) \quad (2)$$

$$ItemPrice(\mathcal{G}) \geq \sum_{j \in [m]} Pr_{t' \sim \mathcal{G}}\{j \in SOLD(t')\} \cdot Q_j \quad (3)$$

Lemma 2 follows immediately from Eqs. 2 and 3 by adding these inequalities together.

Equation 3 is obvious by definition. We now prove Eq. 2. First recall the following result from [3].

Lemma CZ (Cai and Zhao [3])

$$EntryFee(\mathcal{F}) \geq \frac{1}{4} \sum_{j \in [m]} Pr_{t \sim \mathcal{F}}\{j \notin SOLD(t)\} \cdot Q_j - C \cdot RSPM_\xi(v, \mathcal{F}).$$

Note that the righthand side of Eq. 2 and the righthand side of Lemma CZ have exactly the same form, although their numerical values may be quite different. The key insight for proving Eq. 2 is to generalize Lemma CZ to a more relaxed setting: the *Digital Goods*, where the type profile t's distribution can be de-coupled from the distribution of the available sets $\{S_i(t_{<i})\}$. We show that

in this setting, it becomes possible to compare the two entry fees $EntryFee(\mathcal{F})$ and $EntryFee(\mathcal{G})$. We emphasize that we are not casting our original research problem in a new setting. Rather, we merely embed $\mathcal{F}$ and $\mathcal{G}$ in this larger space where we will be able to find a connecting path between them for the purpose of comparing their revenues under $ASPE_{Q,\delta}$. (For more information on Digital Goods, see e.g. Hartline and Karlin [11].)

In what follows, v and $\mathcal{F}$ are fixed, while Q, δ, ξ are determined by v, $\mathcal{F}$ (as specified in [3]).

Digital Goods (DG):
In this setting, each item $j \in [m]$ has an unlimited supply of identical copies, so that j may be assigned to many buyers if necessary. Let $\mathcal{I} = (\mathcal{I}_1, \ldots, \mathcal{I}_n)$ where each $\mathcal{I}_i$ is a distribution over $2^{[m]}$. Consider the following mechanism: given a type profile $t = (t_1, \ldots, t_n)$, the seller generates for each buyer $i \in [n]$ a random $I_i \sim \mathcal{I}_i$ and applies $ASPE_{Q,\delta}$ to buyer i with I_i as the available set of items. Thus, the buyer pays an entry fee $\delta_i(I_i)$ (and the appropriate item prices) if the condition $u_i(t_i, I_i) \geq \delta_i(I_i)$ is satisfied; otherwise buyer i pays nothing and receives no items. For any $\mathcal{I}$ and type profile distribution $\mathcal{H}$, let $EntryFee(\mathcal{H}, \mathcal{I})$ denote the expected total entry fees collected when $t \sim \mathcal{H}$, that is,

$$EntryFee(\mathcal{H}, \mathcal{I}) = \sum_{i \in [n]} E_{t_i \sim H_i} \left[Pr_{I_i \sim \mathcal{I}_i} \{u_i(t_i, I_i) \geq \delta_i(I_i)\} \cdot \delta_i(I_i) \right].$$

Definition 5. (Embedding of $\mathcal{H}$ in DG Space) *For any type profile distribution $\mathcal{H}$, let $\mathcal{I}^{\mathcal{H}} = (\mathcal{I}_1^{\mathcal{H}}, \ldots, \mathcal{I}_n^{\mathcal{H}})$ where each $\mathcal{I}_i^{\mathcal{H}}$ is $\{S_i(t_{<i}) \,|\, t \sim \mathcal{H}\}$. It is obvious that $\mathcal{H} \rightarrow (\mathcal{H}, \mathcal{I}^{\mathcal{H}})$ is an embedding that preserves Entryfee, that is,*

$$EntryFee(\mathcal{H}) = EntryFee(\mathcal{H}, \mathcal{I}^{\mathcal{H}}).$$

We state below the embeddings of our targeted distributions $\mathcal{F}$, $\mathcal{G}$ for easy reference.

Fact 1.

$$EntryFee(\mathcal{F}) = EntryFee(\mathcal{F}, \mathcal{I}^{\mathcal{F}}) \quad \text{where } \mathcal{I}_i^{\mathcal{F}} \text{ is } \{S_i(t_{<i}) \,|\, t \sim \mathcal{F}\},$$
$$EntryFee(\mathcal{G}) = EntryFee(\mathcal{G}, \mathcal{I}^{\mathcal{G}}) \quad \text{where } \mathcal{I}_i^{\mathcal{G}} \text{ is } \{S_i(t'_{<i}) \,|\, t' \sim \mathcal{G}\}.$$

Fact 1 suggests that, $EntryFee(\mathcal{F})$ and $EntryFee(\mathcal{G})$ may be compared in the DG space via a connecting point between $(\mathcal{F}, \mathcal{I}^{\mathcal{F}})$ and $(\mathcal{G}, \mathcal{I}^{\mathcal{G}})$, such as $(\mathcal{F}, \mathcal{I}^{\mathcal{G}})$. This turns out to be indeed the case as seen below.

Lemma 3 (Monotonicity). *For any $\mathcal{I}$ and $\mathcal{G} \succeq_v \mathcal{F}$,*

$$EntryFee(\mathcal{G}, \mathcal{I}) \geq EntryFee(\mathcal{F}, \mathcal{I}).$$

Lemma 4 (Digital Goods Extension of CZ Lemma). *For any $\mathcal{I}$,*

$$EntryFee(\mathcal{F}, \mathcal{I}) \geq \frac{1}{4} \sum_{j \in [m]} B_j \cdot Q_j - C \cdot RSPM_{\xi}(v, \mathcal{F})$$

where $B_j = \min_{i \in [n]} \left[Pr_{I_i \sim \mathcal{I}_i} \{j \in I_i\} \right]$.

Proof of Lemma 3. Write $t_i' \geq_v t_i$ if $v(t_i', S) \geq v(t_i, S)$ for all $S \subseteq [m]$. By definition of v-domination, one can generate a random pair of types (t_i', t_i) such that (a) $t_i' \geq_v t_i$, and (b) marginal distribution of t_i', t_i equals G_i, F_i respectively. Noting that $u_i(t_i', I_i) \geq u_i(t_i, I_i)$ whenever $t_i' \geq_v t_i$, we have

$$\begin{aligned} EntryFee(\mathcal{G}, \mathcal{I}) &= \sum_{i \in [n]} E_{t_i' \sim G_i} \left[Pr_{I_i \sim \mathcal{I}_i} \{ u_i(t_i', I_i) \geq \delta_i(I_i) \} \cdot \delta_i(I_i) \right] \\ &\geq \sum_{i \in [n]} E_{t_i \sim F_i} \left[Pr_{I_i \sim \mathcal{I}_i} \{ u_i(t_i, I_i) \geq \delta_i(I_i) \} \cdot \delta_i(I_i) \right] \\ &= EntryFee(\mathcal{F}, \mathcal{I}) \end{aligned}$$

□

Proof of Lemma 4. This requires a lengthy and complex proof. Fortunately, the proof of Lemma CZ as given in [3] (in the arXiv full paper version, Lemmas 25–28) can be extended line-by-line to our Digital Goods Setting with minimum (and obvious) modifications to yield the present lemma, and hence will not be repeated here. □

Fact 2. For $\mathcal{I}_i = \{ S_i(t'_{<i}) \mid t' \sim \mathcal{G} \}$, we have

$$B_j \geq Pr_{t' \sim \mathcal{G}} \{ j \notin SOLD(t') \}.$$

Proof. Obvious. □

It follows from Fact 1 and Lemma 3 that

$$EntryFee(\mathcal{G}) \geq EntryFee(\mathcal{F}, \mathcal{I})$$

where $\mathcal{I}_i = \{ S_i(t'_{<i}) \mid t' \sim \mathcal{G} \}$.

By Fact 2 and Lemma 4, we have then

$$\begin{aligned} EntryFee(\mathcal{G}) \geq & \frac{1}{4} \sum_{j \in [m]} Pr_{t' \sim \mathcal{G}} \{ j \notin SOLD(t') \} \cdot Q_j \\ & - C \cdot RSPM_{\xi}(v, \mathcal{F}). \end{aligned}$$

This proves Eq. 2 and thus Lemma 2. We have completed the proof of Theorem 3.

References

1. Babaioff, M., Immorlica, N., Lucier, B., Weinberg, S.M.: A simple and approximately optimal mechanism for an additive buyer. In: Proceedings of the 55th Annual IEEE Symposium on Foundations of Computer Science (FOCS) (2014)
2. Cai, Y., Devanur, N.R., Weinberg, S.M.: A duality based unified approach to Bayesian mechanism design. In: Proceedings of the 52th Annual ACM Symposium on Theory of Computing (STOC) (2016)

3. Cai, Y., Zhao, M.: Simple mechanisms for subadditive buyers via duality. In: Proceedings of the 49th Annual ACM Symposium on Theory of Computing (STOC) (2017)
4. Chawla, S., Hartline, J.D., Malec, D.L., Sivan, B.: Multi-parameter mechanism design and sequential posted pricing. In: Proceedings of the 42th Annual ACM Symposium on Theory of Computing (STOC), pp. 311–320 (2010)
5. Chawla, S., Malec, D.L., Sivan, B.: The power of randomness in Bayesian optimal mechanism design. Games Econ. Behav. **91**, 297–317 (2015)
6. Dobzinski, S., Nisan, N., Schapira, M.: Approximation algorithms for combinatorial auctions with complement-free bidders. In: Proceedings of the 37th Annual ACM Symposium on Theory of Computing (STOC) (2005)
7. Feldman, M., Gravin, N., Lucier, B.: Combinatorial auctions via posted price. In: Proceedings of the 26th ACM-SIAM Symposium on Discrete Algorithms (SODA), pp. 123–135 (2015)
8. Gonczarowski, Y., Nisan, N.: Efficient empirical revenue maximization in single-parameter auction environments. In: Proceedings of the 49th Annual ACM Symposium on Theory of Computing (STOC), pp. 856–868 (2017)
9. Hart, S., Nisan, N.: Approximate revenue maximization with multiple items. In: Proceedings of the 13th ACM Conference on Electric Commerce (EC), p. 656 (2012)
10. Hart, S., Reny, P.: Maximizing revenue with multiple goods: nonmonotonicity and other observations. Theor. Econ. **10**, 893–992 (2015)
11. Hartline, J.D., Karlin, A.R.: Profit maximization in mechanism design. In: Nisan, N., Roughgarden, T., Tardos, E., Vazirani, V. (eds.) Algorithmic Game Theory, Chap. 13, pp. 331–361. Cambridge University Press (2007)
12. Kleinberg, R., Weinberg, S.M.: Matroid prophet inequalities. In: Proceedings of the 44th Annual ACM Symposium on Theory of Computing (STOC) (2012)
13. Myerson, R.B.: Optimal auction design. Math. Oper. Res. **6**(1), 58–73 (1981)
14. Rubinstein, A., Weinberg, S.M.: Simple mechanisms for a subadditive buyer and applications to revenue monotonicity. In: Proceedings of the 16th ACM Conference on Electronic Commerce, pp. 377–394 (2015)
15. Yao, A.C.: An n-to-1 bidder reduction for multi-item auctions and its applications. In: Proceedings of the 25th ACM-SIAM Symposium on Discrete Algorithms (SODA), pp. 92–109 (2015)

Restricted Preference Domains in Social Choice: Two Perspectives

Edith Elkind(✉)

University of Oxford, Oxford, UK
eelkind@gmail.com

Abstract. Preference aggregation is a challenging task: Arrow's famous impossibility theorem [1] tells us that there is no perfect voting rule. One of the best-known ways to circumvent this difficulty is to assume that voters' preferences satisfy a structural constraint, such as, e.g., being single-peaked. Indeed, under this assumption many impossibility results in social choice disappear. Restricted preference domains also play an important role in computational social choice: for instance, there are voting rules that are NP-hard to compute in general, but admit efficient winner determination algorithms when voters' preferences belong to a restricted domain. However, restricted domains that have nice social choice-theoretic properties are not necessarily attractive from an algorithmic perspective, and vice versa. In this note, we will discuss some domain restrictions that have proved to be useful from a computational perspective, and compare the use of restricted domains in computational and classic social choice theory.

1 Introduction

A family of three—Alice, Bob, and their daughter Claire—would like to decide what to do on a hot Sunday morning. They are choosing among a bike ride, a trip to the pool, and a visit to the farmers' market, and they only have time for one activity. Bob prefers the bike ride to the pool, and the pool to the farmers' market. Alice prefers the farmers' market to the bike ride, and the bike ride to the pool. Finally, Claire prefers the pool to the farmers' market, and the market to the bike ride.

Aggregating these preferences into a decision that will be accepted by all family members is not easy: whichever option is chosen, there will be another option that would be preferred by a majority of the family members to the original choice. Moreover, the collective preferences exhibit a cyclic structure: a majority (Bob and Alice) prefer the bike ride to the pool, a majority (Bob and Claire) prefer the pool to the farmers' market, and a majority (Alice and Claire) prefer the farmers' market to the bike ride.

This simple example illustrates a fundamental difficulty in preference aggregation, and underpins a classic result in social choice: Arrow's impossibility theorem [1], which tells us that there is no universally acceptable way of aggregating

X. Deng (Ed.): SAGT 2018, LNCS 11059, pp. 12–18, 2018.
https://doi.org/10.1007/978-3-319-99660-8_2

individual preferences into a collective decision. Indeed, social choice theory is unable to help Alice, Bob and Claire in this scenario unless they are willing to use randomization.

Suppose, on the other hand, that Alice, Bob and Claire decide to stop fighting, stay home, turn on the air conditioner and watch a movie. There is still a decision to be made, namely, how to set the room temperature. Alice's preferred option is +21, Bob favors +23, and Claire is happy with +25. It is then safe to assume that Alice also prefers +23 to +25, and Claire prefers +23 to +21. Hence, if we select +23, a majority of the voters would prefer that choice to +21, and a majority of the voters would prefer it to +25. Thus, it makes sense for them to agree on +23.

Why was it easier for Alice, Bob, and Claire to decide on the room temperature, compared to selecting a joint activity? Were they lucky, or was there something fundamental about the structure of the problem that ensured the existence of an acceptable alternative? Perhaps there is always a good solution for a family of three, but a bigger family would struggle to identify a majority-supported alternative? In turns out that there is a fundamental difference between the two examples, which was identified by Black back in 1948 [3]: temperature is, fundamentally, a one-dimensional concept, so anyone whose top choice is x would prefer $x+1$ to $x+2$, and $x-1$ to $x-2$. As demonstrated by Black, this rules out the possibility that collective preferences are cyclic, for any number of voters. Moreover, there always exist a choice that is preferred to any other choice by at least half of the voters.

The preferences studied by Black are formally known as *single-peaked preferences*: the space of alternatives can be viewed as a line, and when we graph each voter's intensity of preferences, the resulting curve has a single peak. There are other preference restrictions with similar properties, though the single-peaked domain is perhaps the most famous one. Restricted preference domains that circumvent Arrow's impossibility result and other similar impossibility theorems have long been a subject of study in social choice theory.

Now, in the examples considered so far, computation was not an issue: with just three voters and three alternatives one can easily implement any reasonable choice mechanism. However, as the number of voters and alternatives increases, algorithmic considerations become important. Consider, for instance, a student computer club that needs to decide on the slate of seminar speakers. There is a limited number of slots, and each club member has ranked all potential speakers; the goal is then to select a subset of speakers so that there is something in the seminar program for every student. One can formalize this goal and define a voting rule that selects an optimal set of speakers according to it; this rule is known as the Chamberlin–Courant rule [4]. However, winner determination under this rule is known to be NP-hard [9]. Inspired by Black's results we can then ask: does this hardness result survive if voters' preferences are single-peaked? It turns out that the answer is no: for single-peaked preferences, the output of the Chamberlin–Courant rule and other similar rules can be computed in polynomial time using a dynamic programming approach [2].

Inspired by this and similar observations, the computational social choice community began to systematically investigate whether existing hardness results in computational social choice still hold when voters' preferences belong to a restricted domain, with a focus on domains defined in the social choice literature. The answer to this question turned out to be more nuanced than one might expect: while classic domain restrictions, such as single-peaked or single-crossing preferences, turn out to be incredibly useful from an algorithmic perspective, there are other domains that have some appealing social choice properties, but do not enable efficient algorithms for important social choice problems, as well as domains that do not rule out cyclic preferences, but are nevertheless algorithmically useful. In this talk, we will consider several such examples, including, in particular, preferences single-peaked on trees [5, 11, 13], preferences single-peaked on a circle [12], and single-peaked preferences in the approval setting [6].

2 Preliminaries and Notation

For every positive integer n, set $[n] = \{1, \dots, n\}$. Let C be a finite set of *alternatives*, or *candidates*, and let $m = |C|$. A *linear order* over C is a binary relation over C that is complete, transitive and antisymmetric. Given a linear order v over C, we denote the top alternative in v by $\mathrm{top}(v)$.

Definition 1. *A* profile $P = (v_1, \dots, v_n)$ *over a set of alternatives* C *is a list of linear orders over* C*. We associate* $P = (v_1, \dots, v_n)$ *with a set of* voters $N = [n]$*; the order* v_i *is called the* vote *of voter* i*. For convenience, we write* $a \succ_i b$ *whenever* $(a, b) \in v_i$*, i.e., when voter* i *strictly prefers* a *to* b*.*

Given a profile C over A, we define its *weak majority relation* $\succeq_{\mathrm{maj}}$ as a binary relation over C such that

$$a \succeq_{\mathrm{maj}} b \iff |\{i \in N : a \succ_i b\}| \geq |\{i \in N : b \succ_i a\}|.$$

We write $a \succ_{\mathrm{maj}} b$ if $a \succeq_{\mathrm{maj}} b$, but not $b \succeq_{\mathrm{maj}} a$. Alternative a is a *weak Condorcet winner* if $a \succeq_{\mathrm{maj}} b$ for all $b \in C$; it is a *Condorcet winner* is $a \succ_{\mathrm{maj}} b$ for all $b \in C$.

Single-Peaked Preferences. Let $\lhd$ be a linear order over the set of alternatives C. A vote v over C is *single-peaked with respect to* $\lhd$ if for every pair of candidates $a, b \in C$ with $\mathrm{top}(v) \lhd b \lhd a$ or $a \lhd b \lhd \mathrm{top}(v)$ it holds that v ranks b above a. A profile P over C is *single-peaked with respect to* $\lhd$ if every vote in P is single-peaked with respect to $\lhd$; P is *single-peaked* if there exists a linear order $\lhd$ over C such that P is single-peaked with respect to $\lhd$. We refer to any such order $\lhd$ as an *axis* for P.

It is known that if voters' preferences are single-peaked then the weak majority relation is transitive; in particular, this means that single-peaked profiles always have weak Condorcet winners.

The Chamberlin–Courant Rule. We will now describe a family of voting rules that will be used to illustrate the algorithmic properties of various restricted preference domains considered in this note.

Rules in this family take a candidate set C, a profile P over this set and a target committee size k as an input, and output a subset of candidates (committee) of size k. Given a candidate set C, $|C| = m$, every vector $\mathbf{s} = (s_1, \dots, s_m)$ of non-negative integers with $0 = s_1 \leq \dots \leq s_m$ defines a *positional misrepresentation function* $\mu_\mathbf{s} : P \times 2^C \to \mathbb{Z}$ as follows: $\mu_\mathbf{s}(v, C') = s_i$ if v ranks her most preferred candidate in C' in position i. The (utilitarian version of the) Chamberlin–Courant rule outputs some committee C' of size k that minimizes the quantity $\sum_{v \in P} \mu_\mathbf{s}(v, C')$ (which we call the $\mathbf{s}$-*score* of C') over all size-k subsets of C. The misrepresentation function associated with the vector $\mathbf{s} = (0, 1, \dots, m-1)$ is known as the *Borda misrepresentation function.*

Finding a winning committee under the Chamberlin–Courant rule is known to be NP-hard, even for the Borda misrepresentation function [9]. In contrast, this problem can be solved in polynomial time for an arbitrary misrepresentation function if the input profile is single-peaked [2]. The algorithm of Betzler et al. [2] proceeds by dynamic programming: it considers the single-peaked ordering $\lhd$ of the set of alternatives C, and, for every $m' \in [m]$ and every $k' \in [k]$, identifies the minimum misrepresentation among all committees of size k' that are contained in the m'-prefix of $\lhd$ and contain the m'-th candidate. This approach works, because it is easy to measure to benefit accomplished by adding candidate m' to a committee whose rightmost member is r, $r < m'$.

In what follows, we will consider several extensions of the concept of single-peaked preferences, and analyze the complexity of computing the Chamberlin–Courant winners for preferences belonging to these domains.

3 Preferences Single-Peaked on a Tree

The domain of preferences single-peaked on a tree was introduced by Demange [5]. Consider a profile P over a set of candidates C and a tree T with vertex set C. The profile P is said to be *single-peaked on* T if it is single-peaked on every path in T. Equivalently, for every vote $v \in P$ and every $s \in [m]$ it holds that the set of top s candidates in v is a connected subset of vertices of T. Clearly, a single-peaked profile is single-peaked on a tree (namely, the path associated with the underlying order $\lhd$), but the converse is not true.

Demange [5] has shown that if a profile is single-peaked on some tree then it has a weak Condorcet winner. However, its weak majority relation need not be transitive. In fact, it is not hard to see that for every tree that is not a path there exists a profile that is single-peaked on that tree, but whose majority relation is not transitive. Indeed, if a tree T is not a path, it has a vertex a with three neighbors b, c, and d. If a voter ranks a first, followed by b, c, and d (in any order), followed by all other candidates, then her preferences are single-peaked on T. Thus, we can construct a profile single-peaked on T where a majority of the voters prefers b to c, a majority prefers c to d and a majority prefers d

to b. Hence, from a social choice perspective profiles that are single-peaked on trees have some desirable properties; however, there is a clear boundary between paths and other kinds of trees.

However, from a computational social choice perspective, the picture is very different. The dynamic programming algorithm of Betzler et al. [2] can be extended to trees, but its running time then becomes exponential in the number of leaves [13]: intuitively, whenever we try to compute an optimal committee of size k' contained in a subtree rooted at a, we have to consider the subtrees rooted at children of a, and, for each child, guess the topmost committee member in its subtree. Indeed, it can be shown that computing Chamberlin–Courant winners remains NP-hard when preferences are single-peaked on a tree, for a large class of misrepresentation functions. The hardness result holds even if the underlying tree is a star or has maximum degree 3. Thus, computing the Chamberlin–Courant winners is easy for long skinny trees, but not for bushy trees. However, this hardness result does not apply to the Borda misrepresentation function; indeed, for this misrepresentation function computing the Chamberlin–Courant winners is easy if the profile is single-peaked on a star, or, more generally, on a tree with a small number of internal vertices [11].

Thus, for a social choice theorist all trees that are not stars have the same properties, whereas a computational social choice theorist distinguishes between 'nice' trees and arbitrary trees.

4 Preferences Single-Peaked on a Circle

Peters and Lackner [12] extend the notion of single-peaked preferences from paths to circles: under their definition, a profile P over a set of candidates C is single-peaked on a circle with vertex set C if for each voter $v \in P$ the circle C could be cut so that v is single-peaked on the resulting path. We immediately note that the preferences of Alice, Bob and Claire over morning activities are single-peaked on a circle, and hence we cannot expect profiles that are single-peaked on a circle to always have weak Condorcet winners or to guarantee the transitivity of weak majority preferences. This explains why this model has not been considered in the social choice literature.

On the other hand, preferences single-peaked on a circle turn out to be simple enough to be algorithmically useful: Peters and Lackner establish that for such preferences one can compute a winning committee under the Chamberlin–Courant rule in polynomial time. However, it is not clear how to use Betzler et al.'s dynamic programming algorithm for this purpose: instead, Peters and Lackner build on the techniques developed by Peters [10] to argue that the problem of finding Chamberlin–Courant winners in this scenario can be encoded as an integer linear program whose constraint matrix is totally unimodular.

5 Single-Peaked Preferences in the Approval Setting

So far, we assumed that voters' preferences are expressed by linear orders over alternatives. Instead, one can consider a simpler scenario: each voter approves a subset of alternatives and disapproves the remaining alternatives. Such preferences are known as approval preferences. The setting of approval preferences is considered to be fairly simple from a social choice perspective: if the goal is to select a single winner, the rule that selects an alternative with the highest number of approvals has a number of attractive normative properties (see, e.g., [8] and the references therein). Thus, domain restrictions in the context of approval preferences have not been considered in the social choice literature. However, voting with approval preferences still presents interesting algorithmic challenges; in particular, a natural adaptation of the Chamberlin–Courant rule for this setting remains NP-hard. On the other hand, for single-peaked approval profiles (where alternatives can be placed on a line so that each voter approves a contiguous segment of this line) Chamberlin–Courant winners can be computed efficiently by dynamic programming [6].

6 Conclusion

We have seen three examples of restricted domains that would be viewed very differently by a social choice theorist and an algorithms researcher: in the first case, the algorithmic approach offers a more fine-grained view of the domain, and in the second and third case, a domain restriction that is seen as not particularly useful from a social choice perspective (either because it does not eliminate violations of the Condorcet principle or because the underlying domain is considered to be simple enough to start with) turns out to be useful from an algorithmic perspective. In this note, we focused on the Chamberlin–Courant rule; one can consider other computationally challenging voting rules, in which case the picture becomes even more complicated. For more details, we refer the reader to the recent survey [7].

Acknowledgments. This work was supported by the European Research Council (ERC) under grant number 639945 (ACCORD).

References

1. Kenneth, J.A.: Social Choice and Individual Values. Wiley, New York (1951)
2. Betzler, N., Slinko, A., Uhlmann, J.: On the computation of fully proportional representation. J. Artif. Intell. Res. **47**(1), 475–519 (2013)
3. Black, D.: On the rationale of group decision-making. J. Polit. Econ. **56**(1), 23–34 (1948)
4. Chamberlin, J.R., Courant, P.N.: Representative deliberations and representative decisions: proportional representation and the Borda rule. Am. Polit. Sci. Rev. **77**(3), 718–733 (1983)
5. Demange, G.: Single-peaked orders on a tree. Math. Soc. Sci. **3**(4), 389–396 (1982)
6. Elkind, E., Lackner, M.: Structure in dichotomous preferences. In: Proceedings of the 24th International Joint Conference on Artificial Intelligence (IJCAI), pp. 2019–2025 (2015)

7. Elkind, E., Lackner, M., Peters, D.: Structured preferences. In: Endriss, U. (ed.) Trends in Computational Social Choice, Chap. 10, pp. 187–207. AI Access (2017)
8. Endriss, U.: Sincerity and manipulation under approval voting. Theory Decis. **74**(3), 335–355 (2013)
9. Lu, T., Boutilier, C.: Budgeted social choice: from consensus to personalized decision making. In: Proceedings of the 22nd International Joint Conference on Artificial Intelligence (IJCAI), pp. 280–286 (2011)
10. Peters, D.: Single-peakedness and total unimodularity: new polynomial-time algorithms for multi-winner elections. In: Proceedings of the 32nd AAAI Conference on Artificial Intelligence (AAAI), pp. 1169–1176 (2016)
11. Peters, D., Elkind, E.: Preferences single-peaked on nice trees. In: Proceedings of the 30th AAAI Conference on Artificial Intelligence (AAAI), pp. 594–600 (2016)
12. Peters, D., Lackner, M.: Preferences single-peaked on a circle. In: Proceedings of the 31st AAAI Conference on Artificial Intelligence (AAAI), pp. 649–655 (2017)
13. Yu, L., Chan, H., Elkind, E.: Multiwinner elections under preferences that are single-peaked on a tree. In: Proceedings of the 23rd International Joint Conference on Artificial Intelligence (IJCAI), pp. 425–431 (2013)

The Complexity of Cake Cutting with Unequal Shares

Ágnes Cseh[1(✉)] and Tamás Fleiner[2]

[1] Institute of Economics, Hungarian Academy of Sciences, Budapest, Hungary
cseh.agnes@krtk.mta.hu
[2] Department of Computer Science and Information Theory,
Budapest University of Technology and Economics, Budapest, Hungary

Abstract. An unceasing problem of our prevailing society is the fair division of goods. The problem of proportional cake cutting focuses on dividing a heterogeneous and divisible resource, the cake, among n players who value pieces according to their own measure function. The goal is to assign each player a not necessarily connected part of the cake that the player evaluates at least as much as her proportional share.

In this paper, we investigate the problem of proportional division with unequal shares, where each player is entitled to receive a predetermined portion of the cake. Our main contribution is threefold. First we present a protocol for integer demands that delivers a proportional solution in fewer queries than all known algorithms. Then we show that our protocol is asymptotically the fastest possible by giving a matching lower bound. Finally, we turn to irrational demands and solve the proportional cake cutting problem by reducing it to the same problem with integer demands only. All results remain valid in a highly general cake cutting model, which can be of independent interest.

1 Introduction

In cake cutting problems, the cake symbolizes a heterogeneous and divisible resource that shall be distributed among n players. Each player has her own measure function, which determines the value of any part of the cake for her. The aim of proportional cake cutting is to allocate each player a piece that is worth at least as much as her proportional share, evaluated with her measure function [21]. The measure functions are not known to the protocol.

The efficiency of a fair division protocol can be measured by the number of queries. In the standard Robertson-Webb model [18], two kinds of queries are

Supported by the Hungarian Academy of Sciences under its Momentum Programme (LP2016-3/2016), its János Bolyai Research Fellowship, Cooperation of Excellences Grant (KEP-6/2017), and OTKA grants K108383, K128611. This work is connected to the scientific program of the "Development of quality-oriented and harmonized R+D+I strategy and functional model at BME" project, supported by the New Hungary Development Plan (Project ID: TÁMOP-4.2.1/B-09/1/KMR-2010-0002).

X. Deng (Ed.): SAGT 2018, LNCS 11059, pp. 19–30, 2018.
https://doi.org/10.1007/978-3-319-99660-8_3

allowed. The first one is the *cut* query, in which a player is asked to mark the cake at a distance from a given starting point so that the piece between these two is worth a given value to her. The second one is the *eval* query, in which a player is asked to evaluate a given piece according to her measure function.

If shares are meant to be *equal* for all players, then the proportional share is defined as $\frac{1}{n}$ of the whole cake. In the *unequal shares* version of the problem (also called cake cutting with entitlements), proportional share is defined as a player-specific demand, summing up to the value of the cake over all players. The aim of this paper is to determine the query complexity of proportional cake cutting in the case of unequal shares. Robertson and Webb [18] write in their seminal book "Nothing approaching general theory of optimal number of cuts for unequal shares division has been given to date. This problem may prove to be very difficult." We now settle the issue for the number of queries, the standard measure of efficiency instead of the number of physical cuts.

1.1 Related Work

Equal Shares. Possibly the most famous cake cutting protocol belongs to the class of Divide and Conquer algorithms. Cut and Choose is a 2-player equal-shares protocol that guarantees proportional shares. It already appeared in the Old Testament, where Abraham divided Canaan to two equally valuable parts and his brother Lot chose the one he valued more for himself. The first n-player variant of this algorithm is attributed to Banach and Knaster [21] and it requires $\mathcal{O}\left(n^2\right)$ cut and eval queries. Other methods include the continuous (but discretizable) Dubins-Spanier [11] and the Even-Paz protocols [13]. The latter show that their method requires $\mathcal{O}\left(n \log n\right)$ queries at most. The complexity of proportional cake cutting in higher dimensions has been studied in several papers [2,3,5,14,15,19], in which cuts are tailored to fit the shape of the cake.

Unequal Shares. The problem of proportional cake cutting with unequal shares is first mentioned by Steinhaus [21]. Motivated by dividing a leftover cake, Robertson and Webb [18] define the problem formally and offer a range of solutions for two players. More precisely, they list cloning players, using Ramsey partitions [17] and most importantly, the Cut Near-Halves protocol [18]. The last method computes a fair solution for 2 players with integer demands d_1 and d_2 in $2\lceil \log_2(d_1 + d_2) \rceil$ queries. Robertson and Webb also show how any 2-player protocol can be generalized to n players in a recursive manner. The number of physical cuts Cut Near-Halves makes for two players can be beaten for certain demands, as Robertson and Webb [18] also note. For some demands, Carney [7] designs such a protocol utilizing a number-theoretic approach.

Irrational Demands. The case of irrational demands in the unequal shares case is interesting from the theoretical point of view, but beyond this, solving it might be necessary, because other protocols might generate instances with irrational demands. E.g., in the maximum-efficient envy-free allocation problem with two players and piecewise linear measure functions, any optimal solution must be specified using irrational numbers, as Cohler et al. [8] show.

Barbanel [1] studies the case of cutting the cake in an irrational ratio between n players and presents an algorithm that constructs a proportional division. Shishido and Zeng [20] solve the same problem with the objective of minimizing the number of resulting pieces. Their protocol is simpler than that of Barbanel [1].

Lower Bounds. The drive towards establishing lower bounds on the complexity of cake cutting protocols is coeval to the cake cutting literature itself [21]. Even and Paz [13] conjectured that their protocol is the best possible, while Robertson and Webb explicitly write that "they would place their money against finding a substantial improvement on the $n \log_2 n$ bound" for proportional cake cutting with equal shares. After approximately 20 years of no breakthrough in the topic, Magdon-Ismail et al. [16] showed that any protocol must make $\Omega(n \log n)$ comparisons – but this was no bound on the number of queries. Essentially simultaneously, Woeginger and Sgall [22] came up with the lower bound $\Omega(n \log n)$ on the number of queries for the case where contiguous pieces are allocated to each player. Not much later, this condition was dropped by Edmonds and Pruhs [12] who completed the query complexity analysis of proportional cake cutting with equal shares by presenting a lower bound of $\Omega(n \log n)$. Brams et al. [6] study the minimum number of actual cuts in the case of unequal shares and prove that $n - 1$ cuts might not suffice – in other words, they show that there is no proportional allocation with contiguous pieces. However, no lower bound on the number of queries has been known in the case of unequal shares.

Generalizations in Higher Dimensions. There are two sets of multiple-dimensional generalizations of the proportional cake cutting problem. The first group focuses on the existence of a proportional division, without any constructive proof. The existence can be shown easily using Lyapunov's theorem, as stated by Dubins and Spanier [11] as Corollary 1.1. Berliant et al. [4] investigate the existence of envy-free divisions. Dall'Aglio [10] considers the case of equal shares and defines a dual optimization problem that allows to compute a proportional solution by minimizing convex functions over a finite dimensional simplex. Complexity issues are not discussed in these papers, in fact, queries are not even mentioned in them.

The second group of multiple-dimensional generalizations considers problems where certain geometric parameters are imposed on the cake and the pieces, see Barbanel et al. [2], Beck [3], Brams et al. [5], Hill [14], Iyer and Huhns [15], Segal-Halevi et al. [19]. Also, some of these have special extra requirements on the output, such as contiguousness or envy-freeness. These works demonstrate the interest in various problems in multi-dimensional cake cutting, for which we define a very general framework.

1.2 Our Contribution

We provide formal definitions for the n-player proportional cake cutting problem with total demand $D \geq n$ in Sect. 2. Then, in Sect. 3 we focus on our protocol for the problem, which is our main contribution in this paper. The idea is that we recursively render the players in two batches so that these batches can simulate two players who aim to cut the cake into two approximately equal halves. Our protocol requires only $2(n-1) \cdot \lceil \log_2 D \rceil$ queries. Other known protocols reach $D \cdot \lceil \log_2 D \rceil$ and $n(n-1) \cdot \lceil \log_2 D \rceil$, thus ours is the fastest procedure that derives a proportional division for the n-player cake cutting problem with unequal shares. Moreover, our protocol also works on a highly general cake (introduced in Sect. 4), extending the notion of the cake to any finite dimension.

We complement our positive result by showing a lower bound of $\Omega(n \log D)$ on the query complexity of the problem in Sect. 5. Our proof generalizes, but does not rely on, the lower bound proof given by Edmonds and Pruhs [12] for the problem of proportional division with equal shares. Moreover, our lower bound remains valid in the generalized cake cutting and query model, allowing a more powerful notion of a query even on the usual, $[0,1]$ interval cake.

In Sect. 6 we turn to irrational demands and solve the proportional cake cutting problem by reducing it to the same problem with integer demands only. By doing so, we provide a novel and simple approach to the problem. Moreover, our method works in the generalized query model as well. The query analysis of known protocols for the n-player proportional cake cutting problem, and missing proofs can be found in the full version of the paper [9].

2 Preliminaries

We begin with formally defining our input. Our setting includes a set of players of cardinality n, denoted by $\{P_1, P_2, \ldots, P_n\}$, and a heterogeneous and divisible good, which we refer to as the cake and project to the unit interval $[0,1]$. Each player P_i has a non-negative, absolutely continuous *measure function* μ_i that is defined on Lebesgue-measurable sets. We remark that absolute continuity implies that every zero-measure set has value 0 according to μ_i as well. In particular, $\mu_i((a,b)) = \mu_i([a,b])$ for any interval $[a,b] \subseteq [0,1]$. Besides measure functions, each player P_i has a *demand* $d_i \in \mathbb{Z}^+$, representing that P_i is entitled to receive $d_i / \sum_{j=1}^{n} d_j \in]0,1[$ part of the whole cake. The value of the whole cake is identical for all players, in particular it is the sum of all demands:

$$\forall 1 \leq i \leq n \quad \mu_i([0,1]) = D = \sum_{j=1}^{n} d_j.$$

We remark that an equivalent formulation is also used sometimes, where the demands are rational numbers that sum up to 1, the value of the full cake. Such an input can be transformed into the above form simply by multiplying all demands by the least common denominator of all demands. As opposed to this, if demands are allowed to be irrational numbers, then no ratio-preserving transformation might be able to transform them to integers. That is why the case of irrational demands is treated separately.

The cake $[0, 1]$ will be partitioned into subintervals in the form $[x, y), 0 \leq x \leq y \leq 1$. A finite union of such subintervals forms a *piece* X_i allocated to player P_i. We would like to stress that a piece is not necessarily connected.

Definition 1. *A set* $\{X_i\}_{1 \leq i \leq n}$ *of pieces is a* division *of the cake* $[0, 1]$ *if* $\bigcup_{1 \leq i \leq n} X_i = [0, 1]$ *and* $X_i \cap X_j = \emptyset$ *for all* $i \neq j$. *We call division* $\{X_i\}_{1 \leq i \leq n}$ proportional *if* $\mu_i(X_i) \geq d_i$ *for all* $1 \leq i \leq n$.

In words, proportionality means that each player receives a piece with which her demand is satisfied. We do not consider Pareto optimality or alternative fairness notions such as envy-freeness in this paper.

We now turn to defining the measure of efficiency in cake cutting. We assume that $1 \leq i \leq n$, $x, y \in [0, 1]$ and $0 \leq \alpha \leq 1$. Oddly enough, the Robertson-Webb query model was not formalized explicitly by Robertson and Webb first, but by Woeginger and Sgall [22], who attribute it to the earlier two. In their query model, a protocol can ask agents the following two types of queries.

- *Cut query* (P_i, α) returns the leftmost point x so that $\mu_i([0, x]) = \alpha$. In this operation x becomes a so-called *cut point*.
- *Eval query* (P_i, x) returns $\mu_i([0, x])$. Here x must be a cut point.

Notice that this definition implies that choosing sides, sorting marks or calculating any other parameter than the value of a piece are not counted as queries and thus they do not influence the efficiency of a protocol.

Definition 2. *The* number of queries *in a protocol is the number of eval and cut queries until termination. We denote the number of queries for a n-player algorithm with total demand D by* $T(n, D)$.

The query definition of Woeginger and Sgall is the strictest formalization of the Robertson-Webb queries. Here we utilize an extension of a query, which has been used in earlier papers [12,13,18,22] and is also referred to as a Robertson-Webb query. The term *proportional cut query* stands for generalized cut queries of the sort "P_i cuts the piece $[x, y]$ in ratio $a : b$", where a, b are integers. It is easy to see that such a query is a concatenation of at most five cut and eval queries. Also, following other papers, we allow queries to start from an arbitrary point of the cake instead of 0 only.

3 Our Protocol

In this section, we present a simple and elegant protocol that beats all three above mentioned protocols in query number. Our main idea is that we recursively render the players in two batches so that these batches can simulate two players who aim to cut the cake into two approximately equal halves. For now we work with the standard cake and query model defined in Sect. 2. Later, in Sect. 4.3 we will show how our protocol can be extended to a more general cake. We remind the reader that cutting near-halves means to cut in ratio $\lfloor \frac{D}{2} \rfloor : \lceil \frac{D}{2} \rceil$.

To ease the notation we assume that the players are indexed so that when they mark the near-half of the cake, the marks appear in an increasing order from 1 to n. In the subsequent rounds, we reindex the players to keep this property intact. Based on these marks, we choose "the middle player", this being the player whose demand reaches the near-half of the cake when summing up the demands in the order of marks from left to right. This player cuts the cake and each player is ordered to the piece her mark falls to. The middle player is cloned if necessary so that she can play on both pieces. The protocol is then repeated on both generated subinstances, with adjusted demands. In the subproblem, the players' demands are according to the ratios listed in the pseudocode.

Proportional division with unequal shares

Each player marks the near-half of the cake X.
Sort the players according to their marks.
Calculate the smallest index j such that $\lfloor \frac{D}{2} \rfloor \leq \sum_{i=1}^{j} d_i =: a$.
Cut the cake in two along P_j's mark.
Define two instances of the same problem and solve them recursively.

1. Players $P_1, P_2, \ldots, P_j$ share piece X_1 on the left. Demands are set to $d_1, d_2 \ldots, d_{j-1}, d_j - a + \lfloor \frac{D}{2} \rfloor$, while measure functions are set to $\mu_i \cdot \lfloor \frac{D}{2} \rfloor / \mu_i(X_1)$, for all $1 \leq i \leq j$.
2. Players $P_j, P_{j+1}, \ldots, P_n$ share piece $X_2 = X \setminus X_1$ on the right. Demands are set to $a - \lfloor \frac{D}{2} \rfloor, d_{j+1}, d_{j+2}, \ldots, d_n$, while measure functions are set to $\mu_i \cdot \lceil \frac{D}{2} \rceil / \mu_i(X_2)$, for all $j \leq i \leq n$.

Example 1. We illustrate our protocol on an example with $n = 3$, depicted in Fig. 1. Let $d_1 = 1, d_2 = 3, d_3 = 1$. Since $D = 5$ is odd, all players mark the near-half of the cake in ratio $2 : 3$. The cake is then cut at P_2's mark, since $d_1 < \lfloor \frac{D}{2} \rfloor$, but $d_1 + d_2 \geq \lfloor \frac{D}{2} \rfloor$. The first subinstance will consist of players P_1 and P_2, both with demand 1, whereas the second subinstance will have the second copy of player P_2 alongside P_3 with demands 2 and 1, respectively. In the first instance, both players mark half of the cake and the one who marked it closer to 0 will receive the leftmost piece, while the other player is allocated the remaining piece. The players in the second instance mark the cake in ratio $1 : 2$. Suppose that the player demanding more marks it closer to 0. The leftmost piece is then allocated to her and the same two players share the remaining piece in ratio $1 : 1$. The player with the mark on the left will be allocated the piece on the left, while the other players takes the remainder of the piece. These rounds require $3 + 2 + 2 + 2 = 9$ proportional cut queries and no eval query.

Theorem 1. *Our "Protocol for proportional division with unequal shares" terminates with a proportional division.*

Theorem 2. *For any $2 \leq n$ and $n < D$, the number of queries in our n-player protocol on a cake of total value D is $T(n, D) \leq 2(n-1) \cdot \lceil \log_2 D \rceil$.*

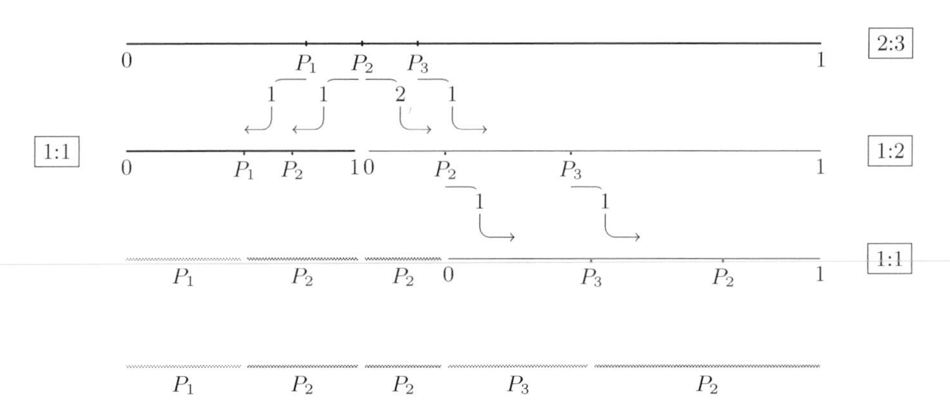

Fig. 1. The steps performed by our algorithm on Example 1. The colored intervals are the pieces already allocated to a player. (Color figure online)

With a query number of $\mathcal{O}(n \log D)$, our protocol is more efficient than all known protocols. We will now point out a further essential difference in fairness when comparing to the fastest known protocol before our result, the generalized Cut Near-Halves. Our protocol treats players equally, while the generalized Cut Near-Halves does not. Equal treatment of players is a clear advantage if one considers the perception of fairness from the point of view of a player.

We remark that our protocol is not truthful, which can be illustrated on a simple example. Take the 2-player equal shares case with nonzero measure functions on any nonzero measure interval. If the player whose mark is at the left knows the measure function of the other player, she can easily manipulate the outcome by marking the half of the cake just before the mark of the other player. As a result, her piece will be larger than what she receives if she reports the truth, unless their measure functions are special.

Remark 1. In the "Protocol for proportional division with unequal shares"

- each player answers the exact same queries as the other players in the same round and same subinstance;
- no player is asked to disclose the outcome of an eval query.

The generalized Cut Near-Halves protocol fails to satisfy both of the above points. It addresses both eval and cut queries to players and treats players differently based on which type of query they got. In the 2-player version of Cut Near-Halves, only one player marks the cake and the other player uses an eval query to choose a side. This enables the second player to have a chance for a piece strictly better than half of the cake, while the first player is only entitled for her exact proportional share and has no chance to receive more than that. Besides this, the player who is asked to evaluate a piece might speculate that she was offered the piece because the other player cut it off the cake—and thus gain information about the measure function of the other player.

4 Generalizations

In this section we introduce a far generalization of cake cutting, where the cake is a measurable set in arbitrary finite dimension and cuts are defined by a monotone function. At the end of the section we prove that even in the generalized setting, $\mathcal{O}(n \log D)$ queries suffice to construct a proportional division.

4.1 A General Cake Definition

Our players remain $\{P_1, P_2, \dots, P_n\}$ with demands $d_i \in \mathbb{Z}^+$, but the cake is now a Lebesgue-measurable subset X of $\mathbb{R}^k$ such that $0 < \lambda(X) < \infty$. Each player P_i has a non-negative, absolutely continuous *measure function* μ_i defined on the Lebesgue-measurable subsets of X. An important consequence of this property is that for every $Z \subseteq X$, $\mu_i(Z) = 0$ if and only if $\lambda(Z) = 0$. The value of the whole cake is identical for all players, in particular it is the sum of all demands:

$$\forall 1 \leq i \leq n \quad \mu_i(X) = D = \sum_{j=1}^{n} d_j.$$

A measurable subset Y of the cake X is called a *piece*. The *volume* of a piece Y is the value $\lambda(Y)$ taken by the Lebesgue-measure on Y. The cake X will be partitioned into pieces $X_1, \dots, X_n$.

Definition 3. *A set $\{X_i\}_{1 \leq i \leq n}$ of pieces is a* division *of X if $\bigcup\limits_{1 \leq i \leq n} X_i = X$ and $X_i \cap X_j = \emptyset$ holds for all $i \neq j$. We call division $\{X_i\}_{1 \leq i \leq n}$* proportional *if $\mu_i(X_i) \geq d_i$ holds for all $1 \leq i \leq n$.*

4.2 A Stronger Query Definition

The more general cake clearly requires a more powerful query notion. Cut and eval queries are defined on an arbitrary piece (i.e. measurable subset) $I \subseteq X$. Beyond this, each cut query specifies a value $\alpha \in \mathbb{R}^+$ and a monotone mapping $f : [0, \lambda(I)] \to 2^I$ (representing a moving knife) such that $f(x) \subseteq f(y)$ and $\lambda(f(x)) = x$ holds for every $0 \leq x \leq y \leq \lambda(I)$.

- *Eval query* (P_i, I) returns $\mu_i(I)$.
- *Cut query* (P_i, I, f, α) returns an $x \leq \lambda(I)$ with $\mu_i(f(x)) = \alpha$ or an error message if such an x does not exist.

As queries involve an arbitrary measurable subset I of X, our generalized queries automatically cover the generalization of the previously discussed Edmonds-Pruhs queries, proportional queries and reindexing. If we restrict our attention to the usual unit interval cake $[0, 1]$, generalized queries open up a number of new possibilities for a query, as Example 2 shows. The new notions also allow us to define cuts on a cake in higher dimensions, see Example 3.

Example 2. On the $[0, 1]$ cake the following rules qualify as generalized queries.

- Evaluate an arbitrary measurable set.
- Cut a piece of value α surrounding a point x so that x is the midpoint of the cut piece.

Example 3. Defined on the generalized cake $X \subseteq \mathbb{R}^k$, the following rules qualify as generalized queries.

- Evaluate an arbitrary measurable set.
- Multiple cut queries on piece $I \subset \mathbb{R}^2$: one player always cuts I along a horizontal line, the other player cuts the same piece along a vertical line.

4.3 The Existence of a Proportional Division

Our algorithm "Proportional division with unequal shares" in Sect. 3 extends to the above described general setting and hence proves that a proportional division always exists.

Theorem 3. *For any $2 \leq n$ and $n < D$, the number of generalized queries in our n-player protocol on the generalized cake of total value D is $T(n, D) \leq 2(n-1) \cdot \lceil \log_2 D \rceil$.*

5 The Lower Bound

In this section, we present our lower bound on the number of queries any deterministic protocol needs to make when solving the proportional cake cutting problem with unequal shares. This result is valid in two relevant settings: 1. on the $[0, 1]$ cake with Robertson-Webb or with generalized queries, 2. on the general cake and queries introduced in Sect. 4.

The lower bound proof is omitted due to space restrictions, but we outline its structure. In the first step, we define a single-player cake-cutting problem where the goal is to identify a piece of small volume and positive value for the sole player. For this problem, we design an adversary strategy and specify the minimum volume of the identified piece as a function of the number of queries asked. In the second step, we turn to the problem of proportional cake cutting with unequal shares. We show that in order to allocate each player a piece of positive value, at least $\Omega(n \log D)$ queries must be addressed to the players—otherwise the allocated pieces overlap.

Theorem 4. *To construct a proportional division in an n-player unequal shares cake cutting problem with total demand D one needs $\Omega(n \log D)$ queries.*

6 Irrational Demands

In this section we consider the case when some demands are irrational numbers. Apart from this, our setting is exactly the same as before. Even though two direct protocols have been presented for the problem of proportional cake cutting with irrational demands [1,20], we feel that our protocol sheds new light to the topic. The complexity of all known protocols for irrational shares falls into the same category: finite but unbounded. Shishido and Zeng [20] present a protocol that is claimed to be simpler than the one of Barbanel [1]. First they present a 2-player protocol, in which one player marks a large number of possibly overlapping intervals that are worth the same for her. The other player then chooses one of these so that it satisfies her demand. The authors then refer to the usual inductive method to the case of n players, in which the n-th player shares each of the $n-1$ pieces the other players have already obtained. This procedure is cumbersome compared to our protocol that reduces the problem to one with rational demands or decreases the number of players. Moreover, our method works on our generalized cake and query model.

Let us choose an arbitrary piece $A \subseteq X$ such that $\mu_i(A) > 0$ for all players P_i. If the players share A and $X \setminus A$ in two separate instances, both in their original ratio $d_1 : d_2 : \ldots : d_n$, then the two proportional divisions will give a proportional $d_1 : d_2 : \ldots : d_n$ division of X itself. Assume now that $\mu_i(A) < \mu_j(A)$ for some players P_i and P_j, and some piece $A \subseteq X$. When generating the two subinstances on A and $X \setminus A$, we reduce d_i on A to 0 and increase it in return on $X \setminus A$ and swap the roles for d_j, increasing it on A and decreasing it on $X \setminus A$. The first generated instance thus has $n-1$ players with irrational demands, while the second instance has n players with irrational demands. We will show in Lemma 2 that if we set the right new demands in these instances, the two proportional divisions deliver a proportional division of X. The key point we prove in Lemma 3, which states that the demands in the second subinstance sum up to slightly below all players' evaluation of $X \setminus A$. Redistributing the slack as extra demand among players gives us the chance to round the demands up to rational numbers in the second subinstance and keep proportionality in the original instance. Iteratively breaking up the instances into an instance with fewer players and an instance with rational demands leads to a set of instances with rational demands only.

We now describe our protocol in detail. Without loss of generality we can assume that $d_1 \leq d_2 \leq \ldots \leq d_n$. As a first step, P_1 answers the cut query with $x = d_1$ and $I = X$. We denote the piece in $f(d_1)$ by A and ask all players to evaluate A. Let P_j be one of the players whose evaluation is the highest. Notice that $\mu_j(A) \geq d_1$, because $\mu_1(A) = d_1$. We distinguish two cases from here.

1. If $\mu_j(A) = d_1$, then $\mu_i(A) \leq d_1$ for all players. We allocate A to P_1 and continue with an instance $\mathcal{I}_1$ with $n-1$ players having the same demands as before. The measure functions need to be normalized to $\frac{D-d_1}{D-\mu_i(A)} \cdot \mu_i$ for all $i \neq 1$ so that all players of $\mathcal{I}_1$ evaluate $X \setminus A$ to $D - d_1$.
2. Otherwise, $\mu_j(A) = d_1 + \varepsilon$, where $\varepsilon > 0$. We generate instances $\mathcal{I}_{2a}$ and $\mathcal{I}_{2b}$.

(a) In the first instance $\mathcal{I}_{2a}$, the cake is A, P_1's demand is 0, P_j's demand is $d_j + d_1$, while all other players keep their original d_i demand. In order to make all players evaluate the full cake to the sum of their demands D, measure functions are modified to $\frac{D}{\mu_i(A)} \cdot \mu_i$.

(b) In the second instance $\mathcal{I}_{2b}$, the cake is $X \setminus A$, P_1's demand is $d_1 + \frac{d_1^2}{D-d_1}$, P_j's demand is $d_j - \frac{d_1(d_1+\varepsilon)}{D-(d_1+\varepsilon)}$, while the original d_i demands are kept for all other players. In order to make all players evaluate the full cake to D, we set $\frac{D}{D-\mu_i(A)} \cdot \mu_i$.

Proportional division with irrational demands

P_1 marks $d_1 \to A$. All players evaluate A. P_j has the highest evaluation.

If $\mu_j(A) = d_1$, then allocate A to P_1 and continue with $n-1$ players on $\mathcal{I}_1$.

Otherwise $\mu_j(A) = d_1 + \varepsilon$. Define two instances $\mathcal{I}_{2a}$ and $\mathcal{I}_{2b}$. While $\mathcal{I}_{2a}$ has $n-1$ players, demands in $\mathcal{I}_{2b}$ sum up to below D and thus can be rationalized.

	$\mathcal{I}_1$	$\mathcal{I}_{2a}$	$\mathcal{I}_{2b}$
cake	$X \setminus A$	A	$X \setminus A$
d_1	0	0	$d_1 + \frac{d_1^2}{D-d_1}$
d_j	d_j	$d_j + d_1$	$d_j - \frac{d_1(d_1+\varepsilon)}{D-(d_1+\varepsilon)}$
d_i	d_i	d_i	d_i
μ_i	$\frac{D-d_1}{D-\mu_i(A)}\mu_i$	$\frac{D}{\mu_i(A)}\mu_i$	$\frac{D}{D-\mu_i(A)}\mu_i$

Lemma 1. *A proportional division in $\mathcal{I}_1$ extends to a proportional division in the original problem once P_1's allocated piece A is added to it.*

Lemma 2. *If each player receives her demanded share in $\mathcal{I}_{2a}$ and $\mathcal{I}_{2b}$, then the union of these pieces gives a proportional division in the original problem.*

Lemma 3. *By slightly increasing all demands, $\mathcal{I}_{2b}$ can be transformed into an instance of proportional cake cutting with rational demands.*

Theorem 5. *Any instance of the proportional cake cutting problem with n players and irrational demands can be transformed into at most $n-1$ proportional cake cutting problems with rational demands and thus can be solved using a finite number of queries.*

We would like to emphasize that even though we have transformed any proportional cake cutting problem with irrational demands into a set of problems with rational demands, we did not show any upper bound on its query complexity. When the problems with rational demands are created, D might grow arbitrarily large, which hugely affects the query number.

Acknowledgment. We thank Simina Brânzei and Erel Segal-Halevi for their generous and insightful advice.

References

1. Barbanel, J.B.: Game-theoretic algorithms for fair and strongly fair cake division with entitlements. Colloq. Math. **69**, 59–73 (1996)
2. Barbanel, J.B., Brams, S.J., Stromquist, W.: Cutting a pie is not a piece of cake. Am. Math. Mon. **116**(6), 496–514 (2009)
3. Beck, A.: Constructing a fair border. Am. Math. Mon. **94**(2), 157–162 (1987)
4. Berliant, M., Thomson, W., Dunz, K.: On the fair division of a heterogeneous commodity. J. Math. Econ. **21**(3), 201–216 (1992)
5. Brams, S.J., Jones, M.A., Klamler, C.: Proportional pie-cutting. Int. J. Game Theory **36**(3), 353–367 (2008)
6. Brams, S.J., Jones, M.A., Klamler, C.: Divide-and-conquer: a proportional, minimal-envy cake-cutting algorithm. SIAM Rev. **53**(2), 291–307 (2011)
7. Carney, E.: A new algorithm for the cake-cutting problem of unequal shares for rational ratios: the divisor reduction method. Sci. Terrapin **3**(2), 15–22 (2012)
8. Cohler, Y.J., Lai, J.K., Parkes, D.C., Procaccia, A.D.: Optimal envy-free cake cutting. In: 25th AAAI Conference on Artificial Intelligence (2011)
9. Cseh, Á., Fleiner, T.: The complexity of cake cutting with unequal shares. CoRR abs/1709.03152 (2018). http://arxiv.org/abs/1709.03152
10. Dall'Aglio, M.: The Dubins-Spanier optimization problem in fair division theory. J. Comput. Appl. Math. **130**(1), 17–40 (2001)
11. Dubins, L.E., Spanier, E.H.: How to cut a cake fairly. Am. Math. Mon. **68**(1), 1–17 (1961)
12. Edmonds, J., Pruhs, K.: Cake cutting really is not a piece of cake. ACM Trans. Algorithms (TALG) **7**(4), 51 (2011)
13. Even, S., Paz, A.: A note on cake cutting. Discrete Appl. Math. **7**(3), 285–296 (1984)
14. Hill, T.P.: Determining a fair border. Am. Math. Mon. **90**(7), 438–442 (1983)
15. Iyer, K., Huhns, M.N.: A procedure for the allocation of two-dimensional resources in a multiagent system. Int. J. Coop. Inf. Syst. **18**(03n04), 381–422 (2009)
16. Magdon-Ismail, M., Busch, C., Krishnamoorthy, M.S.: Cake-cutting is not a piece of cake. In: Alt, H., Habib, M. (eds.) STACS 2003. LNCS, vol. 2607, pp. 596–607. Springer, Heidelberg (2003). https://doi.org/10.1007/3-540-36494-3_52
17. McAvaney, K., Robertson, J., Webb, W.: Ramsey partitions of integers and pair divisions. Combinatorica **12**(2), 193–201 (1992)
18. Robertson, J., Webb, W.: Cake-Cutting Algorithms: Be Fair If You Can. AK Peters, Natick (1998)
19. Segal-Halevi, E., Nitzan, S., Hassidim, A., Aumann, Y.: Fair and square: cake-cutting in two dimensions. J. Math. Econ. **70**, 1–28 (2017)
20. Shishido, H., Zeng, D.Z.: Mark-choose-cut algorithms for fair and strongly fair division. Group Decis. Negot. **8**(2), 125–137 (1999)
21. Steinhaus, H.: The problem of fair division. Econometrica **16**, 101–104 (1948)
22. Woeginger, G.J., Sgall, J.: On the complexity of cake cutting. Discrete Optim. **4**(2), 213–220 (2007)

A Near Optimal Mechanism for Energy Aware Scheduling

Antonios Antoniadis[1] and Andrés Cristi[2(✉)]

[1] Universität des Saarlandes and Max-Planck-Institut für Informatik, Saarbrücken, Germany
antonios.antoniadis@mpi-inf.mpg.de
[2] Universidad de Chile, Santiago, Chile
andres.cristi@ing.uchile.cl

Abstract. With the increased popularity of cloud computing it is of paramount importance to understand energy-efficiency from a game-theoretic perspective. An important question is how the operator of a server should deal with combining energy-efficiency and the particular interests of the users. Consider a cloud server, where clients/agents can submit jobs for processing. The quality of service that each agent perceives is given by a non-decreasing function of the completion time of her job which is private information. The server has to process the jobs and charge each agent while trying to optimize the social cost, defined as the energy expenditure plus the sum of the values of the cost functions of the agents. The operator would like to design a mechanism in order to optimize this objective, which ideally is computationally tractable, charges the users "fairly" and induces a game with an equilibrium.

We describe and analyze one such mechanism called modAVR, which relies on an adaption of the well-known Average Rate (AVR) algorithm for scheduling the jobs. We prove that modAVR combines the aforementioned properties with a constant Price of Anarchy, i.e., despite the fact that it is based on an algorithm designed for optimizing the energy alone, every equilibrium it results in is near-optimal for the total social cost as well. The existence of a Nash equilibrium is proven for both mixed strategies and (in a slightly more restricted setting) pure strategies.

A further interesting feature of modAVR is that it is indirect: each user needs only to declare an upper bound on the completion time of her job, and not the cost function.

Additionally, we prove that for the corresponding mechanism that uses the classical YDS algorithm for scheduling the jobs no pure Nash equilibrium can exist for a very broad and natural class of cost functions. Finally, we are able to extend several of our results for modAVR to a mechanism based on a slight variation of the YDS algorithm. This variation is known also to not admit Nash equilibria in pure strategies.

Antonios Antoniadis was supported by the Deutsche Forschungsgemeinschaft (DFG, German Research Foundation)under AN 1262/1-1, and Andrés Cristi by CONICYT grant PCI PII 20150140 and CONICYTPFCHA/MagísterNacional/2017-22171387.

X. Deng (Ed.): SAGT 2018, LNCS 11059, pp. 31–42, 2018.
https://doi.org/10.1007/978-3-319-99660-8_4

1 Introduction

Chip manufacturers are, at a growing rate, incorporating energy-saving functionalities to their processors as a response to the ever increasing importance of energy-efficiency in computing environments. The most common such functionality is *dynamic speed scaling*, i.e., the capability of the processor to dynamically adjust its operating speed. A higher speed implies a higher performance, but this comes at the cost of a higher energy-consumption per unit of work done. On the other hand, a lower speed although more energy-efficient, leads to a degradation with respect to the Quality of Service (QoS). In practice, it has been observed that the energy consumption of the processor is roughly the cube of its running speed, which results in a nontrivial trade-off between energy and QoS.

Dynamic speed scaling has been extensively studied in the algorithmic literature since the seminal paper by Yao, Demers and Shenker in 1995 [22]. The deadline-based dynamic speed scaling problem that was studied in there, has over the years been considered in the offline, the online, the single and the multiprocessor setting, either alone or in conjunction with other energy-saving functionalities (for some examples see [2,3,5,6,8]). Several variants of the problem where the jobs do not have deadlines and the QoS is flow time have been studied [4,9,11,16]. For two surveys on dynamic speed scaling see [1,15].

We study dynamic speed scaling under a game theoretic point of view. Consider a shared computing system to which users submit jobs to be processed. Each user has a waiting cost function which is non-decreasing in the completion time of her job and is private information. We note that our functions generalize those considered in [13]. This function is modeling the QoS that the user receives. The service operator who manages the computer system would like to simultaneously optimize two opposing objectives: On one hand he wants to keep the total energy consumption for scheduling the jobs as low as possible, while on the other hand he wants to keep the users happy by finishing the processing of their jobs sooner rather than later. However simultaneously optimizing these two objectives is a hard problem [17]. In order to achieve this, the operator would like to design a mechanism, not only for optimizing these objectives, but ideally also so that it is computationally tractable, the payments assigned to the users are proportional to the energy that is required for processing their job, and so that the induced game has at least one Nash equilibrium in pure or mixed strategies.

Outline. We give a formal model of our setting in the next subsection, followed by our contribution in Subsect. 1.2 and discuss connections to previous work in Subsect. 1.3. Then in Sect. 2 we present the modAVR mechanism before showing the existence of Nash equilibria in pure strategies under a slightly restricted setting and in mixed strategies for the more general setting for the induced game. Subsect. 2.4 contains our main result: The game induced by modAVR has a constant Price of Anarchy. In Sect. 3 we study another two very natural choices of scheduling algorithms, one of which was introduced in [13]. After showing that the resulting mechanism does not admit equilibria in pure stratgies even on very simple instances, we prove that, by slightly modifying the algorithm we can extend several results of the previous section to this mechanism. We conclude by discussing generalizations of our work as well as open problems in Sect. 4.

Due to space constraints all of the proofs are deferred to the full version of the paper.

1.1 Formal Problem Statement and Preliminaries

Consider n users, each of which at timepoint 0 presents to the service operator a job i of some particular workload w_i which is assumed to be public information. Each user is also equipped with a non-decreasing waiting cost function $f_i(t)$ which is private information and represents the loss in value that the user faces if her job completes the processing at time t.

The service operator has to schedule the submitted jobs on a single processor equipped with dynamic speed scaling capabilities. Following the speed-scaling literature, we assume that the power consumption of the processor is given by a power function $P(s) = s^{\alpha}$ for some constant value α which is believed to lie in practice between 2 and 3 [7,10], that preemption of jobs is allowed, and that the processor speed is unbounded. The service operator has to decide on the speed $s(t)$ on which the processor resides at each timepoint t as well as on which job to schedule at each timepoint t in order to process all the workload. The total energy consumption of the schedule is given by integrating $P(s)$ over time. We also define the individual energy consumption of a job in the schedule as the integral of $P(s)$ over the timepoints at which the respective job is processed. The problem faced by the operator is offline in the sence that at timepoint 0 all public information is available and the agents know their waiting cost functions.

Ideally the service operator would like to keep the total *social cost* of the solution, which is comprised by the total energy expenditure plus the total waiting costs (which can be seen as the loss on the QoS), as low as possible. The operator would like to design a mechanism in order to achieve this objective. However, besides keeping the total social cost low, there are some desirable properties the mechanism should ideally have, namely:

(i) That the induced game has a Nash equilibrium in mixed strategies,
(ii) To be *strictly* budget balanced: Not only should the sum of the payments be at least the energy consumption, but even more restrictively, the payment from each agent should be exactly the energy consumed processing her job. This is a very natural property which also helps ensure fairness.
(iii) To be computationally tractable for the operator.

We call a mechanism that satisfies the above three properties, a *fairly budget balanced* mechanism.

It is non-trivial to answer whether these conditions can be simultaneously fulfilled. For instance, assuming unlimited computational power from the operator, a VCG type mechanism can be implemented in a way that the payments cover the energy consumption, however individual payments will not equal the consumed energy. Furthermore a VCG mechanism will in general require a larger communication capacity: the agents would have to report their entire waiting cost functions. This paper overcomes these difficulties, and presents a mechanism with the aforementioned desirable properties.

We would also like to stress that the assumption of preemptive jobs can be removed for all of our main results. Indeed, one can use our mechanism for deciding on the payments and on when to return a processed job back to the corresponding agent, but use a different simple mechanism for the actual scheduling of the jobs without incurring any further costs.

Since the algorithms *YDS* and *AVR* by Yao, Demers and Shenker [22] for deadline-based speed scaling play a central role in our results, it is fitting to give a rough description of them here and discuss some of their properties. A more formal description of YDS can be found in the full version of the paper. AVR, and a modification of it are formally defined in Subsect. 2.1.

The YDS Algorithm. The YDS-algorithm takes as input a set of n tasks, each task j with a workload w_j, a release time $r_j = 0$ (in our particular setting) and a deadline d_j that have to be feasibly scheduled on a single, speed-scalable processor with the least possible energy expenditure. The YDS-algorithm then repeatedly and in a greedy fashion identifies a subset of the jobs that *have to* run at the highest possible speed in a particular interval (called *critical interval*), and schedules them at that speed throughout the interval according to the Earliest Deadline First (EDF) principle. It then adapts the remaining instance. YDS produces a solution of minimum energy and runs in polynomial time.

The AVR Algorithm. The AVR-algorithm was originally defined as an online algorithm for the same setting as YDS. At every time t it runs the processor at a speed equal to the sum of the rates $w_i/(d_i - r_i)$ of the available jobs, in any feasible order, for instance EDF. It has a constant competitive ratio.

We note that our results regarding modAVR can be extended to arbitrary r_j. See the full version of the paper.

1.2 Our Contribution

The main contribution of this paper is that we develop a mechanism for the problem that attains all these desired properties. More formally, we show that

Theorem 1. *There exists a* fairly budget balanced *mechanism for the problem such that the social cost of any equilibrium is guaranteed to be within a constant factor of the optimal social cost (bounded Price of Anarchy).*

The first mechanism that we describe is called *modAVR* and is based on a modified version of the well known online algorithm for energy minimization, called AVR which is also due to Yao, Demers and Shenker [22]. We show that the induced game admits a Nash equilibrium in mixed strategies (and even on pure strategies for a slightly more restrictive setting). Furthermore, through a smoothness argument we show a Price of Anarchy of $O(\alpha^{2\alpha-2})$ on the general setting and of $O(\alpha^{\alpha-2})$ when the waiting cost functions are linear. For the practically relevant values of $\alpha \in [2, 3]$ the Price of Anarchy is in the range $[5, 14]$. An interesting feature of our mechanism is that it is indirect (consequently there is no question about truthfulness): each agent reports a requested upper bound on the completion time which is just a positive real number, instead of declaring her entire waiting cost function, which might need the transmission of an unrealistic amount of information in the general setting. We denote that the competitiveness of AVR for the energy part alone and the common release-date setting that we study here is α^α. It is therefore somewhat surprising that a mechanism based solely on AVR performs so well with respect to the total social cost. As a drawback, our result about existence of equilibrium is non-constructive, so even though the implementation is tractable, the equilibrium can be hard to compute.

Furthermore, we show that for the mechanism which uses the classical YDS algorithm for scheduling the jobs, no pure Nash equilibrium can exist. As the mechanism based on a modified version of YDS (modYDS, defined in Sect. 3) is known to also not admit a Nash equilibrium in pure strategies (see [13]), we look into Nash equilbria in mixed strategies and are able to extend several of our results from modAVR and show that such Nash equilibria exist and also give an upper-bound on the Price of Anarchy for modYDS.

In summary, our main contribution is presenting a new mechanism modAVR as well as analyzing modAVR and modYDS. Both satisfy all three properties of Theorem 1. As a side-effect our work also improves upon results from [13].

1.3 Previous Work on the Model

Dürr, Jez and Vasquez [13] investigate the existence of pure Nash equilibria under two different mechanisms. Similar to our model, each player has an internal cost function (however these are restricted to linear), and declares a deadline to the scheduler. The scheduler then implements a modified YDS algorithm which releases each task back to the client at its deadline. Subsequently, the player is charged. They analyze two charging schemes: (i) proportional cost share where every player has to pay exactly the cost generated during the execution of her job (what we call here modYDS), and (ii) marginal cost share where every player pays the marginal increase in energy caused by adding her to the game. They show that the proportional cost share (modYDS) mechanism does not necessarily admit a pure Nash equilibrium, while marginal cost share does, but at the cost of overcharging the players. Furthermore, the convergence time of iterative best-response dynamics to this pure Nash equilibrium may be unbounded. The authors also prove that the social optimum is a pure Nash equilibrium, however as the mechanism is not direct and no general guarantee is provided, it is not clear if it induces a bounded Price of Anarchy.

A similar model is studied by the same set of authors [12]. Here, they consider a mechanism in which the scheduler fixes an arbitrary order of the jobs, and the users have to declare their (again linear) waiting cost functions. The scheduler then efficiently computes the optimal schedule under this fixed order. Their main result states that under a certain payment function this mechanism is truthful and that the sum of the payments covers the energy expenditure.

Unfortunately, fixing the order of the tasks can lead to outcomes with arbitrarily worse social cost than that of an optimal solution that is not restricted to a particular order. Indeed, consider a two-job instance, where $w_1 = n$, $w_2 = 1/n$, $f_1(x) = (1/n)^{\frac{\alpha}{\alpha-1}} x$ and $f_2(x) = n^{\frac{\alpha}{\alpha-1}} x$. By calculating the optimal cost for both possible permutations of the jobs (see also [12, Theorem 1]) one can see that they are a factor n^2 apart.

Finally, our work is related to cost-sharing mechanisms [18,19] which study similar situations in which the cost is only a function of the set of served players. The main concern in that setting is that of finding truthful mechanisms, which are also budget-balanced, economically efficient and computationally tractable.

2 Nash Equilibria and Price of Anarchy for modAVR

We describe a mechanism that we call modAVR, in which each agent reports a requested upper bound d_j on the completion time and the schedule is built using a slight variation of the Average Rate algorithm [22], considering these reports as deadlines. The payments will be exactly the energy consumed by each job, and the additional property that each job j is completed exactly at the corresponding declared deadline d_j will hold.

2.1 Definition of modAVR

We describe modAVR in two parts. In the first part we define the speed profile, i.e., we define the speed function $s(t)$ for each timepoint t. In the second part we define which job is scheduled at each timepoint t at the given speed $s(t)$. Although the first part is identical to the definition of AVR in [22], and therefore the total energy consumed will remain the same, the second part diverges from simply doing earliest deadline first (EDF). This change is key for the continuity of the costs and therefore for the existence of equilibrium, and for our smoothness analysis and therefore for the bounded Price of Anarchy result (whether our results also apply to the mechanism that is induced by classical AVR is an open problem). We note that modAVR here is an offline algorithm, since all jobs are available at timepoint 0.

Part I: The Speed Profile. As in the classical AVR, each job j contributes an amount of w_j/d_j, to the speed throughout the interval $[0, d_j]$. In other words, at each time t, the speed is given by:

$$s(t) := \sum_{j:t \leq d_j} \frac{w_j}{d_j}.$$

We note that since $w_j > 0$ for all j, the speed level will change at exactly the declared deadlines of the jobs. Therefore, we can consider the partition of the time horizon $\mathbb{R}_+$ into intervals $\mathcal{I}(d)$, which is induced by the vector of declared deadlines d. Furthermore, by slightly abusing notation by using $I \leq d_i$ for $I =$

$[a, b]$ and $b \leq d_i$, the speed throughout interval I under modAVR with a vector of declared deadlines d is given by

$$s_I(d) = \sum_{i:I \leq d_i} \frac{w_i}{d_i}.$$

We note that the speed function is a step-wise decreasing function, where each step corresponds to an interval in the partitioning, see also Fig. 1.

Part II: The Job Assignment. It remains to decide which job will be executed at each timepoint. We describe how this is done for a particular interval I in the partitioning. For each job j with $I \leq d_j$ we will process in I a total amount of volume of $w_j^I := w_j \cdot \frac{|I|}{d_j}$.

These amounts of volumes are processed throughout I sequentially in a last deadline first (LDF) order, at speed $s_I(d)$.

It is easy to verify that the resulting schedule is feasible, and has the property that for each job j its completion time C_j is exactly its declared deadline d_j, i.e., $C_j = d_j$.

We note that under the mechanism, the energy expenditure charged to agent i, when d is the vector of declared deadlines, is

$$E_i^{\text{modAVR}}(d) = \sum_{I \in \mathcal{I}(d):I \leq d_i} |I| \cdot s_I(d)^{\alpha} \frac{w_i/d_i}{s_I(d)}.$$

In some proofs where there is no ambiguity we just denote this by $E_i(d)$. Recall that on the induced game, each agent i wishes to minimize $\text{Cost}_i(d) = E_i^{\text{modAVR}}(d) + f_i(C_i) = E_i^{\text{modAVR}}(d) + f_i(d_i)$. This results in the fact that the social cost is exactly the sum of the costs of all agents. We denote this as

$$\text{Cost}(d) = \sum_{i=1}^{n} \text{Cost}_i(d).$$

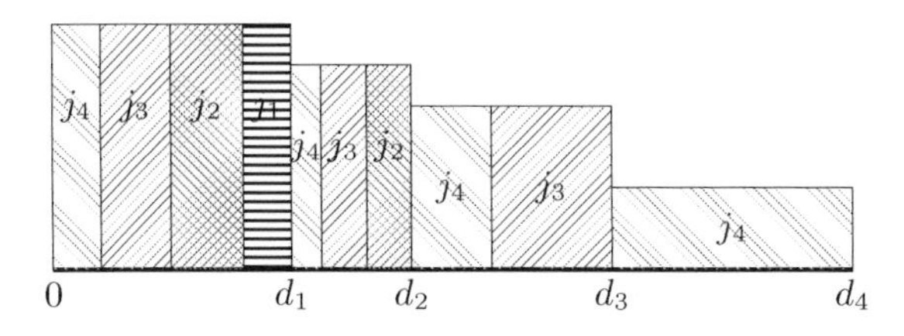

Fig. 1. An example of a modAVR schedule with four jobs. The height of the rectangles represents the speed at which the corresponding job runs.

Corollary 2 *The mechanism modAVR can be implemented in polynomial time, and each agent i is charged* exactly *the energy spent for processing job i.*

In order to prove Theorem 1 it therefore remains to bound the price of Anarchy of modAVR.

2.2 Existence of Nash Equilibria in Pure Strategies for modAVR

We start by proving the existence of Nash equilibria in pure strategies for a natural class of power functions, under some assumptions regarding the waiting cost functions and the workloads of the jobs.

Theorem 3. *For common release times, if $\alpha = 2$, the waiting costs are linear, i.e. $f_i(x) = c_i x$ for positive c_i, and there is an order of the agents such that $\frac{w_i}{c_i} \leq \frac{w_{i+1}}{c_{i+1}}$ and $\frac{w_i^2}{c_i} \leq \frac{w_{i+1}^2}{c_{i+1}}$ for every $i = 1, \ldots, n-1$, then there is a pure strategy N.E. in which the declared deadlines are given by $d_i^* = \sqrt{\frac{w_i}{c_i} \sum_{k=1}^{i} w_k}$.*

Corollary 4 *If $\alpha = 2$, the waiting costs are linear, and all the workloads are the same, then there is a pure strategy N.E. under modAVR.*

Corollary 5 *If $\alpha = 2$ and the waiting costs are all given by the same linear function, then there is a pure strategy N.E. under modAVR.*

We have just proven the existence of Nash equilibria in pure strategies for the mechanism based on modAVR, for a natural and broad class of settings. Whether Nash equilibria in pure strategies without the assumptions on the power function, the waiting cost functions and the workloads exist remains an open question. However, in the next subsection, we will show the existence of Nash equilibria in mixed strategies in the general setting.

2.3 Existence of Nash Equilibria in Mixed Strategies

As the game induced by the mechanism described in Subsect. 2.1 has an infinite strategy space, the existence of equilibria is non-trivial. The goal of this section is to show that it has a Nash equilibrium in mixed strategies. Since we use a smoothness argument for the Price of Anarchy, all the results can be immediately extended to a very broad collection of equilibrium concepts (see Roughgarden [20], [21]).

Theorem 6 ([14]). *Let G be a game played by n agents who have compact Hausdorff pure strategy spaces $A_1, \ldots, A_n$ and real-valued and continuous payoff functions over $A_1 \times \cdots \times A_n$. Then G has a Nash equilibrium in mixed strategies.*

We start by showing that the energy charged to every job is continuous in the vector of declared deadlines.

Lemma 1. *On modAVR the energy consumed by a job, and hence the charging to the agent, is a continuous function in the vector of declared deadlines.*

We now proceed with stating the main theorem of this subsection.

Theorem 7. *Assume that $\alpha \geq 2$ and that for every agent $i \in \{1, \ldots, n\}$ the waiting time cost function $f_i : [0, \infty) \to [0, \infty)$ is continuous, non-decreasing, and satisfies both $f_i(0) = 0$ and $\lim_{x \to \infty} f(x) = \infty$. Then the game induced by the modAVR mechanism has a Nash equilibrium in mixed strategies.*

2.4 Bounded Price of Anarchy

This subsection is devoted to proving a constant (depending on α) upper bound on the *Price of Anarchy* of the modAVR mechanism of Subsect. 2.1. We first state the theorem in its general form, i.e., for any waiting cost functions.

Theorem 8. *The expected social cost of any mixed Nash equilibrium on the game induced by the modAVR mechanism is within a constant factor* $K_\alpha = O(\alpha^{2\alpha-2})$ *of the optimal social cost.*

After proving Theorem 8, we will show an improved bound for the special case when the waiting cost functions are linear, in Theorem 9.

The idea of proving Theorems 8 and 9 is to use a smoothness argument and the Extension Theorem for cost-minimization games due to Roughgarden [20, Proposition 2.3]. We start by stating a sequence of three lemmas.

Lemma 2. *If* $\lambda, \mu > 0$ *are such that* $(x+1)^{\alpha-1} \leq \mu x^\alpha + \lambda, \forall x \geq 0$, *then the inequality*

$$\sum_{i=1}^{n} E_i(d_i^*, d_{-i}) \leq \mu E(d) + \lambda E(d^*)$$

holds for any two vectors of deadlines d *and* d^* *in* $\mathbb{R}_+^n$.

Lemma 3. *Assume* $\alpha \geq 2$. *For any* $\mu \in (0,1)$ *there is a* $\lambda > 0$ *such that* $(x+1)^{\alpha-1} \leq \mu x^\alpha + \lambda$ *for every* $x \geq 0$.

In particular, for $\lambda = \alpha^{\alpha-2} + \frac{2^{\alpha-1}}{\alpha-1}\left(\frac{\alpha}{\alpha-1}\right)^{(\alpha-3)\alpha}$ *and* $\mu = 1/2$, *there holds* $(x+1)^{\alpha-1} \leq \mu x^\alpha + \lambda$, *for all* $x \geq 0$.

The next lemma helps us compare the energy costs of modAVR (AVR) to the energy costs of the optimal (with respect to energy) YDS algorithm.

Lemma 4. *When all the jobs are released at time 0, for any fixed vector of deadlines* $d \in \mathbb{R}_+^n$, *the energy cost for modAVR (AVR) is at most* α^α *times the energy cost for YDS.*

Linear Waiting Cost Functions. We now look a bit more closely to the special case when the waiting cost functions are of the form $f_i(t) = c_i \cdot t$, for some constants c_i, in which we are able to find a better *Price of Anarchy*. The key idea behind the proof is that the linearity of the waiting cost implies a balance between it and the energy expenditure.

Theorem 9. *If the waiting cost functions are linear, the expected social cost of any mixed Nash equilibrium on the game induced by the modAVR mechanism is within a constant factor* $O(\alpha^{\alpha-2})$ *of the optimal social cost.*

3 Analysis of YDS and modYDS

In this section we (re)define and analyze the mechanisms YDS and modYDS.

The YDS Mechanism. We call YDS mechanism the one that asks for deadlines, implements the YDS algorithm, releases the jobs when completed and charges each agent the individual energy consumption.

The modYDS Mechanism. Similar to the definition of modAVR, we define modYDS, as a modified version of the YDS mechanism that releases each job exactly at its deadline, instead of the completion time.

The modYDS mechanism has already been shown in [13] (where it is called proportional cost share mechanism) to not admit a pure Nash equilibrium. We prove that this is also the case for the YDS mechanism. More specifically, with respect to the YDS mechanism there can be no pure Nash equilibrium for $n \geq n_0$ many agents. Here, n_0 is a constant depending on α and the maximum degree of the polynomial functions f_j. This holds even for the special case when all agents share the same w, and have the same waiting cost function f. We then extend our results about the existence of mixed Nash equilibria and the bounded PoA from modAVR to modYDS, with a small loss in the factor of the latter.

Theorem 10. *The YDS mechanism does not always admit a pure Nash equilibrium, even in the case when all the jobs have the same weight and every agent has the same waiting cost function $f(x) = cx^k$, for any $k > 1$, and any $\alpha > 1$.*

3.1 Nash Equilibria in Mixed Strategies for ModYDS

Analogously to our results for modAVR, we start by showing the existence of Nash equilibria in mixed strategies for modYDS.

Theorem 11. *Assume that $\alpha \geq 2$ and that for every agent $i \in \{1, \ldots, n\}$ the waiting time cost function $f_i : [0, \infty) \to [0, \infty)$ is continuous, non-decreasing, and satisfies both $f_i(0) = 0$ and $\lim_{x \to \infty} f(x) = \infty$. Then the game induced by the modYDS mechanism has a Nash equilibrium in mixed strategies.*

3.2 Bounded Price of Anarchy for ModYDS

Regarding the smoothness argument for modYDS, we use the following two lemmas to build upon the smoothness of modAVR.

Lemma 5. *Let d be a fixed vector of valid deadlines. If the release dates are all zero and $\alpha \geq 2$, then for any job i, $E_i^{YDS}(d) \leq E_i^{AVR}(d)$.*

Using this lemma in combination with Lemma 3 and the smoothness for the energy in modAVR, we can prove the smoothness for the energy in modYDS. As we do for modAVR, we use the smoothness for the energy to prove a bound on the social cost:

Theorem 12. *The expected social cost of any mixed Nash equilibrium on the game induced by the modYDS mechanism is within a constant factor K'_α of the optimal social cost.*

4 Discussion

We have studied two mechanisms based on variations of the AVR and YDS algorithm respectively. We have shown that these mechanisms (i) can be implemented efficiently, (ii) have the property that the payment from each agent is exactly the energy spent for processing her job, and (iii) the induced games have an equilibrium and a bounded Price of Anarchy.

An interesting question would be whether modAVR admits a pure Nash equilibrium for any $\alpha > 2$. We conjecture that this is indeed the case, at least when the remaining conditions in Corollaries 4 and 5 hold.

We note that Theorems 7 and 8 (see the full version of the paper for more details) can be extended, at a slight loss in the constant factors involved, to the case where each job i is associated with a release-time r_i before which it cannot be processed. However Lemma 5 does not hold in this setting (see the full version of the paper for a counterexample). It would therefore be interesting to study whether one can nevertheless extend Theorem 12 to the setting with release-times.

Furthermore, our mechanism assumes that the players are forced to participate. A natural question to investigate, is whether it is possible to extend the above results to a voluntary participation model, namely, when the waiting cost functions do not tend to infinity.

Finally, it would be very interesting to study whether the online nature of the AVR algorithm can be exploited, i.e., to study the context in which the agents can submit their jobs for scheduling in an online fashion. This would however require using a different solution concept, such as *subgame perfect Nash equilibria*.

Acknowledgments. We would like to thank José Correa, Dimitris Fotakis, Martin Hoefer, Ruben Hoeksma, Minming Li, and Sebastian Ott for interesting discussions related to this work.

References

1. Albers, S.: Energy-efficient algorithms. Commun. ACM **53**(5), 86–96 (2010)
2. Albers, S., Antoniadis, A.: Race to idle: new algorithms for speed scaling with a sleep state. ACM Trans. Algorithms **10**(2), 9:1–9:31 (2014)
3. Albers, S., Antoniadis, A., Greiner, G.: On multi-processor speed scaling with migration. J. Comput. Syst. Sci. **81**(7), 1194–1209 (2015)
4. Albers, S., Fujiwara, H.: Energy-efficient algorithms for flow time minimization. ACM Trans. Algorithms **3**(4), 49 (2007)
5. Angel, E., Bampis, E., Chau, V., Thang, N.K.: Throughput maximization in multiprocessor speed-scaling. In: Ahn, H.-K., Shin, C.-S. (eds.) ISAAC 2014. LNCS, vol. 8889, pp. 247–258. Springer, Cham (2014). https://doi.org/10.1007/978-3-319-13075-0_20
6. Antoniadis, A., Huang, C., Ott, S.: A fully polynomial-time approximation scheme for speed scaling with sleep state. In: Proceedings of the Twenty-Sixth Annual ACM-SIAM Symposium on Discrete Algorithms, SODA 2015, pp. 1102–1113. SIAM (2015)

7. Bansal, N., Chan, H.-L., Lam, T.-W., Lee, L.-K.: Scheduling for speed bounded processors. In: Aceto, L., Damgård, I., Goldberg, L.A., Halldórsson, M.M., Ingólfsdóttir, A., Walukiewicz, I. (eds.) ICALP 2008. LNCS, vol. 5125, pp. 409–420. Springer, Heidelberg (2008). https://doi.org/10.1007/978-3-540-70575-8_34
8. Bansal, N., Kimbrel, T., Pruhs, K.: Speed scaling to manage energy and temperature. J. ACM **54**(1), 3:1–3:39 (2007)
9. Bansal, N., Pruhs, K., Stein, C.: Speed scaling for weighted flow time. SIAM J. Comput. **39**(4), 1294–1308 (2009)
10. Chan, H., Chan, W., Lam, T.W., Lee, L., Mak, K., Wong, P.W.H.: Energy efficient online deadline scheduling. In: Proceedings of the Eighteenth Annual ACM-SIAM Symposium on Discrete Algorithms, SODA 2007, pp. 795–804. SIAM (2007)
11. Chan, H., Edmonds, J., Lam, T.W., Lee, L., Marchetti-Spaccamela, A., Pruhs, K.: Nonclairvoyant speed scaling for flow and energy. Algorithmica **61**(3), 507–517 (2011)
12. Dürr, C., Jez, L., Vásquez, O.C.: Scheduling under dynamic speed-scaling for minimizing weighted completion time and energy consumption. Discrete Appl. Math. **196**, 20–27 (2015)
13. Dürr, C., Jez, L., Vásquez, O.C.: Mechanism design for aggregating energy consumption and quality of service in speed scaling scheduling. Theor. Comput. Sci. **695**, 28–41 (2017)
14. Glicksberg, I.L.: A further generalization of the kakutani fixed theorem, with application to nash equilibrium points. Proc. Am. Math. Soc. **3**(1), 170 (1952)
15. Irani, S., Pruhs, K.: Algorithmic problems in power management. SIGACT News **36**(2), 63–76 (2005)
16. Lam, T.W., Lee, L., To, I.K., Wong, P.W.H.: Online speed scaling based on active job count to minimize flow plus energy. Algorithmica **65**(3), 605–633 (2013)
17. Megow, N., Verschae, J.: Dual techniques for scheduling on a machine with varying speed. In: Fomin, F.V., Freivalds, R., Kwiatkowska, M., Peleg, D. (eds.) ICALP 2013. LNCS, vol. 7965, pp. 745–756. Springer, Heidelberg (2013). https://doi.org/10.1007/978-3-642-39206-1_63
18. Mehta, A., Roughgarden, T., Sundararajan, M.: Beyond moulin mechanisms. In: Proceedings of the 8th ACM Conference on Electronic Commerce, EC 2007, pp. 1–10 (2007)
19. Moulin, H.: Incremental cost sharing: characterization by coalition strategy-proofness. Soc. Choice Welfare **16**(2), 279–320 (1999)
20. Roughgarden, T.: Intrinsic robustness of the price of anarchy. In: Proceedings of the Forty-first Annual ACM Symposium on Theory of Computing, STOC 2009, pp. 513–522. ACM, New York (2009)
21. Roughgarden, T.: The price of anarchy in games of incomplete information. ACM Trans. Econ. Comput. **3**(1), 6:1–6:20 (2015)
22. Yao, F.F., Demers, A.J., Shenker, S.: A scheduling model for reduced CPU energy. In: 36th Annual Symposium on Foundations of Computer Science, Milwaukee, Wisconsin, 23–25 October 1995, pp. 374–382. IEEE Computer Society (1995)

Information Elicitation for Bayesian Auctions

Jing Chen, Bo Li(✉), and Yingkai Li

Department of Computer Science, Stony Brook University, Stony Brook, NY 11794, USA
{jingchen,boli2,yingkli}@cs.stonybrook.edu

Abstract. In this paper we design information elicitation mechanisms for Bayesian auctions. While in Bayesian mechanism design the distributions of the players' private types are often assumed to be common knowledge, information elicitation considers the situation where the players know the distributions better than the decision maker. To weaken the information assumption in Bayesian auctions, we consider an information structure where the knowledge about the distributions is *arbitrarily scattered* among the players. In such an unstructured information setting, we design mechanisms for auctions with unit-demand or additive valuation functions that *aggregate* the players' knowledge, generating revenue that are constant approximations to the optimal Bayesian mechanisms with a common prior. Our mechanisms are 2-step dominant-strategy truthful and the revenue increases gracefully with the amount of knowledge the players collectively have.

Keywords: Information elicitation · Distributed knowledge
Removing common prior

1 Introduction

Bayesian auction design has been extremely flourishing since the seminal work of [23]. One of the main focuses is to generate revenue, by selling m heterogeneous items to n players. Each player has a private valuation function describing how much he values each subset of the items, and the valuations are drawn from prior distributions. An important assumption in Bayesian mechanism design is that the distributions are commonly known by the seller and the players—the

The first author thanks Matt Weinberg for reading a draft of this paper and for helpful discussions. The authors thank Constantinos Daskalakis, János Flesch, Hu Fu, Pinyan Lu, Silvio Micali, Rafael Pass, Andrés Perea, Elias Tsakas, several anonymous reviewers, and the participants of seminars at Stony Brook University, Shanghai Jiaotong University, Shanghai University of Finance and Economics, Maastricht University, MIT, and IBM Thomas J. Watson Research Center for helpful comments. This work is partially supported by NSF CAREER Award No. 1553385. A full version of this extended abstract is available at https://arxiv.org/abs/1702.01416.

X. Deng (Ed.): SAGT 2018, LNCS 11059, pp. 43–55, 2018.
https://doi.org/10.1007/978-3-319-99660-8_5

common prior assumption. However, as pointed out by another seminal work [25], such common knowledge is "rarely present in experiments and never in practice", and "only by repeated weakening of common knowledge assumptions will the theory approximate reality."

In this paper, we weaken the information assumption about the seller and the players by adopting an information elicitation approach [22]. We consider a framework for auctions where the knowledge about the players' value distributions are *arbitrarily scattered* among the players and the seller. The seller must *aggregate* pieces of information from all players to gain a good understanding about the distributions, so as to decide how to sell the items.

As in information elicitation, the players get rewards for reporting their knowledge. However, different from classic information elicitation where a player's utility is exactly his reward, in our model a player's utility comes not only from his knowledge, but also from participating in the auction. Moreover, information elicitation usually assumes the prior distribution is correlated: each player observes a private signal and reports the corresponding posterior distribution. This means every player has information about every other player. In our model, following the convention in multi-item auctions, the players' value distributions for individual items are assumed to be independent. A player may be totally ignorant about some players and only partially knows some other players.

We focus on auctions with unit-demand valuation functions and auctions with additive valuation functions—two valuation types widely studied in the literature [7,16]. In such auctions, a player's valuation function is specified by m values, one for each item. For each player i and item j, the value v_{ij} is independently drawn from a distribution $\mathcal{D}_{ij}$. When all players are unit-demand (respectively, additive), we call such an auction a *unit-demand auction* (respectively, *an additive auction*) for short. Each player privately knows his own values and some (or none) of the distributions of some other players for some items, like long-time competitors in the market. There is no constraint about who knows which distributions. The seller may also know some of the distributions, but he does not know which player knows what.

We introduce directed *knowledge graphs* to succinctly describe the players' knowledge. Each player knows the distributions of his neighbors, different items' knowledge graphs may be totally different, and the structures of the graphs are *not known* by anybody. Interestingly, the intuition behind such an information structure has long been considered by philosophers. In [19], the author discussed a world where "everything in the world might be known by somebody, yet not everything by the same knower." Due to lack of space, most proofs and extensions of our results are given in the full version [9].

1.1 Main Results

Under Arbitrary Knowledge Graphs. Our goal is to design *2-step dominant strategy truthful* (2-DST) information elicitation mechanisms whose expected revenue approximates that of the optimal Bayesian incentive compatible (BIC) mechanism, denoted by OPT. [1] In order for the seller to aggregate the players' knowledge about the distributions, it is natural for the mechanism to ask each player to report his knowledge to the seller, together with his own values. A 2-DST mechanism [2] is such that, *(1) no matter what knowledge the players may report about each other, it is* dominant *for each player to report his true values; and (2) given that all players report their true values, it is* dominant *for each player to report his true knowledge about others.*

When the knowledge graphs are such that some distributions are not known by anybody, it is easy to see that no information elicitation mechanism can be a bounded approximation to OPT. Thus it is natural to consider the following benchmark: the optimal BIC mechanism applied to players and items for whom the distributions are indeed known by somebody, denoted by OPT_K. This is a natural benchmark when considering players with limited knowledge and, if every distribution is known by somebody, then it is exactly OPT. We have the following, formalized in Sect. 3.

Theorems 1 and 3 (sketched). *For any knowledge graph, there is a 2-DST information elicitation mechanism for unit-demand auctions with revenue* $\geq \frac{OPT_K}{96}$, *and such a mechanism for additive auctions with revenue* $\geq \frac{OPT_K}{70}$.

To prove Theorem 1, we actually show a general result: any Bayesian mechanism for unit-demand auctions that is a good approximation in the COPIES setting (formally defined in Sect. 3.2) can be converted to information elicitation mechanisms; see Theorem 2. This applies to a large class of Bayesian mechanisms, including the ones in [7,8,20].

To prove Theorem 3, we have developed a novel approach for using the *adjusted revenue* [27]. Although this concept is very useful in Bayesian auctions, it was unexpected that we found an interesting and highly non-trivial way of using it to analyze information elicitation mechanisms.

When Everything Is Known by Somebody. When the knowledge graphs become denser, the amount of knowledge increases and the seller may generate more revenue. Indeed, if every distribution is known by somebody, $OPT_K = OPT$. We show the revenue that can be generated by information elicitation mechanisms *increases* gracefully together with the amount of knowledge. More precisely, for any integer $k \geq 1$, let $\tau_k = \frac{k}{(k+1)^{\frac{k+1}{k}}}$. Note $\tau_1 = \frac{1}{4}$ and $\tau_k \to 1$ when k gets larger. We have the following theorems, formalized in [9].

Theorems 5 and 6 (sketched). $\forall k \in [n-1]$, *when each distribution is known by at least k players, there is a 2-DST information elicitation mechanism for unit-demand auctions with revenue* $\geq \frac{\tau_k}{24} \cdot OPT$, *and such a mechanism for additive auctions with revenue* $\geq \max\{\frac{1}{11}, \frac{\tau_k}{6+2\tau_k}\}OPT$.

Finally, by exploring the knowledge graph's combinatorial structure, we have the following for single-good auctions, formalized in Sect. 4.

[1] A Bayesian mechanism is BIC if it is a Bayesian Nash equilibrium for all players to report their true values.

Theorem 4 (sketched). *When the knowledge graph is 2-connected,*[2] *there is a 2-DST information elicitation mechanism for single-good auctions with revenue* $\geq (1 - \frac{1}{n})OPT$.

1.2 Discussions

The Use of Scoring Rules. Since our mechanisms elicit the players' knowledge about each other's value distributions, we will use *scoring rules* (see, e.g., [4]) to reward the players for their reported knowledge, as typical in information elicitation. However, the use of scoring rules does not solve the main problems in our auctions. Indeed, because a player's utility comes both from the reward and from participating in the auction, the difficulties in designing information elicitation mechanisms are to guarantee that, even without rewarding the players for their knowledge, (1) it is dominant for each player to report his true values, (2) reporting his true knowledge *never hurts him*, and (3) the resulting revenue approximates the desired benchmark.

Accordingly, in Sects. 3 and 4 we focus on designing information elicitation mechanisms without rewarding the players. Scoring rules are used later solely to break the utility-ties and make it *strictly better* for a player to report his true knowledge. In [9], we show how to add scoring rules to our mechanisms.

Extensions of Our Results. In our main results, the seller asks the players to report the distributions in their entirety, without being concerned with the communication complexity for doing so. This is common in information elicitation and allows us to focus on the main difficulties in aggregating the players' knowledge. In [9], we show how to modify our mechanisms so that the players only report a small amount of information about the distributions.

Furthermore, in the main body of this paper we consider auction settings where a player i's knowledge about another player i' for an item j is exactly the prior distribution $\mathcal{D}_{i'j}$. This simplifies the description of the knowledge graphs. In [9], we consider settings where a player may observe private signals about other players and can further refine the prior.

Future Directions. As Bayesian auctions require the seller (and the players under common-prior assumption) has correct knowledge about *all* distributions, in our main results we do not consider scenarios where players have "insider" knowledge. If the insider knowledge is correct, then our mechanisms' revenue increases. Still, how to aggregate the incorrect information that the players may have about each other is a very interesting question for future studies. Another direction is to elicit players' information for BIC mechanisms. For example, the BIC mechanisms in [5,13] are optimal in their own settings, and it is unclear how to convert them to information elicitation mechanisms.

[2] A directed graph is 2-connected if for any node i, the graph with i and all adjacent edges removed is still strongly connected..

1.3 Related Work

Information Elicitation. Following [22], information elicitation has become an important research area in the past decade [24,28]. A mechanism asks each player to report his private knowledge about the prior distribution. The decision maker wants the mechanism to be BIC, and a player is rewarded based on his reported information and other players' report. Different from auctions, there are no allocations or prices, and a player's utility equals his reward. Most studies on information elicitation require a common prior. Mechanisms without this assumption are considered by [26], and our work is information elicitation in auctions without a common prior. Moreover, information elicitation does not consider the players to have any cost for revealing their knowledge. It would be interesting to include such costs in the general model as well as in ours, to see how the mechanisms will change accordingly.

Bayesian Auction Design. In his seminal work [23], Myerson introduced the first optimal Bayesian mechanism for single-good auctions, which also applies to single-parameter settings [1]. Since then, there has been a huge literature on designing (approximately) optimal Bayesian mechanisms that are either BIC or dominant-strategy truthful (DST); see [17] for an introduction. Mechanisms for multi-parameter settings have been constructed recently. In [5], the authors characterize optimal BIC mechanisms for combinatorial auctions. For unit-demand auctions, [6–8,20] construct DST Bayesian mechanisms that are constant approximations. For additive auctions, [6,16,21,27] provide logarithmic or constant approximations under different conditions.

Removing the Common Prior Assumption. Following [25], a lot of effort has been made to remove the common prior assumption. In DST Bayesian mechanisms it suffices to assume that the seller knows the prior distribution [20,27]. In prior-free mechanisms [14,18] the distribution is unknown and the seller learns it from the values of randomly selected players. In [12,15] the seller observes independent samples from the distribution before the auction begins. In [10,11] the players have arbitrary possibilistic belief hierarchies about each other. In robust mechanism design [3] the players have arbitrary probabilistic belief hierarchies. In crowdsourced Bayesian auctions [2] each player privately knows *all* the distributions (or their refinements), which is a special case of our model. Indeed, all knowledge graphs will be complete graphs under their setting (that is, everybody knows everything), while we allow arbitrary knowledge graphs. In [9], we further discuss how to elicit the players' knowledge refinements, and how to handle correlated distributions in a setting that is a special case of our model but is still more general than that of [2].

2 Preliminaries

In this work, we focus on multi-item auctions with n players (denoted by N) and m items (denoted by M). A player i's value for an item j, v_{ij}, is independently drawn from a distribution $\mathcal{D}_{ij}$. Let $v_i = (v_{ij})_{j \in M}$, $\mathcal{D}_i = \times_{j \in M} \mathcal{D}_{ij}$ and $\mathcal{D} =$

$\times_{i \in N} \mathcal{D}_i$. Player i's value for a subset S of items is $\max_{j \in S} v_{ij}$ in *unit-demand* auctions, and is $\sum_{j \in S} v_{ij}$ in *additive* auctions. The players' utilities, denoted by u_i, are quasi-linear, and the players are risk-neutral.

Knowledge Graphs. It is illustrative to model the players' knowledge graphically. We consider a vector of *knowledge graphs*, $G = (G_j)_{i \in M}$, one for each item. Each G_j is a directed graph with n nodes, one for each player. For any $i \neq i'$, an edge (i, i') is in G_j if and only if player i knows $\mathcal{D}_{i'j}$. There is no constraint about the knowledge graphs: the same player's distributions for different items may be known by different players, different players' distributions for the same item may also be known by different players, and some distributions may not be known by anybody. Each player knows his own out-going edges, and neither the players nor the seller knows the whole graph.

We measure the amount of knowledge in the system by the number of players knowing each distribution. For any $k \in \{0, 1 \ldots, n-1\}$, a knowledge graph is *k-informed* if each node has in-degree at least k: a player's distribution is known by at least k other players. The vector G is k-informed if all knowledge graphs are so. Note that every knowledge graph is 0-informed, and "everything is known by somebody" when $k \geq 1$. A common prior would imply all knowledge graphs are complete directed graphs, or $(n-1)$-informed, which is the strongest condition in our model. The seller's knowledge can be naturally incorporated into the knowledge graphs by considering him as a special "player 0". All our mechanisms can easily utilize the seller's knowledge, and we will not further discuss this issue.

Information Elicitation Mechanisms. Let $\hat{\mathcal{I}} = (N, M, \mathcal{D})$ be a Bayesian auction instance and $\mathcal{I} = (N, M, \mathcal{D}, G)$ a corresponding information elicitation instance, where G is a knowledge graph vector. Different from Bayesian mechanisms, which has $\mathcal{D}$ as input, an information elicitation mechanism has neither $\mathcal{D}$ nor G as input. Instead, it asks each player i to report a valuation $b_i = (b_{ij})_{j \in M}$ and a *knowledge* $K_i = \times_{i' \neq i, j \in M} \mathcal{D}^i_{i'j}$ —a distribution for the valuation subprofile v_{-i}. K_i may contain "$\perp$" at some places, indicating i does not know the corresponding distributions. K_i is i's *true knowledge* if $\mathcal{D}^i_{i'j} = \mathcal{D}_{i'j}$ whenever $(i, i') \in G_j$, and $\mathcal{D}^i_{i'j} = \perp$ otherwise. An information elicitation mechanism maps a strategy profile $(b_i, K_i)_{i \in N}$ to an allocation and a price profile, and may be randomized. To distinguish whether a mechanism $\mathcal{M}$ is a Bayesian or an information elicitation mechanism, we may explicitly write $\mathcal{M}(\hat{\mathcal{I}})$ or $\mathcal{M}(\mathcal{I})$. The (expected) revenue of $\mathcal{M}$ is denoted by $Rev(\mathcal{M})$, and sometimes by $\mathbb{E}_{\mathcal{D}} Rev(\mathcal{M})$ to emphasize the distribution.

3 Under Arbitrary Knowledge Graphs

3.1 Knowledge-Based Revenue Benchmark

When the knowledge graphs can be totally arbitrary, some distributions may not be known by anybody. It is not hard to see that in this case, no information elicitation mechanism can be a bounded approximation to OPT. Indeed, if all but one value distributions of the players are constantly 0, and if the only non-zero distribution, denoted by $\mathcal{D}_{ij}$, is unknown by anybody, then a Bayesian mechanism can find the optimal reserve price based on $\mathcal{D}_{ij}$, while an information elicitation mechanism can only set the price for player i based on the reported values of the other players, which are all 0.

Thus, for arbitrary knowledge graphs, we define a natural revenue benchmark: the optimal Bayesian revenue *on players and items for which the distributions are known in the information elicitation setting.* More precisely, let $\hat{\mathcal{I}} = (N, M, \mathcal{D})$ be a Bayesian instance and $\mathcal{I} = (N, M, \mathcal{D}, G)$ a corresponding information elicitation instance. Let $\mathcal{D}' = \times_{i \in N, j \in M} \mathcal{D}'_{ij}$ be such that $\mathcal{D}'_{ij} = \mathcal{D}_{ij}$ if there exists a player i' with $(i', i) \in G_j$, and $\mathcal{D}'_{ij}$ is constantly 0 otherwise. We refer to $\mathcal{D}'$ as $\mathcal{D}$ *projected on* G. Letting $\mathcal{I}' = (N, M, \mathcal{D}')$ be the resulting Bayesian instance, the *knowledge-based* revenue benchmark is $OPT_K(\mathcal{I}) \triangleq OPT(\mathcal{I}')$, the optimal BIC revenue on $\mathcal{I}'$. This is a demanding benchmark in information elicitation settings: it takes into consideration the knowledge of *all* players, no matter who knows what. When everything is known by somebody, even if G is only 1-informed, we will have $\mathcal{I}' = \hat{\mathcal{I}}$ and $OPT_K(\mathcal{I}) = OPT(\hat{\mathcal{I}})$.

3.2 Unit-Demand Auctions

For unit-demand auctions, sequential post-price Bayesian mechanisms have been constructed by [8,20]. For information elicitation, if the seller asks the players to report both their values and knowledge, and directly uses the reported distributions in these mechanisms, then a player may want to withhold his knowledge about the other players. By doing so, a player may prevent the seller from selling the items to the others, so the items are still available when it is his turn to buy.

A simple idea is to partition the players into two groups: a set of *reporters* who will not receive any item and is only asked to report their knowledge; and a set of *potential buyers* whose knowledge is never used. It is possible that the reported knowledge may not contain a potential buyer's value distributions on all items, thus the technical part is to prove that the seller generates a good revenue even though the players' knowledge is only partially recovered.

Our mechanism $\mathcal{M}_{IEUD}$ is simple and intuitive; see Mechanism 1, where $\mathcal{M}_{UD}$ is the Bayesian mechanism of [20]. It's worth pointing out that, although mechanism $\mathcal{M}_{UD}$ is used as a black-box, Mechanism 1 is not a reduction from arbitrary Bayesian mechanisms. Instead, we will prove a *projection lemma* that allows such a reduction from an important class of Bayesian mechanisms, where mechanism $\mathcal{M}_{UD}$ is an important example. We have the following theorem.

Theorem 1. *Mechanism $\mathcal{M}_{IEUD}$ for unit-demand auctions is 2-DST and, for any instances $\hat{\mathcal{I}} = (N, M, \mathcal{D})$ and $\mathcal{I} = (N, M, \mathcal{D}, G)$, $Rev(\mathcal{M}_{IEUD}(\mathcal{I})) \geq \frac{OPT_K(\mathcal{I})}{96}$.*

The key of the 2-DSTness is that the use of the players' values and the use of their knowledge are disentangled. In [9], we add scoring rules to mechanism

Mechanism 1. $\mathcal{M}_{IEUD}$

1: Each player i reports to the seller a valuation $b_i = (b_{ij})_{j \in M}$ and a knowledge $K_i = (\mathcal{D}^i_{i'j})_{i' \neq i, j \in M}$.
2: Randomly partition the players into two sets, N_1 and N_2, where each player is independently put in each set with probability $\frac{1}{2}$.
3: Set $N_3 = \emptyset$.
4: **for** players $i \in N_1$ lexicographically **do**
5: For each player $i' \in N_2$ and item $j \in M$, if $\mathcal{D}'_{i'j}$ has not been defined yet and $\mathcal{D}^i_{i'j} \neq \bot$, then set $\mathcal{D}'_{i'j} = \mathcal{D}^i_{i'j}$ and add player i' to N_3.
6: **end for**
7: For each $i \in N_3$ and $j \in M$ such that $\mathcal{D}'_{ij}$ is not defined, set $\mathcal{D}'_{ij} \equiv 0$ (i.e., 0 with probability 1) and $b_{ij} = 0$.
8: Run mechanism $\mathcal{M}_{UD}$ on the unit-demand Bayesian auction $(N_3, M, (\mathcal{D}'_{ij})_{i \in N_3, j \in M})$, with the players' values being $(b_{ij})_{i \in N_3, j \in M}$. Let $x' = (x'_{ij})_{i \in N_3, j \in M}$ be the resulting allocation where $x'_{ij} \in \{0, 1\}$, and let $p' = (p'_i)_{i \in N_3}$ be the prices. Without loss of generality, $x'_{ij} = 0$ if $\mathcal{D}'_{ij} \equiv 0$.
9: For each player $i \notin N_3$, i gets no item and his price is $p_i = 0$.
10: For each player $i \in N_3$, i gets item j if $x'_{ij} = 1$, and his price is $p_i = p'_i$.

$\mathcal{M}_{IEUD}$ to reward the players' knowledge, so that a player's utility will be strictly larger when he reports his true knowledge than when he lies.

To analyze the revenue of $\mathcal{M}_{IEUD}$, note that it runs the Bayesian mechanism on a smaller (randomized) Bayesian instance: $\hat{\mathcal{I}}$ projected to the player-item pairs (i, j) such that $i \in N_3$ and $\mathcal{D}_{ij}$ has been reported. To understand how much revenue is lost by the projection, we consider the COPIES instance [8], $\hat{\mathcal{I}}^{CP} = (N^{CP}, M^{CP}, \mathcal{D}^{CP})$, as a bridge between the original Bayesian instance and the information elicitation instance. $\hat{\mathcal{I}}^{CP}$ is obtained from $\hat{\mathcal{I}}$ by replacing each player with m copies and each item with n copies, where a player i's copy j only wants item j's copy i, with the value distributed according to $\mathcal{D}_{ij}$. Thus $\hat{\mathcal{I}}^{CP}$ is a single-parameter auction, with $N^{CP} = N \times M$, $M^{CP} = M \times N$, and $\mathcal{D}^{CP} = \times_{(i,j) \in N^{CP}} \mathcal{D}_{ij}$. We now lower-bound the optimal BIC revenue in the *projected COPIES instance.* For any subset $NM \subseteq N \times M$, let $\hat{\mathcal{I}}^{CP}_{NM}$ be $\hat{\mathcal{I}}^{CP}$ projected to NM. By definition, $OPT(\hat{\mathcal{I}}^{CP}_{NM})$ is the optimal BIC revenue for $\hat{\mathcal{I}}^{CP}_{NM}$. Moreover, let $OPT(\hat{\mathcal{I}}^{CP})_{NM}$ be the revenue of the optimal BIC mechanism for $\hat{\mathcal{I}}^{CP}$ *obtained from players in* NM.

Lemma 1 ***(The projection lemma).*** *For any* $\hat{\mathcal{I}}$ *and* $NM \subseteq N \times M$,

$$OPT(\hat{\mathcal{I}}^{CP}_{NM}) \geq OPT(\hat{\mathcal{I}}^{CP})_{NM}.$$

We elaborate the related definitions and prove Lemma 1 in [9]. Given mechanism $\mathcal{M}_{IEUD}$, the subset NM is the set of player-item pairs (i, j) such that $i \in N_3$ and $\mathcal{D}_{ij}$ is reported. Theorem 1 holds by combining the projection lemma, the randomized partition in $\mathcal{M}_{IEUD}$, and the results on COPIES setting in Bayesian auctions [6,20].

Note that Lemma 1 is only concerned with COPIES instances. Using this lemma and similar to our proof of Theorem 1, any Bayesian mechanism $\mathcal{M}$ whose revenue can be properly lower-bounded by the COPIES instance can be converted to an information elicitation mechanism in a black-box way. We have the following theorem, with the proof omitted.

Theorem 2. *Let $\mathcal{M}$ be any DST Bayesian mechanism such that $Rev(\mathcal{M}(\hat{\mathcal{I}})) \geq \alpha OPT(\hat{\mathcal{I}}^{CP})$ for some $\alpha > 0$. There exists a 2-DST information elicitation mechanism that uses $\mathcal{M}$ as a black-box and is a $\frac{\alpha}{16}$-approximation to OPT_K.*

By Theorem 2, the mechanisms in [7,8] automatically imply information elicitation mechanisms. For single-good auctions, replacing $\mathcal{M}_{UD}$ with Myerson's mechanism, the resulting mechanism is a 4-approximation to OPT_K.

3.3 Additive Auctions

Information elicitation mechanisms for additive auctions are harder to construct and analyze. First, randomly partitioning the players as before may cause a significant revenue loss, as the revenue of additive auctions may come from selling a subset of items as a *bundle* to a player i. Even when i's value distribution for each item is reported with constant probability, the probability that his distributions for all items in the bundle are reported may be very low, thus the mechanism may rarely sell the bundle to i at the optimal price. Second, the seller can no longer "throw away" player-item pairs whose distributions are not reported and focus on the projected instance. When the players are not partitioned into reporters and potential buyers, doing so may cause a player to lie and withhold his knowledge about others, so that they are thrown away.

To simultaneously achieve truthfulness and a good revenue guarantee, our mechanism is very stingy and never throws away any information. If a player i's value distribution for an item j is reported by others, then j may be sold to i via the *β-Bundling* mechanism of [27], denoted by *Bund*. If i's distribution for j is not reported, then j may still be sold to i via the second-price mechanism. Indeed, our mechanism handles the players neither solely based on the original Bayesian instance $\hat{\mathcal{I}}$ nor solely based on the projected instance $\mathcal{I}'$. Rather, it works on a *hybrid* of the two.

Our mechanism $\mathcal{M}_{IEA}$ is still simple; see Mechanism 2. However, significant effort is needed to analyze its revenue. Indeed, note that in Mechanism 2, each M_i is defined according to the original Bayesian instance $\hat{\mathcal{I}}$, while the partition of M is done according to the knowledge graphs in the information elicitation instance $\mathcal{I}$. The mechanism *Bund* is run on a hybrid instance, where β_i is based on $\hat{\mathcal{I}}$ and $\mathcal{D}'_i$ is based on $\mathcal{I}$. Finally, part of player i's winning set is sold according to mechanism *Bund* and part of it is sold using second-price.

To bound the revenue of $\mathcal{M}_{IEA}$, in [9], we develop a novel way to use the *adjusted revenue* [27] in our analysis. As we show there, the adjusted revenue in a hybrid information setting, combined with the revenue of the second-price sale, eventually provides a desirable lower-bound to the revenue of $\mathcal{M}_{IEA}$. We have the following theorem.

Mechanism 2. $\mathcal{M}_{IEA}$

1: Each player i reports a valuation $b_i = (b_{ij})_{j \in M}$ and a knowledge $K_i = (\mathcal{D}^i_{i'j})_{i' \neq i, j \in M}$.
2: For each item j, set $i^*(j) = \arg\max_i b_{ij}$ (ties broken lexicographically) and $p_j = \max_{i \neq i^*} b_{ij}$.
3: **for** each player i **do**
4: Let $M_i = \{j \mid i^*(j) = i\}$ be player i's *winning set*.
5: Partition M into M_i^1 and M_i^2 as follows: $\forall j \in M_i^1$, some i' has reported $\mathcal{D}^{i'}_{ij} \neq \bot$ (if there are more than one reporters, take the lexicographically first); and $\forall j \in M_i^2$, $\mathcal{D}^{i'}_{ij} = \bot$ for all i'.
6: $\forall j \in M_i^1$, set $\mathcal{D}'_{ij} = \mathcal{D}^{i'}_{ij}$; and $\forall j \in M_i^2$, set $\mathcal{D}'_{ij} \equiv 0$.
7: Compute the optimal entry fee e_i and reserve prices $(p'_j)_{j \in M_i^1}$ according to mechanism *Bund* with respect to $(\mathcal{D}'_i, \beta_i)$, where $\beta_{ij} = \max_{i' \neq i} b_{i'j}$ $\forall j \in M$. By the definition of *Bund*, we always have $p'_j \geq \beta_{ij}$ for each j. If $e_i = 0$ then it is possible that $p'_j > \beta_{ij}$ for some j; while if $e_i > 0$ then $p'_j = \beta_{ij}$ for every j.
8: Sell $M_i^1 \cap M_i$ to player i according to *Bund*. That is, if $e_i > 0$ then do the following: if $\sum_{j \in M_i^1 \cap M_i} b_{ij} \geq e_i + \sum_{j \in M_i^1 \cap M_i} p'_j$, player i gets $M_i^1 \cap M_i$ with price $e_i + \sum_{j \in M_i^1 \cap M_i} p'_j$; otherwise the items in $M_i^1 \cap M_i$ are not sold. If $e_i = 0$ then do the following: for each item $j \in M_i^1 \cap M_i$, if $b_{ij} \geq p'_j$, player i gets item j with price p'_j; otherwise item j is not sold.
9: In addition, sell each item j in $M_i^2 \cap M_i$ to player i with price $p_j (= \beta_{ij})$.
10: **end for**

Theorem 3. *Mechanism $\mathcal{M}_{IEA}$ for additive auctions is 2-DST and, for any instances $\hat{\mathcal{I}} = (N, M, \mathcal{D})$ and $\mathcal{I} = (N, M, \mathcal{D}, G)$, $Rev(\mathcal{M}_{IEA}(\mathcal{I})) \geq \frac{OPT_K(\mathcal{I})}{70}$.*

4 When Everything Is Known by Somebody

When the knowledge graph vector G is k-informed with $k \geq 1$, both mechanisms in Sect. 3 of course apply, but we can do better when k gets larger. In fact, in [9], we show that, for both unit-demand auctions and additive auctions, as k gets larger, the approximation ratio approaches the best known approximation to OPT by DST Bayesian mechanisms [6,20,27].

Below, we show that for single-good auctions, if the knowledge graph is only k-informed for some small k, but has nice combinatorial structures, then *nearly optimal* revenue can be generated by leveraging such structures.

Single-Good Auctions with 2-Connected Knowledge Graphs. If the knowledge graph is only k-informed for some small k but has certain combinatorial structures, then good revenue may be generated by leveraging such structures. For single-good auctions, we show a *nearly optimal* information elicitation mechanism under a natural combinatorial structure.

Recall that a directed graph is *strongly connected* if there is a directed path from any node i to any other node i'. A directed graph is *2-connected* if it remains strongly connected after removing any single node and the adjacent edges. For a knowledge graph G, being strongly connected means that for any two players i and i', "i knows a guy who knows a guy ... who knows i'". G being 2-connected means there does not exist a crucial player as an "information hub", without whom the players will split into two parts, with one part having no information about the other. Note that being 2-connected implies being 2-informed.

We define the *information elicitation Myerson* mechanism $\mathcal{M}_{IEM}$ in Mechanism 3. Recall that Myerson's mechanism maps each player i's reported value b_i to the *(ironed) virtual value*, $\phi_i(b_i; \mathcal{D}_i)$. It runs the second-price mechanism with reserve price 0 on virtual values and maps the resulting "virtual price" back to the winner's value space, as his price.

Theorem 4. *For any single-good auction instances $\hat{\mathcal{I}} = (N, M, \mathcal{D})$ and $\mathcal{I} = (N, M, \mathcal{D}, G)$ where G is 2-connected, $\mathcal{M}_{IEM}$ is 2-DST and $Rev(\mathcal{M}_{IEM}(\mathcal{I})) \geq (1 - \frac{1}{n})OPT(\hat{\mathcal{I}})$.*

The mechanism disentangles the use of the players' values and the use of their knowledge, but in a more subtle and stingy way than randomized partition. When computing a player's virtual value in Step 10, his knowledge has not been used yet. Only when a player is removed from S—that is, when it is guaranteed that he will not get the item, will his knowledge be used.

It is important for G to be 2-connected, so that mechanism $\mathcal{M}_{IEM}$ does not stop until $N' = \emptyset$, and *all* players' distributions are reported. Indeed, by 2-connectedness, there is always an edge from $N \setminus (N' \cup \{i^*\})$ to N'. Therefore $\mathcal{M}_{IEM}$ recovers $\hat{\mathcal{I}}$ and runs Myerson's mechanism on it after randomly excluding a player a, and the revenue guarantee follows. Note that if the seller knows at least two distributions, then the mechanism can use him as the starting point and the revenue will be exactly OPT.

Mechanism 3. $\mathcal{M}_{IEM}$

1: Each player i reports a value b_i and a knowledge $K_i = (\mathcal{D}^i_j)_{j \in N \setminus \{i\}}$.
2: Randomly choose a player a, let $S = \{j \mid \mathcal{D}^a_j \neq \bot\}$, $N' = N \setminus (\{a\} \cup S)$, and $\mathcal{D}'_j = \mathcal{D}^a_j \ \forall j \in S$.
3: If $S = \emptyset$, the item is unsold, price p_i is 0 $\forall i$, and halt.
4: Set $i^* = \arg\max_{j \in S} \phi_j(b_j; \mathcal{D}'_j)$. (Ties are broken lexicographically.)
5: **while** $N' \neq \emptyset$ **do**
6: Set $S' = \{j \mid j \in N', \exists i' \in S \setminus \{i^*\} \text{ s.t. } \mathcal{D}^{i'}_j \neq \bot\}$.
7: If $S' = \emptyset$ then go to Step 12.
8: For each $j \in S'$, set $\mathcal{D}'_j = \mathcal{D}^{i'}_j$, where i' is the first player in $S \setminus \{i^*\}$ with $\mathcal{D}^{i'}_j \neq \bot$.
9: Set $S = \{i^*\} \cup S'$ and $N' = N' \setminus S'$.
10: Set $i^* = \arg\max_{j \in S} \phi_j(b_j; \mathcal{D}'_j)$.
11: **end while**
12: Set $\phi_{second} = \max_{j \in N \setminus (\{a, i^*\} \cup N')} \phi_j(b_j; \mathcal{D}'_j)$ and $p_i = 0$ for each player i.
13: If $\phi_{i^*}(b_{i^*}; \mathcal{D}'_{i^*}) < 0$ then the item is unsold; otherwise, the item is sold to player i^* and $p_{i^*} = \phi^{-1}_{i^*}(\max\{\phi_{second}, 0\}; \mathcal{D}'_{i^*})$.

References

1. Archer, A., Tardos, É.: Truthful mechanisms for one-parameter agents. In: 42nd Symposium on Foundations of Computer Science (FOCS 2001), pp. 482–491 (2001)
2. Azar, P., Chen, J., Micali, S.: Crowdsourced Bayesian auctions. In: 3rd Innovations in Theoretical Computer Science Conference (ITCS 2012), pp. 236–248 (2012)
3. Bergemann, D., Morris, S.: Robust Mechanism Design. World Scientific, Hackensack (2012)
4. Brier, G.W.: Verification of forecasts expressed in terms of probability. Mon. Weather Rev. **78**(1), 1–3 (1950)
5. Cai, Y., Daskalakis, C., Weinberg, M.: Optimal multi-dimensional mechanism design: reducing revenue to welfare maximization. In: 53rd Symposium on Foundations of Computer Science (FOCS 2012), pp. 130–139 (2012)
6. Cai, Y., Devanur, N., Weinberg, M.: A duality based unified approach to Bayesian mechanism design. In: 48th Symposium on Theory of Computing (STOC 2016), pp. 926–939 (2016)
7. Chawla, S., Hartline, J., Kleinberg, R.: Algorithmic pricing via virtual valuations. In: 8th Conference on Electronic Commerce (EC 2007), pp. 243–251 (2007)
8. Chawla, S., Hartline, J., Malec, D., Sivan, B.: Multi-parameter mechanism design and sequential posted pricing. In: 43th Symposium on Theory of Computing (STOC 2010), pp. 311–320 (2010)
9. Chen, J., Li, B., Li, Y.: Information elicitation for Bayesian auctions, full version. arXiv:1702.01416 (2017)
10. Chen, J., Micali, S.: Mechanism design with possibilistic beliefs. J. Econ. Theory **156**, 77–102 (2013)
11. Chen, J., Micali, S., Pass, R.: Tight revenue bounds with possibilistic beliefs and level-k rationality. Econometrica **83**(4), 1619–1639 (2015)
12. Cole, R., Roughgarden, T.: The sample complexity of revenue maximization. In: 46th Symposium on Theory of Computing (STOC 2014), pp. 243–252 (2014)
13. Cremer, J., McLean, R.: Full extraction of the surplus in Bayesian and dominant strategy auctions. Econometrica **56**(6), 1247–1257 (1988)
14. Devanur, N., Hartline, J.: Limited and online supply and the Bayesian foundations of prior-free mechanism design. In: 10th Conference on Electronic Commerce (EC 2009), pp. 41–50 (2009)
15. Devanur, N., Huang, Z., Psomas, C.A.: The sample complexity of auctions with side information. In: 48th Symposium on Theory of Computing (STOC 2016), pp. 426–439 (2016)
16. Hart, S., Nisan, N.: Approximate revenue maximization with multiple items. In: 13th Conference on Electronic Commerce (EC 2012), pp. 656–656 (2012)
17. Hartline, J., Karlin, A.: Profit maximization in mechanism design. In: Nisan, N., Roughgarden, T., Tardos, É., Vazirani, V. (eds.) Algorithmic Game Theory, pp. 331–361. Cambridge University Press, Cambridge (2007)
18. Hartline, J., Roughgarden, T.: Optimal mechanism design and money burning. In: 40th Symposium on Theory of Computing (STOC 2008), pp. 75–84 (2008)
19. James, W.: Some Problems of Philosophy. Harvard University Press, Cambridge (1979)
20. Kleinberg, R., Weinberg, M.: Matroid prophet inequalities. In: 44th Symposium on Theory of Computing (STOC 2012), pp. 123–136 (2012)
21. Li, X., Yao, A.C.C.: On revenue maximization for selling multiple independently distributed items. Proc. Natl. Acad. Sci. **110**(28), 11232–11237 (2013)

22. Miller, N., Resnick, P., Zeckhauser, R.: Eliciting informative feedback: the peer-prediction method. Manag. Sci. **51**(9), 1359–1373 (2005)
23. Myerson, R.: Optimal auction design. Math. Oper. Res. **6**(1), 58–73 (1981)
24. Radanovic, G., Faltings, B.: A robust Bayesian truth serum for non-binary signals. In: 27th AAAI Conference on Artificial Intelligence (AAAI 2013), pp. 833–839 (2013)
25. Wilson, R.: Game-theoretic analysis of trading processes. In: Bewley, T. (ed.) Advances in Economic Theory: Fifth World Congress, pp. 33–77 (1987)
26. Witkowski, J., Parkes, D.: Peer prediction without a common prior. In: 13th Conference on Electronic Commerce (EC 2012), pp. 964–981 (2012)
27. Yao, A.: An n-to-1 bidder reduction for multi-item auctions and its applications. In: 26th Symposium on Discrete Algorithms (SODA), pp. 92–109 (2015)
28. Zhang, P., Chen, Y.: Elicitability and knowledge-free elicitation with peer prediction. In: 13th International Conference on Autonomous Agents and Multigent Systems (AAMAS 2014), pp. 245–252 (2014)

Coreness of Cooperative Games with Truncated Submodular Profit Functions

Wei Chen[1], Xiaohan Shan[2,3](✉), Xiaoming Sun[2,3], and Jialin Zhang[2,3]

[1] Microsoft, Beijing, China
weic@microsoft.com
[2] CAS Key Lab of Network Data Science and Technology, Institute of Computing Technology, Chinese Academy of Sciences, Beijing, China
{shanxiaohan,sunxiaoming,zhangjialin}@ict.ac.cn
[3] University of Chinese Academy of Sciences, Beijing, China

Abstract. *Coreness* represents solution concepts related to core in cooperative games, which captures the stability of players. Motivated by the scale effect in social networks, economics and other scenario, we study the coreness of cooperative game with truncated submodular profit functions. Specifically, the profit function $f(\cdot)$ is defined by a truncation of a submodular function $\sigma(\cdot)$: $f(\cdot) = \sigma(\cdot)$ if $\sigma(\cdot) \geq \eta$ and $f(\cdot) = 0$ otherwise, where η is a given threshold. In this paper, we study the core and three core-related concepts of truncated submodular profit cooperative game. We first prove that whether core is empty can be decided in polynomial time and an allocation in core also can be found in polynomial time when core is not empty. When core is empty, we show hardness results and approximation algorithms for computing other core-related concepts including *relative* least-core value, *absolute* least-core value and least *average dissatisfaction* value.

1 Introduction

With the wide popularity of social media and social network sites such as Facebook, Twitter, WeChat, etc., social networks have become a powerful platform for spreading information among individuals. Thus, influential users always play important role in a social network. Motivated by this background, influence diffusion in social networks has been extensively studied [3,9,15]. Most of previous works focus on exploring influential nodes. To the best of our knowledge, there is no study about the "stability" of influential nodes (seed set) when they are treated as a coalition.

Consider the following scenario. A group of influential people in a social network are considering forming a coalition so that they can better serve many advertisers through viral marketing in the social network. To make the coalition

This work is supported in part by the National Natural Science Foundation of China Grant 61433014, 61502449, 61602440, the 973 Program of China Grants No. 2016YFB1000201.

X. Deng (Ed.): SAGT 2018, LNCS 11059, pp. 56–68, 2018.
https://doi.org/10.1007/978-3-319-99660-8_6

stable, we need to design a fair profit allocation scheme among the members of the coalition, such that no individual or a subset of people have incentive to deviate from this coalition, thinking that the allocation to them is unfair and they could earn more by the deviation and forming an alliance by themselves. A useful and mature framework of studying such incentives for stable coalition formation is the cooperative game theory, and in particular the coreness (core and its related concepts) of the cooperative games [7,17].

In the above social influence scenario, the typical way of measuring the contribution of any set S of influential people is by its influence spread function $\sigma(S)$, which measures the expected number of people in the social network that could be influenced by S under some stochastic diffusion model. Extensive researches have been done on stochastic diffusion models, and it has been shown that under a large class of models $\sigma(S)$ is both monotone and submodular[1] [3,15,18]. However, the advertisers would only be interested in the coalition as a viral marketing platform when the influence spread reaches certain scale level. In other words, the coalition can only receive profit after the influence spread is above a certain scale threshold η. Therefore, the true profit function for the coalition is $f(S) = \sigma(S)$ when $\sigma(S) \geq \eta$, and $f(S) = 0$ otherwise. We call such f truncated submodular functions. This motivate us to study the coreness of the cooperative games with truncated submodular profit functions.

Both submodularity and scale effect are common in economic behaviors beyond the above example of viral marketing in social networks. Therefore, considering truncated submodular functions as the profit functions is reasonable. In this paper, we study the computational issues related to the coreness of cooperative games with truncated submodular profit functions.

Solution Concepts in Cooperative Games. A cooperative game $\Gamma = (V, \gamma)$ consists of a player set $V = \{1, 2, \cdots, n\}$ and a profit function $\gamma : 2^V \to \mathbb{R}$ with $\gamma(\emptyset) = 0$. A subset of players $S \subseteq V$ is called a *coalition* and V is called the *grand coalition*. For each coalition S, $\gamma(S)$ represents the profit obtained by S without help of other players. An allocation over the players is denoted by a vector $x = (x_1, x_2, \cdots, x_n) \in \mathbb{R}^n$ whose components are one-to-one associated with players in V, where $x_i \in \mathbb{R}$ is the value received by player $i \in V$ under allocation x. For any player set $S \subseteq V$, we use the shorthand notation $x(S) = \sum_{i \in S} x_i$. A set of all allocations satisfying some specific requirements is called a *solution concept*.

The *core* [11,21] is one of the earliest and most attractive solution concepts that directly addresses the issue of stability. The core of a game is the set of allocations ensuring that no coalition would have an incentive to split from the grand coalition, and do better on its own. More precisely, the core of a game Γ (denoted by $\mathcal{C}(\Gamma)$), is the following set of allocations: $\mathcal{C}(\Gamma) = \{x \in \mathbb{R}^n : x(V) = \gamma(V), x(S) \geq \gamma(S),\ \forall\ S \subseteq V\}$. Intuitively, the requirement of $x(S) \geq \gamma(S)$ means that the coalition S receives profit allocation $x(S)$ that is at least their profit contribution $\gamma(S)$, so they would prefer to stay with the grand coalition.

[1] A set function f is monotone if $f(S) \leq f(T)$ for all $S \subseteq T$, and is submodular if $f(S \cup \{u\}) - f(S) \geq f(T \cup \{u\}) - f(T)$ for all $S \subseteq T$ and $u \notin T$.

In practice, core is very strict and may be even empty in some cases. When $\mathcal{C}(\Gamma)$ is empty, there must be some coalition becoming dissatisfied since they can obtain more benefits if they leave the grand coalition and work as a separated team. In this case, we use the dissatisfaction degree (or dissatisfaction value), defined as $dv(S,x) = \max\{\gamma(S) - x(S), 0\}$, to capture the instability of player set S with respect to the allocation x. Then, the overall stability of the game can be measured as either the worst-case or average-case dissatisfaction degree, for which we consider the following three versions.

The first one is the *relative least-core value* ($\mathcal{RLCV}$) [10], which reflects the relative stability, i.e. the minimum value of the maximum proportional difference between the profits and the payoffs among all coalitions.

Definition 1. *Given a cooperative game Γ, the* relative least-core value *of Γ ($\mathcal{RLCV}(\Gamma)$) is* $\min_x \max_S \frac{dv(S,x)}{\gamma(S)}$. *Technically, $\mathcal{RLCV}(\Gamma)$ is the optimal solution of the following linear programming:*

$$\begin{aligned} &\min \ r \\ &s.t. \begin{cases} x(V) = \gamma(V) & \\ x(S) \geq (1-r)\gamma(S) & \forall\ S \subseteq V \\ x(\{i\}) \geq 0 & \forall\ i \in V \end{cases} \end{aligned} \tag{1}$$

The second one is the *absolute least-core value* ($\mathcal{ALCV}$) [16], which reflects the absolute stability, i.e. the minimum value of the maximum difference between the profits and the payoffs among all coalitions. The formal definition is as following.

Definition 2. *Given a cooperative game Γ, the* absolute least-core value *of Γ ($\mathcal{ALCV}(\Gamma)$) is* $\min_x \max_S dv(S,x)$. *Technically, $\mathcal{ALCV}(\Gamma)$ is the optimal solution of the following linear programming:*

$$\begin{aligned} &\min \ \varepsilon \\ &s.t. \begin{cases} x(V) = \gamma(V) & \\ x(S) \geq \gamma(S) - \varepsilon & \forall\ S \subseteq V \\ x(\{i\}) \geq 0 & \forall\ i \in V \end{cases} \end{aligned} \tag{2}$$

The above two classical least-core values capture the stability from the perspective of the most dissatisfied coalition i.e. the worst-case stability. Sometimes the worst case is too extreme to reflect the real stability. Thus, we introduce the *least average dissatisfaction value* ($\mathcal{LADV}$), which reflects the minimum value of average dissatisfaction degree among all coalitions.

Definition 3. *Given a cooperative game Γ, the* least average dissatisfaction value *of Γ ($\mathcal{LADV}(\Gamma)$) is* $\min_x \mathbb{E}_S(dv(S,x))$. *Technically, $\mathcal{LADV}(\Gamma)$ is the optimal value of the following linear programming:*

$$\begin{aligned} &\min \ \tfrac{1}{2^n} \textstyle\sum_{S \subseteq V} \max\{\gamma(S) - x(S), 0\} \\ &s.t. \begin{cases} x(V) = \gamma(V) & \\ x(\{i\}) \geq 0 & \forall\ i \in V \end{cases} \end{aligned} \tag{3}$$

In this paper, we consider the following computational problems in the context of truncated submodular functions: (a) Whether the core of a given cooperative game is empty? (b) How to find an allocation in core if the core is not empty? (c) If the core is empty, how to compute the relative least-core value, the absolute least-core value and the least average dissatisfaction value of a cooperative game?

Contributions. We study the coreness of truncated submodular profit cooperative game Γ_f. We consider computational properties of the core, the relative least-core value, the absolute least-core value and the least average dissatisfaction value of Γ_f, which are denoted by $\mathcal{C}(\Gamma_f)$, $\mathcal{RLCV}(\Gamma_f)$, $\mathcal{ALCV}(\Gamma_f)$ and $\mathcal{LADV}(\Gamma_f)$, respectively.

We first prove that checking the non-emptiness of $\mathcal{C}(\Gamma_f)$ can be done in polynomial time. Moreover, we can find an allocation in the core if the core is not empty. Next, we consider the case when the core is empty. For the problem of computing the relative least-core value ($\mathcal{RLCV}(\Gamma_f)$), we show that it is in general NP-hard, but when truncation threshold $\eta = 0$, there is a polynomial time algorithm. Along the way, we also find an interesting partial result showing that there is no polynomial time separation oracle for the $\mathcal{RLCV}(\Gamma_f)$'s linear program unless P = NP, which is of independent interest since it reveals close connections with a new class of combinatorial problems. For the absolute least-core value problem $\mathcal{ALCV}(\Gamma_f)$, we prove that finding $\mathcal{ALCV}(\Gamma_f)$ is APX-hard even when $\sigma(\cdot)$ is defined as the influence spread under the classical independent cascade (IC) model in social network. We also prove that there exists a polynomial time algorithm which can guarantee an additive term approximation. Finally, for the least average dissatisfaction value problem $\mathcal{LADV}(\Gamma_f)$, we show that we can use the stochastic gradient descent algorithm to compute $\mathcal{LADV}(\Gamma_f)$ to an arbitrary small additive error.

Related Work. Cooperative game theory is a branch of (micro-)economics that studies the behavior of self-interested agents in strategic settings where binding agreements between agents are possible [2]. Numerous classical studies about cooperative game provide rich mathematical framework to solve issues related to cooperation in multi-agent systems [6,8]. Schulz and Uhan study the approximation of the absolute least core value of supermodular cost cooperative games [19], the results of which can be generalized to submodular profit cooperative games. An important application of our study is to analyze the stability of influential people in social networks. Almost all the existing studies focus on selecting the seed set [5,12,22]. To the best of our knowledge, there is no study considering the stability of the selected seed set. We utilize cooperative game theory to analyse the stability of seed set, and generalize it to a generic cooperative game with truncated submodular functions. The truncated operation represents the "threshold effect" which has been studied widely in literature [1,13].

2 Model and Problems

2.1 Cooperative Games with Truncated Submodular Profit Functions

A truncated submodular profit cooperative game is denoted by $\Gamma_f = (V, f(\cdot))$. In Γ_f, V is the player set and $f(\cdot)$ is the profit function which is defined as follows:

$$f(S) = \begin{cases} \sigma(S), & \text{if } \sigma(S) \geq \eta \\ 0, & \text{if } \sigma(S) < \eta \end{cases}$$

Note that $\sigma(\cdot)$ is a nonnegative monotone increasing submodular function with $\sigma(\emptyset) = 0$ and $0 \leq \eta \leq \sigma(V)$ is a nonnegative threshold. To make it explicit, henceforth, a truncated submodular profit cooperative game is denoted by a triple $(V, \sigma(\cdot), \eta)$. Note that the explicit representation of $\sigma(\cdot)$ might be exponential in the size of V. The standard way to bypass this difficulty is to assume that $\sigma(\cdot)$ is given as a value oracle.

2.2 Computational Problems on the Coreness

Given a truncated submodular profit cooperative game Γ_f, we focus on the following problems:
CORE: Is $\mathcal{C}(\Gamma_f) \neq \emptyset$ and how to find an allocation in $\mathcal{C}(\Gamma_f)$ when $\mathcal{C}(\Gamma_f) \neq \emptyset$?
ALCV: When $\mathcal{C}(\Gamma_f) = \emptyset$, how to compute $\mathcal{ALCV}(\Gamma_f)$?
RLCV: When $\mathcal{C}(\Gamma_f) = \emptyset$, how to compute $\mathcal{RLCV}(\Gamma_f)$?
LADV: When $\mathcal{C}(\Gamma_f) = \emptyset$, how to compute $\mathcal{LADV}(\Gamma_f)$?

Before we analyze the above problems, we introduce a specific instance of truncated submodular profit cooperative game (see Sect. 2.3).

2.3 Influence Cooperative Game (Γ_{inf})

As the description in our introduction, an important motivation of our model is influence in social networks. In this section, we introduce a specific instance of truncated submodular profit cooperative game, *influence cooperative game.*

Social Graph. A social graph is a directed graph $G = (V \cup U, E; P)$, where $V \cup U$ is the vertex set and E is the edge set. $P = \{p_e\}_{e \in E}$ and p_e is the influence probability on each edge $e \in E$. Note that, V and U denote the vertex set of influential people and target people in G, respectively.

Influence Diffusion Model. The information diffusion process follows the independent cascade (IC) model proposed by [15]. In the IC model, discrete time steps $t = 0, 1, 2, \cdots$ are used to model the diffusion process. Each node in G has two states: inactive or active. At step 0, nodes in seed set S are active and other nodes are inactive. For any step $t \geq 1$, if a node u is newly active at step $t - 1$, u has a single chance to influence each of its inactive out-neighbor v with independent probability p_{uv} to make v active. Once a node becomes active, it will never return to the inactive state. The diffusion process stops when there is no new active nodes at a time step. For any $S \subseteq V$, we use $\sigma^{\mathrm{IC}}(S)$ to denote the influence spread of S, the expected number of activated nodes in U from seed set $S \subseteq V$, at the end of an IC diffusion. According to [15], $\sigma^{\mathrm{IC}}(\cdot)$ is a monotone submodular function.

Definition 4. *An* influence cooperative game $\Gamma_{\mathrm{inf}} = (V, \sigma^{\mathrm{IC}}(\cdot), \eta)$ *is a special form of the truncated cooperative game, with V as the player set, and the truncation of influence spread function $\sigma^{\mathrm{IC}}(\cdot)$ as the profit function.*

In the rest of this paper, we analyze problems defined in Sect. 2.2 one by one. Note that our positive results (properties and algorithms) could apply to all truncated submodular profit cooperative games including influence cooperative game. Our hardness results are established for the influence cooperative games, so it is stronger than the hardness results for general truncated submodular cooperative games.

3 Computing Core

We start by considering the core of Γ_f ($\mathcal{C}(\Gamma_f)$). In Γ_f, we say a player $i \in V$ is a *veto player* if $\sigma(S) < \eta$ for any $S \subseteq V \setminus \{i\}$. That is to say, a successful coalition must include all veto players.

Lemma 1. $\mathcal{C}(\Gamma_f) \neq \emptyset$ *if and only if:*

(i) There exists at least one veto player in Γ_f, or
(ii) $\sigma(S) = \sum_{i \in S} \sigma(\{i\})$, for any $S \subseteq V$.

Proof. Suppose the player set of Γ_f is $V = \{1, 2, \cdots, n\}$. We first prove the sufficiency of Lemma 1. On one hand, suppose i is a veto player of Γ_f, then we can find a trivial allocation x in $\mathcal{C}(\Gamma_f)$: $x(\{i\}) = \sigma(V)$ and $x(\{j\}) = 0$, $\forall\, j \in V \setminus \{i\}$. On the other hand, $x(\{i\}) = \sigma(\{i\})$ ($\forall i \in V$) is an allocation in $\mathcal{C}(\Gamma_f)$ if $\sigma(S) = \sum_{i \in S} \sigma(\{i\})$.

Now we prove the necessity. Suppose $\mathcal{C}(\Gamma_f) \neq \emptyset$ and $x \in \mathcal{C}(\Gamma_f)$. Let $\sigma(V) = \sum_{i=1}^{n} M_i$, where $M_i = \sigma(\{1, 2, \cdots, i\}) - \sigma(\{1, 2, \cdots, i-1\})$ is the marginal increasing of player i. If there is no veto player, then for any $i \in V$, $\sigma(V \setminus \{i\}) \geq \eta$ since $\sigma(S)$ is monotone. Thus, $f(V \setminus \{i\}) = \sigma(V \setminus \{i\})$, $\forall\, i \in V$. Suppose $\sigma(V \setminus \{i\}) = \sum_{j=1}^{i-1} M_j + \sum_{j=i+1}^{n} M'_{ij}$, where $M'_{ij} = \sigma(\{1, 2, \cdots, i-1, i+1, \cdots, j\}) - \sigma(\{1, 2, \cdots, i-1, i+1, \cdots, j-1\})$. Note that $M'_{ij} \geq M_j$ since $\sigma(S)$ is submodular. By the definition of the core, for any $i \in \{1, 2, \cdots, n\}$, we have: $x(V \setminus \{i\}) \geq f(V \setminus \{i\}) = \sigma(\{V \setminus \{i\}\})$. That is, $x(V) - x(\{i\}) \geq \sum_{j=1}^{i-1} M_j + \sum_{j=i+1}^{n} M'_{ij}$, $\forall\, i \in V$.

Summing up these inequalities for all $i \in V$, we have, $(n-1)\sum_{i=1}^{n} x(\{i\}) \geq \sum_{i=1}^{n}(\sum_{j=1}^{i-1} M_j + \sum_{j=i+1}^{n} M'_{ij}) \geq \sum_{i=1}^{n}(\sum_{j=1}^{i-1} M_j + \sum_{j=i+1}^{n} M_j) = \sum_{i=1}^{n}(\sigma(V) - M_i) = (n-1)\sigma(V)$.

We have known that $\sum_{i=1}^{n} x(\{i\}) = \sum_{j=1}^{n} M_j = \sigma(V)$ and then $M_j = M'_{ij}$, $\forall i, j \in V$. Thus, $\sigma(S) = \sum_{i \in S} \sigma(\{i\})$.

An important application of Lemma 1 is Theorem 1.

Theorem 1. *Deciding whether $\mathcal{C}(\Gamma_f)$ is empty can be done in polynomial time and an allocation in $\mathcal{C}(\Gamma_f)$ can be computed in polynomial time if $\mathcal{C}(\Gamma_f)$ is not empty.*

Proof (Sketch). First, it takes polynomial time to check the non-emptiness of $\mathcal{C}(\Gamma_f)$. When $\mathcal{C}(\Gamma_f)$ is not empty, then $(x_j = \sigma(V), \mathbf{0}_{\{i:i\neq j\}}) \in \mathcal{C}(\Gamma_f)$ when j is a veto player and $(\sigma(\{1\}), \cdots, \sigma(\{n\})) \in \mathcal{C}(\Gamma_f)$ when (ii) is satisfied.

The detail proof of Theorem 1 is shown in our full version [4].

4 Computing Relative Least-Core Value

From Lemma 1, $\mathcal{C}(\Gamma_f)$ may be empty in many cases. It is obvious that $\mathcal{RLCV}(\Gamma_f) > 0$ if $\mathcal{C}(\Gamma_f) = \emptyset$ and $\mathcal{RLCV}(\Gamma_f) = 0$ otherwise. In this section, we study computational properties of the RLCV problem. The linear programming corresponding to $\mathcal{RLCV}(\Gamma_f)$ (LP-RLCV) is as follows:

$$\min r \\ \text{s.t.} \begin{cases} x(V) = \sigma(V) \\ x(S) \geq (1-r)\sigma(S) \; \forall \; S \subseteq V, \; \sigma(S) \geq \eta \\ x(\{i\}) \geq 0 \qquad\qquad \forall \; i \in V \end{cases} \tag{4}$$

A special case of computing $\mathcal{RLCV}(\Gamma_f)$ is when $\eta = 0$. It captures the scenario that the profit of any coalition exactly equals to its influence spread under influence cooperative game. In Theorem 2 we show that, although there are exponential number of constraints, LP-RLCV can be solved in polynomial time by providing a polynomial time separation oracle when $\eta = 0$. A separation oracle for a linear program is an algorithm that, given a putative feasible solution, checks whether it is indeed feasible, and if not, outputs a violated constraint. It is known that a linear program can be solved in polynomial time by the ellipsoid method as long as it has a polynomial time separation oracle [14].

Theorem 2. *There exists a polynomial time separation oracle of LP-RLCV when $\eta = 0$. Therefore, RLCV can be solved in polynomial time when $\eta = 0$.*

Proof. Given any solution candidate of LP-RLCV (x', r'), we need to either assert (x', r') is a feasible solution or find a constraint in LP-RLCV such that (x', r') violates it. Note that, checking $x'(V) = \sigma(V)$ and $x'(\{i\}) \geq 0$ ($\forall\; i \in V$) can be done in polynomial time. Thus, we only need to check whether $g(S) \triangleq 1 - x'(S)/\sigma(S) \leq r'$, $\forall S \subseteq V$.

An important property is that $g(S)$ achieves its maximum value when S contains only one single player. This is because $g(S) = 1 - \frac{x'(S)}{\sigma(S)} \leq 1 - \frac{\sum_{i\in S} x'_i}{\sum_{i\in S}\sigma(\{i\})} \leq 1 - \min_{i:i\in S}\{\frac{x'_i}{\sigma(\{i\})}\} = \max_{i:i\in S}\{g(\{i\})\}$. The first inequality is due to the submodularity of $\sigma(S)$ and the second inequality is due to $\min_{i:i\in[n]}\{\frac{a_i}{b_i}\} \leq \frac{\sum_{i=1}^n a_i}{\sum_{i=1}^n b_i}$,

$\forall a_i, b_i \in \mathbb{R}$. Thus, the exponential number of constraints can be simplified to n constraints on all single players. Then, we can find a polynomial time separation oracle of LP-RLCV directly.

When $\eta = 0$, RLCV can be solved in polynomial time is mainly because the most dissatisfaction coalition is a single player. However, when $\eta \neq 0$, it becomes intractable to find the most dissatisfaction coalition.

Theorem 3. *There is no polynomial time separation oracle of LP-RLCV for some $\eta > 0$, unless P = NP.*

Theorem 3 can not imply the NP-hardness of RLCV. However, the proof of Theorem 3 reveals an interesting connection between RLCV problem and a series of well defined combinatorial problems. The proof of Theorem 3 and the generalized combinatorial problem is shown in our full version [4].

In the left of this section, we prove the NP-hardness of RLCV, a stronger hardness result than which in Theorem 3.

Theorem 4. *It is NP-hard to compute $\mathcal{RLCV}(\Gamma_f)$, even under influence cooperative game.*

Proof (Sketch). We construct a reduction from the SAT problem. A boolean formula is in conjunctive normal form (CNF) if it is expressed as an AND of clauses, each of which is the OR of one or more literals. The SAT problem is defined as follows: given a CNF formula F, determine whether F has a satisfiable assignment. Let F be a CNF formula with m clauses $C_1, C_2, \cdots, C_m$, over n literals $z_1, z_2, \cdots, z_n$. Without loss of generality, we set $m > 4n$.

We construct a social graph G as follows: $G = (V_1 \cup V_2 \cup V_3, E)$ is a tripartite graph (see the sketch graph in Fig. 1). In the first layer (V_1), there are two nodes S_i and T_i corresponding to each $i \in \{1, 2, \cdots, n\}$, $n + 1$ dummy nodes labeled as $u_1, u_2, \cdots, u_{n+1}$ and n dummy nodes labeled as $v_1, v_2, \cdots, v_n$. In the second layer (V_2), there are two nodes x_i and $\overline{x}_i$ corresponding to each $i \in \{1, 2, \cdots, n\}$, one node c_j for each $j \in \{1, 2, \cdots, m\}$ and a dummy node w. The third layer (V_3) contains only node Q. Edges exist only between the

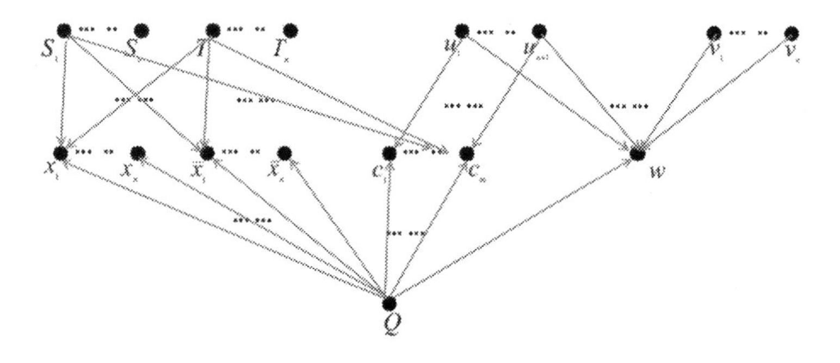

Fig. 1. The reduction from SAT to $\mathcal{RLCV}(\Gamma_f)$

adjacent layers. For each $i \in \{1, 2, \cdots, n\}$, S_i sends an edge to every node in $\{x_i, \overline{x}_i\} \cup \{c_j$: clause C_j contains literal $z_i, j \in \{1, 2, \cdots, m\}\}$. Similarly, for each $i \in \{1, 2, \cdots, n\}$, T_i sends an edge to every node in $\{x_i, \overline{x}_i\} \cup \{c_j$: clause C_j contains literal $\overline{z}_i, j \in \{1, 2, \cdots, m\}\}$. The probabilities on edges sent from S_i and T_i are 1. There is an edge with influence probability 1 from u_i to c_i for any $i \in \{1, 2, \cdots, n\}$ and $m - n$ edges form u_{n+1} to $c_{n+1}, c_{n+2}, \cdots, c_m$. There is an edge from u_i to w with influence probability $1 - \sqrt[n+1]{1/2}$ for any $i \in \{1, 2, \cdots, n+1\}$. There is also exists an edge from v_i to w with influence probability $1 - \sqrt[n]{1/2}$ for any $i \in \{1, 2, \cdots, n\}$. The left edges are from Q to all nodes in the second layer. The influence probability on edge (Q, w) is $1/2$ and all other probabilities on edges sent from Q is 1. The influence cooperative game defined on G is $\Gamma(G) = (V = V_1 \cup V_3, \sigma^{\mathrm{IC}}(\cdot), \eta = 2n + m + 1/2)$. For convenient, we set $N = 2n + m$.

Suppose r^* is the optimal solution of the relative least-core value of $\Gamma(G)$. We can prove that $r^* \geq 1 - \frac{1}{3}(N + \frac{7}{8})/(N + \frac{1}{2})$ if F is satisfiable and $r < 1 - \frac{1}{3}(N + \frac{7}{8})/(N + \frac{1}{2})$ if F is un-satisfiable. The proof of this part is shown in the full version [4].

5 Computing Absolute Least-Core Value

5.1 Hardness of ALCV

Theorem 5. *ALCV problem of influence cooperative game cannot be approximated within 1.139 under the unique games conjecture.*

Proof (Sketch). We construct a reduction from MAX-CUT problem. Under our construction, for any instance of MAX-CUT problem, we can construct an instance of ALCV problem such that the optimal solution of these two instances are equal. The detail proof is shown in our full version [4].

5.2 Approximating $\mathcal{ALCV}(\Gamma_f)$

In this section, we approximate $\mathcal{ALCV}(\Gamma_f)$ by approximating the following linear programming (LP-PRIME):

$$\min \ \varepsilon$$
$$\text{s.t.} \begin{cases} x(V) = \sigma(V) & \\ x(\{S\}) \geq \sigma(\{S\}) - \varepsilon & \forall S \subseteq V, \sigma(S) \geq \eta \\ x(\{u\}) \geq 0 & \forall u \in V \end{cases}$$

The intractability of LP-PRIME lies on the exponential number of constraints and the hardness of identifying all successful coalitions. We use a relaxed version LP-RE and a strengthen version LP-STR of LP-PRIME to design an approximation algorithm of $\mathcal{ALCV}(\Gamma_f)$. (5) and (6) are formal definitions of LP-RE and LP-STR, respectively.

$$\begin{aligned} &\min\ \varepsilon \\ &\text{s.t.} \begin{cases} x(V) = \sigma(V) \\ x(S) \geq \eta - \varepsilon & \forall\, S \subseteq V, \sigma(S) \geq \eta \\ x(\{u\}) \geq 0 & \forall\, u \in V \end{cases} \end{aligned} \tag{5}$$

$$\begin{aligned} &\min\ \varepsilon \\ &\text{s.t.} \begin{cases} x(V) = \sigma(V) \\ x(S) \geq \sigma(S) - \varepsilon & \forall\, S \subseteq V \\ x(\{u\}) \geq 0 & \forall\, u \in V \end{cases} \end{aligned} \tag{6}$$

Intuitively, LP-RE and LP-STR denote absolute least-core values of two cooperative games with new profit functions. Specifically, LP-RE relaxes the constraints in LP-PRIME by reducing the profits of all successful coalitions excepting V to η. Formally, the profit function in LP-RE is $g(S)$: $g(V) = \sigma(V)$, $\forall\, S \subset V$, $g(S) = \eta$ if $\sigma(S) \geq \eta$ and $g(S) = 0$ otherwise. The profit function in LP-STR is $h(S) = \sigma(S)$, $\forall S \subseteq V$. Clearly, LP-STR strengthens LP-PRIME by increasing the profits of all unsuccessful coalitions.

Our main result in this section is shown in Theorem 6.

Theorem 6. $\forall\, \delta > 0$, *there exists an approximate algorithm $\mathcal{A}$ of the $\mathcal{ALCV}(\Gamma_f)$ problem with running time in poly$(n, 1/\delta, \log\sigma(V))$, $\mathcal{A}$ outputs ε'_p such that $\varepsilon^*_p \leq \varepsilon'_p \leq \min\{\varepsilon^*_p + \sigma(V) - \eta + 2\delta,\ \max\{3\varepsilon^*_p, \eta\}\}$.*

We prove Theorem 6 by show Lemmas 2, 3 and 4 in order.

Lemma 2. *Suppose the optimal value of LP-PRIME, LP-RE and LP-STR are ε^*_p, ε^*_r and ε^*_s, respectively. Then, we have*

$$\varepsilon^*_p \leq \varepsilon^*_r + (\sigma(V) - \eta) \leq \varepsilon^*_p + (\sigma(V) - \eta), \tag{7}$$

$$\varepsilon^*_p \leq \varepsilon^*_s \leq \max\{\varepsilon^*_p, \eta\}. \tag{8}$$

Lemma 3. *There exists a polynomial time approximate algorithm of LP-STR outputting ε'_s such that $\varepsilon^*_s \leq \varepsilon'_s \leq 3\varepsilon^*_s$.*

Lemma 4. $\forall\, \delta > 0$, *there exists an algorithm of LP-RE outputting ε'_r such that $\varepsilon^*_r \leq \varepsilon'_r \leq \varepsilon^*_r + 2\delta$, with runs time in poly$(n, 1/\delta, \log\sigma(V))$.*

The proofs of Lemmas 2–4 rely heavily on mathematical computation and we report them in our full version [4].

6 Computing Least Average Dissatisfaction Value

Based on Definition 3, $\mathcal{LADV}(\Gamma_f)$ equals the optimal value of the following linear program:

$$\begin{aligned} &\min F(x) = \tfrac{1}{2^n} \textstyle\sum_{S \subseteq V} \max\{f(S) - x(S), 0\} \\ &\text{s.t.} \begin{cases} x(V) = \sigma(V) \\ x(\{i\}) \geq 0 & \forall\, i \in V \end{cases} \end{aligned} \tag{9}$$

where $f(S) = \sigma(S)$ if $\sigma(S) \geq \eta$ and $f(S) = 0$ otherwise. There are exponential terms in $F(x)$, however, we can utilize stochastic gradient algorithm to approximate the optimal solution of (9). This is because the object function $F(x)$ is a convex function (Lemma 5) and the feasible solution area in (9) is a convex set.

Lemma 5. *$F(x)$ is a convex function.*

The proof of Lemma 5 is shown in our full version [4]. The stochastic gradient descent algorithm (SGD, cf. [20]) can be used to compute $\mathcal{LADV}(\Gamma_f)$ (see Algorithm 1).

Algorithm 1. Stochastic gradient descent for LADV

1: **Parameters:** Scaler $\alpha > 0$, integer $T > 0$
2: **Initialize:** $\mathbf{X}^1 = \mathbf{0}$, $t = 0$.
3: Set $D = \{\mathbf{X} : \mathbf{X}_i \geq 0 (\forall\, i \in V),\ \sum_{i \in V} \mathbf{X}_i = \sigma(V)\}$.
4: **for** $t = 1$ to T **do**
5: /*choose a random $\mathbf{Y}^t$ such that $\mathbb{E}[\mathbf{Y}^t | \mathbf{X}^t]$ is a subgradient of F.*/
6: Uniformly at random choose a set $S \in 2^V$.
7: **if** $f(S) \geq \mathbf{X}^t(S)$ **then**
8: Set $\mathbf{Y}^t = (-\mathbf{1}_S, \mathbf{0}_{V \setminus S})$.
9: **else**
10: Set $\mathbf{Y}^t = \mathbf{0}$.
11: **end if**
12: update $\mathbf{X}^{t+\frac{1}{2}} = \mathbf{X}^t - \alpha \mathbf{Y}^t$.
13: /*Project $\mathbf{X}^{t+\frac{1}{2}}$ to D*/
14: $\mathbf{X}^{t+1} = \arg\min_{\mathbf{X} \in D} \|\mathbf{X} - \mathbf{X}^{t+\frac{1}{2}}\|^2$.
15: **end for**
16: **return** $\hat{F} = \min\{F(\mathbf{X}^t)\}_{t \in \{1,2,\cdots,T\}}$.

Let F^* be the optimal solution of $\mathcal{LADV}(\Gamma_f)$, $\hat{F}$ be the output of Algorithm 1 and the profit of grand coalition $\sigma(V) = V$. Then, the performance of Algorithm 1 can be formalized in the following theorem.

Theorem 7. *$\forall\ \varepsilon > 0$, $\mathbb{E}[\hat{F}] - F^* \leq \varepsilon$ if $T \geq \frac{\sigma(V)^4 n^4}{\varepsilon^2}$ and $\alpha = \sqrt{\frac{\sigma(V)^4}{Tn^4}}$ in Algorithm 1.*

Following the standard analysis of SGD (e.g. in Chap. 14 of [20]), Theorem 7 holds since it is easy to check that $\mathbb{E}[\mathbf{Y}^t | \mathbf{X}^t]$ is a subgradient of $F(\mathbf{X})$ at node $\mathbf{X}^t$, for any $t \in [T]$ (lines 6–11 in Algorithm 1).

7 Conclusion and Future Work

In this paper, we study the core related solution concepts of truncated submodular profit cooperative game. One possible future work is to change the way of truncating a function. For example, we can set $f(S) = \sigma(S)$ if $|S| \geq k$ and

$f(S) = 0$ otherwise. This setting is a special case of the setting in our paper and thus it may allow efficient algorithms. In this paper, we prove that computing the relative least-core value is NP-hard. We also prove that the relative least-core value can be solved in polynomial time in a special case. A directly future work is to design an approximate algorithm of RLCV under general case.

References

1. Bemmaor, A.C.: Testing alternative econometric models on the existence of advertising threshold effect. J. Mark. Res. **21**(3), 298–308 (1984)
2. Chalkiadakis, G., Elkind, E., Wooldridge, M.: Computational aspects of cooperative game theory. Synth. Lect. Artif. Intell. Mach. Learn. **5**(6), 1–168 (2011)
3. Chen, W., Lakshmanan, L.V.S., Castillo, C.: Information and Influence Propagation in Social Networks. Morgan & Claypool Publishers, San Francisco (2013)
4. Chen, W., Shan, X., Sun, X., Zhang, J.: Coreness of cooperative games with truncated submodular profit functions. arXiv:1806.10833 (2018)
5. Chen, W., Wang, Y., Yang, S.: Efficient influence maximization in social networks. In: KDD, pp. 199–208. ACM (2009)
6. Conitzer, V., Sandholm, T.: Complexity of constructing solutions in the core based on synergies among coalitions. Artif. Intell. **170**(6–7), 607–619 (2006)
7. Demange, G.: On group stability in hierarchies and networks. J. Polit. Econ. **112**, 754–778 (2004)
8. Deng, X., Papadimitriou, C.H.: On the complexity of cooperative solution concepts. Math. Oper. Res. **19**(2), 257–266 (1994)
9. Domingos, P., Richardson, M.: Mining the network value of customers. In: KDD, pp. 57–66. ACM (2001)
10. Faigle, U., Kern, W.: On some approximately balanced combinatorial cooperative games. Math. Methods Oper. Res. **38**, 141–152 (1993)
11. Gillies, D.: Some theorems on n-Person games. Ph.D. thesis, Princeton University (1953)
12. Goyal, A., Bonchi, F., Lakshmanan, L.V., Venkatasubramanian, S.: On minimizing budget and time in influence propagation over social networks. Soc. Netw. Anal. Min. **3**(2), 1–14 (2012)
13. Granovetter, M.: Threshold models of collective behavior. Am. J. Sociol. **83**, 489–515 (1978)
14. Grötschel, M., Lovász, L., Schrijver, A.: Geometric Algorithms and Combinatorial Optimization, vol. 2. Springer, Heidelberg (2012)
15. Kempe, D., Kleinberg, J., Tardos, É.: Maximizing the spread of influence through a social network. In: KDD, pp. 137–146. ACM (2003)
16. Maschler, M., Peleg, B., Shapley, L.S.: Geometric properties of the Kernel, nucleolus, and related solution concepts. Math. Oper. Res. **4**(4), 303–338 (1979)
17. Meir, R., Rosenschein, J.S., Malizia, E.: Subsidies, stability, and restricted cooperation in coalitional games. In: IJCAI, pp. 301–306 (2011)
18. Mossel, E., Roch, S.: Submodularity of influence in social networks: from local to global. SIAM J. Comput. **39**(6), 2176–2188 (2010)
19. Schulz, A.S., Uhan, N.A.: Approximating the least core value and least core of cooperative games with supermodular costs. Discrete Optim. **10**(2), 163–180 (2013)
20. Shalev-Shwartz, S., Ben-David, S.: Understanding Machine Learning: From Theory to Algorithms. Cambridge University Press, New York (2014)

21. Shapley, L.S.: Markets as cooperative games. In: IJCAIR and Corporation Memorandum (1955)
22. Tang, Y., Shi, Y., Xiao, X.: Influence maximization in near-linear time: a martingale approach. In: SIGMOD, pp. 1539–1554. ACM (2015)

Simple Games Versus Weighted Voting Games

Frits Hof[1], Walter Kern[1], Sascha Kurz[2], and Daniël Paulusma[3](✉)

[1] University of Twente, Enschede, The Netherlands
f.hof@home.nl, w.kern@math.utwente.nl
[2] University of Bayreuth, Bayreuth, Germany
sascha.kurz@uni-bayreuth.de
[3] Durham University, Durham, UK
daniel.paulusma@durham.ac.uk

Abstract. A simple game (N, v) is given by a set N of n players and a partition of 2^N into a set $\mathcal{L}$ of losing coalitions L with value $v(L) = 0$ that is closed under taking subsets and a set $\mathcal{W}$ of winning coalitions W with $v(W) = 1$. Simple games with $\alpha = \min_{p \geq 0} \max_{W \in \mathcal{W}, L \in \mathcal{L}} \frac{p(L)}{p(W)} < 1$ are exactly the weighted voting games. Freixas and Kurz (IJGT, 2014) conjectured that $\alpha \leq \frac{1}{4}n$ for every simple game (N, v). We confirm this conjecture for two complementary cases, namely when all minimal winning coalitions have size 3 and when no minimal winning coalition has size 3. As a general bound we prove that $\alpha \leq \frac{2}{7}n$ for every simple game (N, v). For complete simple games, Freixas and Kurz conjectured that $\alpha = O(\sqrt{n})$. We prove this conjecture up to a $\ln n$ factor. We also prove that for graphic simple games, that is, simple games in which every minimal winning coalition has size 2, computing α is NP-hard, but polynomial-time solvable if the underlying graph is bipartite. Moreover, we show that for every graphic simple game, deciding if $\alpha < a$ is polynomial-time solvable for every fixed $a > 0$.

1 Introduction

Cooperative Game Theory provides a mathematical framework for capturing situations where subsets of agents may form a coalition in order to obtain some collective profit or share some collective cost. Formally, a *cooperative game (with transferable utilities)* consists of a pair (N, v), where N is a set of n agents called *players* and $v : 2^N \to \mathbb{R}_+$ is a *value function* that satisfies $v(\emptyset) = 0$. In our context, the value $v(S)$ of a *coalition* $S \subseteq N$ represents the profit for S if all players in S choose to collaborate with (only) each other. The central problem in cooperative game theory is to allocate the total profit $v(N)$ of the *grand coalition* N to the individual players $i \in N$ in a "fair" way. To this end various *solution concepts* such as the core, Shapley value or nucleolus have been designed; see [24] for an overview. For example, core solutions try to allocate the total profit such that every coalition $S \subseteq N$ gets at least $v(S)$. This is of course not always possible, *e.g.*, the core might be empty. This leads to related

X. Deng (Ed.): SAGT 2018, LNCS 11059, pp. 69–81, 2018.
https://doi.org/10.1007/978-3-319-99660-8_7

questions like: "How much do we need to spend in total if we want to give at least $v(S)$ to each coalition $S \subseteq N$?" In the specific case of simple games (*cf.* below) where v takes only values 0 and 1, classifying coalitions into "losing" and "winning" coalitions *resp.*, one may also ask: "How much do we have to give in the worst case to a losing coalition if we want to give at least $v(S) = 1$ to each winning coalition?"

As mentioned above, we study simple games. Simple games form a classical class of games, which are well studied; see also the book of Taylor and Zwicker [29]. The notion of being simple means that every coalition either has some equal amount of power or no power at all. Formally, a cooperative game (N, v) is *simple* if v is a monotone 0–1 function with $v(\emptyset) = 0$ and $v(N) = 1$, so $v(S) \in \{0, 1\}$ for all $S \subseteq N$ and $v(S) \leq v(T)$ whenever $S \subseteq T$. In other words, if v is simple, then there is a set $\mathcal{W} \subseteq 2^N$ of *winning coalitions* W that have value $v(W) = 1$ and a set $\mathcal{L} \subseteq 2^N$ of *losing coalitions* L that have value $v(L) = 0$. Note that $N \in \mathcal{W}$, $\emptyset \in \mathcal{L}$ and $\mathcal{W} \cup \mathcal{L} = 2^N$. The monotonicity of v implies that subsets of losing coalitions are losing and supersets of winning coalitions are winning. A winning coalition W is *minimal* if every proper subset of W is losing, and a losing coalition L is *maximal* if every proper superset of L is winning.

A simple game is a *weighted voting game* if there exists a payoff vector $p \in \mathbb{R}^n_+$ such that a coalition S is winning if $p(S) \geq 1$ and losing if $p(S) < 1$. Weighted voting games are also known as *weighted majority games* and form one of the most popular classes of simple games.

However, it is easy to construct simple games that are not weighted voting games. We give an example below, but in fact there are many important simple games that are not weighted voting games, and the relationship between weighted voting games and simple games is not yet fully understood. Therefore, Gvozdeva, Hemaspaandra, and Slinko [16] introduced a parameter α, called the *critical threshold value*, to measure the "distance" of a simple game to the class of weighted voting games:

$$\alpha \;=\; \alpha(N, v) \;= \min_{p \geq 0} \; \max_{W, L} \; \frac{p(L)}{p(W)}, \tag{1}$$

where the maximum is taken over all winning coalitions in $\mathcal{W}$ and all losing coalitions in $\mathcal{L}$. A simple game (N, v) is a weighted voting game if and only if $\alpha < 1$. This follows from observing that each optimal solution p of (1) can be scaled to satisfy $p(W) \geq 1$ for all winning coalitions W.

A concrete example of a simple game (N, v) that is not a weighted voting game and that has in fact a large value of α was given in [12]:

Example. Let $N = \{1, \ldots, n\}$ for some even integer $n \geq 4$, and let the minimal winning coalitions be the pairs $\{1, 2\}, \{2, 3\}, \ldots \{n-1, n\}, \{n, 1\}$. Consider any payoff $p \geq 0$ satisfying $p(W) \geq 1$ for every winning coalition W. Then $p_i + p_{i+1} \geq 1$ for $i = 1, \ldots, n$ (where $n + 1 = 1$). This means that $p(N) \geq \frac{1}{2}n$. Then, for at least one of $L = \{2, 4, 6, \ldots, n\}$ and $L = \{1, 3, 5, \ldots, n-1\}$, we have $p(L) \geq \frac{1}{4}n$, showing that $\alpha \geq \frac{1}{4}n$. On the other hand, it is easily seen that $p \equiv \frac{1}{2}$ satisfies

$p(W) \geq 1$ for all winning coalitions and $p(L) \leq \frac{1}{4}n$ for all losing coalitions, showing that $\alpha \leq \frac{1}{4}n$. Thus $\alpha = \frac{1}{4}n$.

This example led the authors of [12] to the following conjecture:
Conjecture 1 [12]. For every simple game (N, v), it holds that $\alpha \leq \frac{1}{4}n$.

Our Results. In Sect. 2 we prove that Conjecture 1 holds for the case where all minimal winning coalitions have size 3 and for its complementary case where no minimal winning collection has size 3. We were not able to prove Conjecture 1 for all simple games. However, in Sect. 3 we show that $\alpha \leq \frac{2}{7}n \approx 0.2858n$.

In Sect. 4 we consider a subclass of simple games based on a natural desirability order [25]. A simple game (N, v) is *complete* if the players can be ordered by a complete, transitive ordering $\succeq$, say, $1 \succeq 2 \succeq \cdots \succeq n$, indicating that higher ranked players have more "power" than lower ranked players. More precisely, $i \succeq j$ means that $v(S \cup i) \geq v(S \cup j)$ for any coalition $S \subseteq N \backslash \{i, j\}$. The class of complete simple games properly contains all weighted voting games [14]. For complete simple games, we show an asymptotically lower bound on α, namely $\alpha = O(\sqrt{n} \ln n)$. This bound matches, up to a $\ln n$ factor, the lower bound of $\Omega(\sqrt{n})$ in [12] (conjectured to be tight in [12]). Intuitively, complete simple games are much closer to weighted voting games than arbitrary simple games. So, from this perspective, our result seems to support the hypothesis that α is indeed a sensible measure for the distance to weighted voting games.

In Sect. 5 we discuss some algorithmic and complexity issues. We focus on instances where all minimal winning coalitions have size 2. We say that such simple games are *graphic*, as they can conveniently be described by a graph $G = (N, E)$ with vertex set N and edge set $E = \{ij \mid \{i, j\} \text{ is winning}\}$. For graphic simple games we show that computing α is NP-hard in general, but polynomial-time solvable if the underlying graph $G = (N, E)$ is bipartite, or if α is known to be small (less than a fixed number a).

Related Work. Due to their practical applications in voting systems, computer operating systems and model resource allocation (see e.g. [3, 7]), structural and computational complexity aspects for solution concepts for weighted voting games have been thoroughly investigated [9, 10, 13, 16].

Another way to measure the distance of a simple game to the class of weighted voting games is to use the *dimension* of a simple game [28], which is the smallest number of weighted voting games whose intersection equals a given simple game. However, computing the dimension of a simple game is NP-hard [8], and the largest dimension of a simple game with n players is $2^{n-o(n)}$ [21]. Moreover, α may be arbitrarily large for simple games with dimension larger than 1. Hence there is no direct relation between the two distance measures. Gvozdeva, Hemaspaandra, and Slinko [16] introduced two other distance parameters as well. One measures the power balance between small and large coalitions. The other one allows multiple thresholds instead of threshold 1 only.

For graphic simple games, it is natural to take the number of players n as the input size for answering complexity questions, but in general simple games may have different representations. For instance, one can list all minimal winning coalitions or all maximal losing coalitions. Under these two representations the problem of deciding if $\alpha < 1$, that is, if a given simple game is a weighted voting game, is also polynomial-time solvable. This follows from results of [17,23], as shown in [13]. The latter paper also showed that the same result holds if the representation is given by listing all winning coalitions or all losing coalitions.

As mentioned, a crucial case in our study is when the simple game is graphic, that is, defined on some graph $G = (N, E)$. In the corresponding *matching game* a coalition $S \subseteq N$ has value $v(S)$ equal to the maximum size of a matching in the subgraph of G induced by S. One of the most prominent solution concepts is the *core* of a game, defined by $core(N, v) := \{p \in \mathbb{R}^n \mid p(N) = v(N),\ p(S) \geq v(S)\ \forall S \subseteq N\}$. Matching games are not simple games. Yet their core constraints are readily seen to simplify to $p \geq 0$ and $p_i + p_j \geq 1$ for all $ij \in E$. Classical solution concepts, such as the core and core-related ones like least core, nucleolus or nucleon are well studied for matching games, see, for example, [4,5,11,19,20,27]. However, for graphic simple games we aim to bound $p(L)$ over all losing coalitions, subject to $p \geq 0$, $p_i + p_j \geq 1$ for all $ij \in E$, whereas for matching games with an empty core we wish to bound $p(N)$, subject to $p \geq 0$, $p_i + p_j \geq 1$ for all $ij \in E$. Nevertheless, basic tools from matching theory like the Gallai-Edmonds decomposition play a role in both cases.

2 Two Complementary Cases

We will treat the following two "complementary" cases: when all winning coalitions have size equal to 3, and when no winning coalition has size equal to 3. First observe that winning coalitions of size 1 do not cause any problems. If $\{i\}$ is a winning coalition of size 1, we satisfy it by setting $p_i = 1$. Since no losing coalition L contains i, we may remove i from the game and solve (1) with respect to the resulting subgame. A similar argument applies if some $i \in N$ is not contained in any minimal winning coalition. We then simply define $p_i = 0$ and remove i from the game. Thus, we may assume without loss of generality that all minimal winning coalitions have size at least 2 and that they cover all of N.

We first investigate the case where all minimal winning coalitions have size exactly 2. This case (which is a crucial case in our study) can conveniently be translated to a graph-theoretic problem. Let $G = (N, E)$ be the graph with vertex set N whose edges are exactly the minimal winning coalitions of size 2 in our game (N, v). Our assumption that N is completely covered by minimal winning coalitions means that G has no isolated vertices. Losing coalitions correspond to independent sets of vertices $L \subseteq N$. Then the min max problem (1) becomes

$$\alpha := \alpha_G := \min_p \max_L p(L), \tag{2}$$

where the minimum is taken over all *feasible* pay-off vectors p, that is, $p \in \mathbb{R}^n_+$ with $p_i + p_j \geq 1$ for every $ij \in E$, and the maximum is taken over all independent sets $L \subseteq N$.

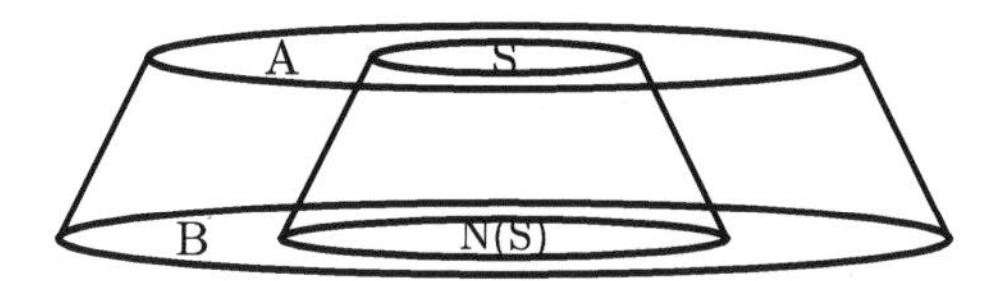

Fig. 1. A well-spread bipartite graph.

We first consider the case where $G = (A \cup B, E)$ is bipartite. To explain the basic idea, we introduce the following concept (illustrated in Fig. 1).

Definition. Let $G = (A \cup B, E)$ be a bipartite graph of order $n = |A| + |B|$ without isolated notes and assume without loss of generality that $|A| \leq |B|$. Let $\lambda \leq \frac{1}{2}$ such that $|A| = \lambda n$ (and $|B| = (1 - \lambda)n$). We say that G is *well-spread* with parameter λ if for all $S \subseteq A$ we have

$$\frac{|S|}{|N(S)|} \leq \frac{|A|}{|B|} = \frac{\lambda}{1 - \lambda}.$$

(Here, as usual, $N(S) \subseteq B$ denotes the set of neighbors of S in B.)

Examples of well-spread bipartite graphs are biregular graphs or biregular graphs minus an edge. Note that if G is well-spread with parameter $\lambda \leq \frac{1}{2}$, then Hall's condition $|N(S)| \geq |S|$ for all $S \subseteq A$ is satisfied, implying that A can be completely matched to B (see, for example, [22]). The following lemma is the key observation.

Lemma 1. *Let $G = (A \cup B, E)$ be well-spread with parameter $\lambda \leq \frac{1}{2}$. Then $p \equiv \lambda$ on B and $p \equiv 1 - \lambda$ on A yields $\alpha_G \leq \frac{1}{4}n$.*

Proof. Assume $L \subseteq N$ is an independent set. Let $\rho \leq 1$ such that $|L \cap A| = \rho\lambda n$. Since G is well-spread, we get $|N(L \cap A)| \geq \rho(1 - \lambda)n$, so that $|L \cap B| \leq (1 - \rho)(1 - \lambda)n$. Thus $p(L) = |L \cap A|(1 - \lambda) + |L \cap B|\lambda \leq \rho\lambda n(1 - \lambda) + (1 - \rho)(1 - \lambda)n\lambda \leq \rho\frac{1}{4}n + (1 - \rho)\frac{1}{4}n \leq \frac{1}{4}n$. □

In general, when $G = (A \cup B, E)$ is not well-spread, we seek to decompose G into well-spread induced subgraphs $G_i = (A_i \cup B_i, E_i)$ with $A = \bigcup A_i$ and $B = \bigcup B_i$. Of course, this can only work if $G = (A \cup B, E)$ is such that A can be matched to B in G.

Proposition 1. *Let $G = (A \cup B, E)$ be a bipartite graph without isolated vertices and assume that A can be matched into B. Then G decomposes into well-spread induced subgraphs $G_i = (A_i \cup B_i, E_i)$, with $A = \bigcup A_i$ and $B = \bigcup B_i$ in such a way that for all i, j with $i < j$, $\lambda_i \geq \lambda_j$ and no edges join A_i to B_j.*

Proof. Let $S \subseteq A$ maximize $|S|/|N(S)$. Set $A_1 := S$ and $B_1 := N(S)$. Let G' be the subgraph of G induced by $A \backslash A_1$ and $B' := B \backslash B_1$. Then G' satisfies the assumption of the Proposition. Indeed, if A' cannot be matched into B' in G', then there must be some $S' \subseteq A'$ with $|S'| > |N'(S')|$, where $N'(S') = N(S') \backslash B_1$ is the neighborhood of S' in G'. But then $|S \cup S'| = |S| + |S'|$ and $|N(S \cup S')| \leq |N(S)| + |N'(S)|$ shows that S cannot maximize $|S|/|N(S)|$, a contradiction. Thus, by induction, we may assume that G' decomposes in the desired way into well-spread subgraphs $G_2, \ldots, G_k$ with parameters $\lambda_2 \geq \cdots \geq \lambda_k$. The claim then follows by observing that (i) no edges join B_1 to A'; and (ii) $\lambda_1 \geq \lambda_2$ (otherwise $S \cup A_2$ would contradict the choice of S maximizing $|S|/|N(S)|$). □

We combine the last two results.

Corollary 1. *For every bipartite graph $G = (A \cup B, E)$ of order n satisfying the assumption of Proposition 1, there exists a payoff vector $p \geq 0$ such that $p_i + p_j \geq 1$ for $ij \in E$ and $p(L) \leq \frac{1}{4}n$ for any independent set $L \subseteq A \cup B$. In addition, p can be chosen so as to satisfy $p \geq \frac{1}{2}$ on A.*

Proof. The result follows immediately from Lemma 1 and Proposition 1. Note that if p is chosen as $p \equiv 1 - \lambda_i$ on A_i, then it holds that $p \geq \frac{1}{2}$ indeed. □

As we will see, the assumption of Proposition 1 is not really restrictive for our purposes. A (connected) component C of a graph G is *even* (*odd*) if C has an even (odd) number of vertices. A graph $G = (N, E)$ is *factor-critical* if for every vertex $v \in V(G)$, the graph $G - v$ has a perfect matching. We recall the well-known Gallai–Edmonds Theorem (see [22]) for characterizing the structure of maximum matchings in G; see also Fig. 2. There exists a (unique) subset $A \subseteq N$, called a *Tutte set*, such that (i) every even component of $G - A$ has a perfect matching; (ii) every odd component of $G - A$ is factor-critical; and (iii) every maximum matching in G is the union of a perfect matching in each even component, a nearly perfect matching in each odd component and a matching that matches A (completely) to the odd components.

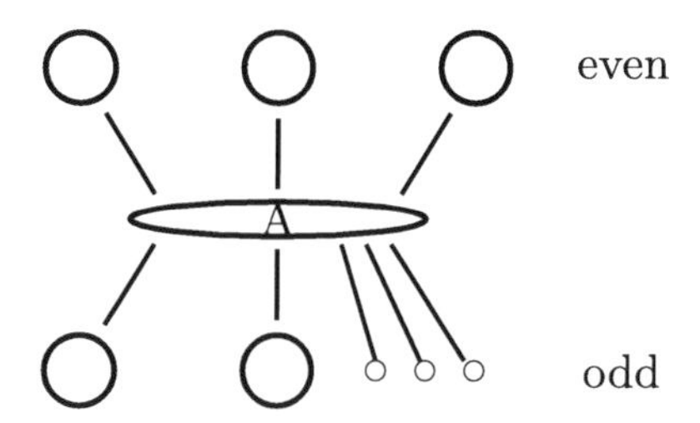

Fig. 2. Tutte set A splitting G into even and odd components (possibly single nodes).

We are now ready to derive our first main result.[1]

[1] For n is odd, the upper bound in Theorem 1 can be slightly strengthened to $\frac{n^2-1}{4n}$ [18].

Theorem 1. *Let $G = (N, E)$ be a graph of order n. Then $\alpha_G \leq \frac{1}{4}n$.*

Proof. Let $A \subseteq N$ be a Tutte set. Contract each odd component in $G - A$ to a single vertex and let B denote the resulting set of vertices. The subgraph $\bar{G}$ induced by $A \cup B$ then satisfies the assumption of Corollary 1. Let $\bar{p} \in \mathbb{R}^{|A|+|B|}$ be the corresponding payoff vector. We define $p \in \mathbb{R}^n$ by setting $p_i = \bar{p}_i$ for every vertex $i \in A$ and every vertex i that corresponds to an odd component of size 1 in $G - A$. All other vertices get $p_j = \frac{1}{2}$.

It is straightforward to check that $p \geq 0$ and $p_i + p_j \geq 1$. Indeed, it holds that $\bar{p} \geq \frac{1}{2}$ everywhere except on B, so the only critical edges ij have $i \in A$ and j a singleton odd component. But in this case $p_i + p_j = \bar{p}_i + \bar{p}_j \geq 1$. Thus we are left to prove that for every independent set $L \subseteq N$, $p(L) \leq \frac{1}{4}n$. Let B_0 denote the set of singleton odd components $i \in B$, $L_0 := (L \cap A) \cup (L \cap B_0)$ and $n_0 := |A| + |B|$. Clearly, L_0 is an independent set in the bipartite graph $\bar{G}$, and $p \equiv \bar{p}$ on L_0. We thus conclude that $p(L_0) \leq \frac{1}{4}n_0$.

Next let us analyze $L \cap C$ where $C \subseteq N \backslash A$ is an even component. As C is perfectly matchable, L contains at most $|C|/2$ vertices of C. So $p(L \cap C) \leq \frac{1}{4}|C|$. A similar argument applies to the odd components. Let C be an odd component in $G - A$ of size at least 3. Then certainly L cannot contain all vertices of C, so there exists some $i \in C \backslash L$. Since C is factor-critical, $C \backslash i$ is perfectly matchable, implying that L can contain at most half of $C \backslash i$. Thus $|L \cap C| \leq (|C| - 1)/2$ and $p(L \cap C) \leq (|C| - 1)/4$.

Summarizing, $n - n_0 = |N| - (|A| + |B|)$ is the sum over all $|C|$, where C is an even component plus the sum over all $|C| - 1$ where C is an odd component, and $p(L \backslash L_0)$ is at most a $\frac{1}{4}$ fraction of this, finishing the proof. □

We note that both decompositions that we use to define the payoff p can be computed efficiently. For the Edmonds–Gallai decomposition, this is a well-known fact (see, for example, [22]). For the decomposition into well-spread subgraphs, this follows from the observation that deciding whether $\max_S \frac{|S|}{|N(S)|} \leq r$ is equivalent to $\min_S r|N(S)| - |S| \geq 0$, which amounts to minimizing the submodular function $f(S) = r|N(S)| - |S|$; see, for example, [26] for a strongly polynomial-time algorithm.

We now deal with the more general case where there are, in addition, minimal winning coalitions of size 4 or larger. First recall how the payoff p that we proposed in Corollary 1 works. For a bipartite graph $G = (A \cup B, E)$ that is split into well-spread subgraphs $G_i = (A_i \cup B_i, E_i)$ with parameter λ_i, we let $p \equiv \lambda_i$ on B_i. So for $\lambda_i < \frac{1}{4}$, p may be infeasible, that is, we may encounter winning coalitions W of size 4 or larger with $p(W) < 1$. This problem can easily be remedied by raising p a bit on each B_i and decreasing it accordingly on A_i. Indeed, the standard $(\lambda, 1 - \lambda)$ allocation rule proposed in Lemma 1 is based on the simple fact that $\lambda(1 - \lambda) \leq \frac{1}{4}$, which gives us some flexibility for modification in the case where λ is small. More precisely, defining the payoff to be $p :\equiv \frac{1}{4(1-\lambda)} > \frac{1}{4}$ on B and $1 - p < \frac{3}{4}$ on A for a bipartite graph $(G = (A \cup B, E)$, well-spread with parameter λ, would work as well and thus solve the problem.

Indeed, the unique independent set L that maximizes $p(L)$ is $L = B$ in this case, which gives $p(L) = p(B) = |B|/(4(1-\lambda)) = \frac{1}{4}n$.

There is one thing that needs to be taken care of. Namely, in Proposition 1 we assumed that $G = (A \cup B, E)$ has no isolated vertices, an assumption that can be made without loss of generality if we only have 2-element winning coalitions. Now we may have isolated vertices that are part of winning coalitions of size 4 or larger. But this does not cause any problems either. We simply assign $p := \frac{1}{4}$ to these isolated vertices to ensure that indeed all winning coalitions W have $p(W) \geq 1$. Formally, this can also be seen as an extension of our decomposition: if $G = (A \cup B, E)$ contains isolated vertices, then they are all contained in B (once we assume that A can be completely matched into B). So the set of isolated vertices can be seen as a "degenerate" well-spread final subgraph $(A_k \cup B_k, E_k)$ with $A_k = \emptyset$ and parameter $\lambda_k = 0$. Our proposed payoff $p \equiv \frac{1}{4(1-\lambda_k)}$ would then indeed assign $p = \frac{1}{4}$ to all isolated vertices.

It remains to observe that when we pass to general graphs, no further problems arise. Indeed, all that happens is that vertices in even and odd components get payoffs $p = \frac{1}{2}$ which certainly does no harm to the feasibility of p. Thus we have proved the following result.

Corollary 2. *Let (N, v) be a simple game with no minimal winning coalition of size 3. Then $\alpha(N, v) \leq \frac{1}{4}n$.*

We end this section with the complementary case where all minimal winning coalitions have size 3.

Proposition 2. *Let (N, v) be a simple game with all minimal winning coalitions of size 3. Then $\alpha(N, v) \leq \frac{1}{4}n$.*

Proof. We try $p :\equiv \frac{1}{3}$, which is certainly feasible. If this yields $\max p(L) \leq \frac{1}{4}n$, then we are done. Otherwise, there exists a losing coalition $L \subseteq N$ with $p(L) = \frac{1}{3}|L| > \frac{1}{4}n$, or equivalently, $|L| > \frac{3}{4}n$. In this case we use an alternative payoff $\tilde{p}$ given by $\tilde{p} \equiv 1$ on $N \backslash L$ and $\tilde{p} \equiv 0$ on L. Since $|N \setminus L| < \frac{1}{4}n$, this ensures $\tilde{p}(\tilde{L}) < \frac{1}{4}n$ for any losing coalition $\tilde{L}$. On the other hand, $\tilde{p}$ is feasible, since a winning coalition W cannot be completely contained in L, that is, there exists a player $i \in W$ with $\tilde{p}_i = 1$ and hence $\tilde{p}(W) \geq 1$. □

We note that Proposition 2 is a pure existence result. To compute $\tilde{p}$ it requires to solve a maximum independent set problem in 3-uniform hypergraphs, which is NP-hard. This can be seen from a reduction from the maximum independent set problem in graphs, which is well known to be NP-hard (see [15]). Given a graph $G = (V, E)$, construct a 3-uniform hypergraph $\bar{G}$ as follows. Add $n = |V|$ new vertices labeled $1, \ldots, n$ and extend each edge $e = ij \in E$ to n edges $\{i, j, 1\}, \ldots, \{i, j, n\}$ in $\bar{G}$. It is readily seen that a maximum independent set of vertices in $\bar{G}$ (that is, a set of vertices that does not contain any hyperedge) consists of the n new vertices plus a maximum independent set in G.

3 Minimal Winning Coalitions of Arbitrary Size

In this section we try to combine the ideas for the two complementary cases to derive an upper bound $\alpha \leq \frac{2}{7}$ for the general case. The payoffs p that we consider will all satisfy $p \geq \frac{1}{4}$ so that only winning coalitions of size 2 and 3 are of interest. The basic idea is to start with a bipartite graph $(A \cup B, E)$ representing the size 2 winning coalitions and a payoff satisfying all these. Standard payoffs that we use satisfy $p \geq \frac{1}{4}$ on B and $p \geq \frac{1}{2}$ on A. Hence we have to worry only about 3-element winning coalitions contained in B. We seek to satisfy these by raising the payoff of some vertices in B without spending too much in total.

More precisely, consider a bipartite graph $G = (A \cup B, E)$ representing the winning coalitions of size 2. As before, we assume that A can be completely matched into B, so that our decomposition into well-spread subgraphs $G_i = (A_i \cup B_i, E_i)$ applies (with possibly the last subgraph $G_k = (A_k \cup B_k, E_k)$ having $A_k = \emptyset$ and B_k consisting of isolated points, as explained at the end of the previous section). Recall the payoff $\bar{\lambda}_i :\equiv \frac{1}{4(1-\lambda_i)}$ on B_i and $1 - \bar{\lambda}_i$ on A_i defined for the proof of Corollary 2. We first consider the following payoff $\bar{p} :\equiv 1 - \bar{\lambda}_i$ on A_i and $\bar{p} :\equiv \bar{\lambda}_i$ on B_i for $\lambda_i \geq \frac{1}{4}$, so $\bar{\lambda}_i \geq \frac{1}{3}$. For subgraphs with $\lambda_i < \frac{1}{4}$ (including possibly a final $\lambda_k = 0$) we define $\bar{p} \equiv \frac{2}{3}$ on A_i and $\bar{p} \equiv \frac{1}{3}$ on B_i. Thus it holds that $\bar{p} \geq \frac{1}{3}$ everywhere, in particular, $\bar{p}$ is feasible with respect to all winning coalitions of size at least 3.

Let $\bar{L}$ be a losing coalition with maximum $\bar{p}(L)$. We define an alternative payoff $\tilde{p}$ as follows: For $\lambda_i \geq \frac{1}{4}$ we set $\tilde{p} :\equiv 1 - \bar{\lambda}_i$ on A_i, $\tilde{p} :\equiv \bar{\lambda}_i$ on $B \cap \bar{L}$ and $\tilde{p} :\equiv \frac{1}{2}$ on $B_i \backslash \bar{L}$. For $\lambda_i < \frac{1}{4}$ we set $\tilde{p} :\equiv \frac{3}{4}$ on A_i, $\tilde{p} :\equiv \frac{1}{4}$ on $B_i \cap \bar{L}$ and $\tilde{p} :\equiv \frac{1}{2}$ on $B_i \backslash \bar{L}$. Clearly, both $\bar{p}$ and $\tilde{p}$ are feasible. We claim that a suitable combination of these two yields the desired upper bound (proof omitted) yielding Theorem 2.

Lemma 2. *For* $p := \frac{3}{7}\bar{p} + \frac{4}{7}\tilde{p}$ *we get* $\alpha = \max_L \ p(L) \leq \frac{2}{7}n$.

Theorem 2. *For every simple game* (N, v), $\alpha(N, v) \leq \frac{2}{7}n$.

4 Complete Simple Games

Intuitively, the class of complete simple games is "closer" to weighted voting games than general simple games. The next result quantifies this expectation.

Theorem 3. *A complete simple game* (N, v) *has* $\alpha \leq \sqrt{n} \ln n$.

Proof. Let $N = \{1, \ldots, n\}$ be the set of players and assume without loss of generality that $1 \succeq 2 \succeq \cdots \succeq n$. Let $k \in N$ be the largest number such that $\{k, \ldots, n\}$ is winning. For $i = 1, \ldots, k$, let s_i denote the smallest size of a winning coalition in $\{i, \ldots, n\}$. Define $p_i := 1/s_i$ for $i = 1, \ldots, k$ and $p_i := p_k$ for $i = k+1, \ldots, n$. Thus, obviously, $p_1 \geq \cdots \geq p_k = \cdots = p_n$.

Consider a winning coalition $W \subseteq N$ and let i be the first player in W (with respect to $\succeq$). If $|W| \leq \sqrt{n}$, then $s_i \leq |W| \leq \sqrt{n}$ and hence $p(W) \geq p_i = \frac{1}{s_i} \geq \frac{1}{\sqrt{n}}$. On the other hand, if $|W| > \sqrt{n}$, then $p(W) > \sqrt{n} p_k \geq \sqrt{n}\frac{1}{n} = \frac{1}{\sqrt{n}}$.

For a losing coalition $L \subseteq N$, we conclude that $|L \cap \{1, \dots, i\}| \leq s_i - 1$ (otherwise L would dominate the winning coalition of size s_i in $\{i, \dots, n\}$). So $p(L)$ is bounded by $\max \sum_{i=1}^{k} x_i \frac{1}{s_i}$ subject to $\sum_{j=1}^{i} x_j \leq s_i - 1$, $i = 1, \dots, k$. The optimal solution of this maximization problem is $x_1 = s_1 - 1, x_i = s_i - s_{i-1}$ for $2 \leq i \leq k$. Hence $p(L) \leq (s_1 - 1)\frac{1}{s_1} + (s_2 - s_1)\frac{1}{s_2} + \cdots + (s_k - s_{k-1})\frac{1}{s_k} \leq \frac{1}{2} + \cdots + \frac{1}{s_k} \leq \ln n$. Summarizing, we obtain $p(L)/p(W) \leq \sqrt{n} \ln n$. □

In [12] it is conjectured that $\alpha = O(\sqrt{n})$ holds for complete simple games. In the same paper a lower bound of order $\sqrt{n}$ is given, as well as specific subclasses of complete simple games for which $\alpha = O(\sqrt{n})$ can be proven.

5 Algorithmic Aspects

A fundamental question concerns the complexity of our original problem (1). For general simple games this depends on how the game in question is given, and we refer to Sect. 1 for a discussion. Here we concentrate on the graphic" case.

Proposition 3. *Computing α_G for bipartite graphs G can be done in polynomial time.*

Proof. Let $P \subseteq \mathbb{R}^n$ Be the set of feasible payoffs (satisfying $p \geq 0$ and $p_i + p_j \geq 1$ for $ij \in E$). For $\alpha \in \mathbb{R}$, let $P_\alpha := \{p \in P \mid p(L) \leq \alpha \text{ for all independent } L \subseteq N\}$. Thus $\alpha_G = \min\{\alpha \mid P_\alpha \neq \emptyset\}$. The separation problem for P_α (for any given α) is efficiently solvable. Given $p \in \mathbb{R}^n$, we can check feasibility and whether $\max\{p(L) \mid L \subseteq N \text{ independent}\} \leq \alpha$ by solving a corresponding maximum weight independent set problem in the bipartite graph G. Thus we can, for any given $\alpha \in \mathbb{R}$, apply the ellipsoid method to either compute some $p \in P_\alpha$ or conclude that $P_\alpha = \emptyset$. Binary search then exhibits the minimum value for which P_α is non-empty; binary search works indeed in polynomial time as the optimal α has size polynomially bounded in n, which follows from observing that

$$\alpha = \min\{a \mid p_i + p_j \geq 1 \ \forall ij \in E,\ p(L) - a \leq 0 \ \forall L \subseteq N \text{ independent}, p \geq 0\} \quad (3)$$

can be computed by solving a linear system of n constraints defining an optimal basic solution of the above linear program. □

The proof of Proposition 3 also applies to other classes of graphs, such as claw-free graphs (see [6]) in which finding a weighted maximum independent set is polynomial-time solvable. In general, the problem is NP-hard.

Proposition 4. *Computing α_G for arbitrary graphs G is* NP*-hard.*

Proof. Let $G' = (N', E')$ and $G'' = (N'', E'')$ be two disjoint copies of a graph $G = (N, E)$ with independence number k. For each $i' \in N'$ and $j'' \in N''$ add an edge $i'j''$ if and only if $i = j$ or $ij \in E$ and call the resulting graph $G^* = (N^*, E^*)$. We claim that $\alpha_{G^*} = k/2$ (thus computing α_{G^*} is as difficult as computing k).

First note that the independent sets in G^* are exactly the sets $L^* \subseteq N^*$ that arise from an independent set $L \subseteq N$ in G by splitting L into two complementary sets L_1 and L_2 and defining $L^* := L_1' \cup L_2''$. Hence, $p \equiv \frac{1}{2}$ on N^* yields $\max p(L^*) = k/2$ where the maximum is taken over all independent sets $L^* \subseteq N^*$ in G^*. This shows that $\alpha_{G^*} \leq k/2$.

Conversely, let p^* be any feasible payoff in G^*, that is, $p^* \geq 0$ and $p_i^* + p_j^* \geq 1$ for all $ij \in E^*$. Let $L \subseteq N$ be a maximum independent set of size k in G and construct L^* by including for each $i \in L$ either i' or i'' in L^*, whichever has p-value at least $\frac{1}{2}$. Then, by construction, L^* is an independent set in G^* with $p^*(L^*) \geq k/2$, showing that $\alpha_{G^*} \geq k/2$. □

Summarizing, for graphic simple games, computing α_G is as least as hard as computing the size of a maximum independent in G. For our last result we assume that a is a fixed integer, that is, a is not part of the input.

Proposition 5. *For every fixed $a > 0$, it is possible to decide if $\alpha_G \leq a$ in polynomial time for an arbitrary graph $G = (N, E)$.*

Proof. Let $k = 2\lceil a + \epsilon \rceil$ for some $\epsilon > 0$. By brute-force, we can check in $O(n^{2k})$ time if N contains $2k$ vertices $\{u_1, \ldots, u_k\} \cup \{v_1, \ldots, v_k\}$ that induce k disjoint copies of P_2, that is, paths $P_i = u_i v_i$ of length 2 for $i = 1, \ldots, k$ with no edges joining any two of these paths. If so, then the condition $p(u_i) + p(v_i) \geq 1$ implies that one of u_i, v_i, say u_i, must receive a payoff $p(u_i) \geq \frac{1}{2}$, and hence $U = \{u_1, \ldots, u_k\}$ has $p(U) \geq k/2 > a$. As U is an independent set, $\alpha(G) > a$.

Now assume that G does not contain k disjoint copies of P_2 as an induced subgraph, that is, G is kP_2-free. For every $s \geq 1$, the number of maximal independent sets in a sP_2-free graphs is $n^{O(s)}$ due to a result of Balas and Yu [2]. Tsukiyama, Ide, Ariyoshi, and Shirakawa [30] show how to enumerate all maximal independent sets of a graph G on n vertices and m edges using time $O(nm)$ per independent set. Hence we can find all maximal independent sets of G and thus solve, in polynomial time, the linear program (3). Then it remains to check if the solution found satisfies $\alpha \leq a$. □

6 Conclusions

After our paper appeared, Kanstantsin Pashkovich [1] found a proof of Conjecture 1. Hence it remains to tighten the upper bound for complete simple games to $O(\sqrt{n})$. In order to classify simple games, many more subclasses of simple games have been identified in the literature. Besides the two open problems, no optimal bounds for α are known for other subclasses of simple games, such as *strong*, *proper*, or *constant-sum* games, that is, where $v(S) + v(N \backslash S) \geq 1$, $v(S) + v(N \backslash S) \leq 1$, or $v(S) + v(N \backslash S) = 1$ for all $S \subseteq N$, respectively.

Acknowledgments. The second and fourth author thank Péter Biró and Hajo Broersma for fruitful discussions on the topic of the paper.

References

1. Pashkovich, K.: On critical threshold value for simple games. arXiv:1806.03170v2, 11 June 2018
2. Balas, E., Yu, C.S.: On graphs with polynomially solvable maximum-weight clique problem. Networks **19**(2), 247–253 (1989)
3. Bilbao, J.M., García, J.R.F., Jiménez, N., López, J.J.: Voting power in the European Union enlargement. Eur. J. Oper. Res. **143**(1), 181–196 (2002)
4. Biro, P., Kern, W., Paulusma, D.: Computing solutions for matching games. Int. J. Game Theory **41**, 75–90 (2012)
5. Bock, A., Chandrasekaran, K., Könemann, J., Peis, B., Sanitá, L.: Finding small stabilizers for unstable graphs. Math. Program. **154**, 173–196 (2015)
6. Brandstaett, A., Mosca, R.: Maximum weight independent set in $l$claw-free graphs in polynomial time. Discrete Appl. Math. **237**, 57–64 (2018)
7. Chalkiadakis, G., Elkind, E., Wooldridge, M.: Computational Aspects of Cooperative Game Theory. Morgan and Claypool Publishers (2011)
8. Deineko, V.G., Woeginger, G.J.: On the dimension of simple monotonic games. Eur. J. Oper. Res. **170**(1), 315–318 (2006)
9. Elkind, E., Chalkiadakis, G., Jennings, N.R.: Coalition structures in weighted voting games, vol. 178, pp. 393–397 (2008)
10. Elkind, E., Goldberg, L.A., Goldberg, P.W., Wooldridge, M.: On the computational complexity of weighted voting games. Ann. Math. Artif. Intell. **56**(2), 109–131 (2009)
11. Faigle, U., Kern, W., Fekete, S., Hochstaettler, W.: The nucleon of cooperative games and an algorithm for matching games. Math. Program. **83**, 195–211 (1998)
12. Freixas, J., Kurz, S.: On α-roughly weighted games. Int. J. Game Theory **43**(3), 659–692 (2014)
13. Freixas, J., Molinero, X., Olsen, M., Serna, M.: On the complexity of problems on simple games. RAIRO Oper. Res. **45**(4), 295–314 (2011)
14. Freixas, J., Puente, M.A.: Dimension of complete simple games with minimum. Eur. J. Oper. Res. **188**(2), 555–568 (2008)
15. Garey, M.R., Johnson, D.S.: Computers and Intractability: A Guide to the Theory of NP-Completeness. W. H. Freeman & Co., New York (1979)
16. Gvozdeva, T., Hemaspaandra, L.A., Slinko, A.: Three hierarchies of simple games parameterized by "resource" parameters. Int. J. Game Theory **42**(1), 1–17 (2013)
17. Hegedüs, T., Megiddo, N.: On the geometric separability of Boolean functions. Discrete Appl. Math. **66**(3), 205–218 (1996)
18. Hof, F.: Weight distribution in matching games. MSc thesis, University of Twente (2016)
19. Kern, W., Paulusma, D.: Matching games: the least core and the nucleolus. Math. Oper. Res. **28**, 294–308 (2003)
20. Koenemann, J., Pashkovich, K., Toth, J.: Computing the nucleolus of weighted cooperative matching games in polynomial time arXiv:1803.03249v2, 9 March 2018
21. Kurz, S., Molinero, X., Olsen, M.: On the construction of high dimensional simple games. In: Proceedings ECAI 2016, New York, pp. 880–885 (2016)
22. Lovász, L., Plummer, M.D.: Matching Theory, vol. 367. American Mathematical Society (2009)
23. Peled, U.N., Simeone, B.: Polynomial-time algorithms for regular set-covering and threshold synthesis. Discrete Appl. Math. **12**(1), 57–69 (1985)

24. Peters, H.: Game Theory: A Multi-Leveled Approach. Springer, Heidelberg (2008). https://doi.org/10.1007/978-3-540-69291-1
25. Isbell, J.R.: A class of majority games. Q. J. Math. **7**, 183–187 (1956)
26. Schrijver, A.: A combinatorial algorithm minimizing submodular functions in strongly polynomial time. J. Comb. Theory, Ser. B **80**(2), 346–355 (2000)
27. Solymosi, T., Raghavan, T.E.: An algorithm for finding the nucleolus of assignment games. Int. J. Game Theory **23**, 119–143 (1994)
28. Taylor, A.D., Zwicker, W.S.: Weighted voting, multicameral representation, and power. Games Econ. Behav. **5**, 170–181 (1993)
29. Taylor, A.D., Zwicker, W.S.: Simple Games: Desirability Relations, Trading, Pseudoweightings. Princeton University Press (1999)
30. Tsukiyama, S., Ide, M., Ariyoshi, H., Shirakawa, I.: A new algorithm for generating all the maximal independent sets. SIAM J. Comput. **6**(3), 505–517 (1977)

Hide and Seek Game with Multiple Resources

Marcin Dziubiński[1(✉)] and Jaideep Roy[2]

[1] Institute of Informatics, University of Warsaw, Banacha 2, 02-097 Warsaw, Poland
m.dziubinski@mimuw.edu.pl

[2] Department of Economics, School of Social Science, University of Bath, Claverton Down, Bath BA2 7AY, UK
j.roy@bath.ac.uk

Abstract. We study a generalization of hide and seek game of von Neumann [14], where each player has one or more resources. We characterize the value and Nash equilibria of such games in terms of their unidimensional marginal distributions. We propose a $\mathcal{O}(n \log(n))$ time algorithm for computing unidimensional marginal distributions of equilibrium strategies and a quadratic time algorithm for computing mixed strategies given the margins. The characterization allows us to establish a number of interesting qualitative features of equilibria.

1 Introduction

The *hide and seek* game is defined as follows. There are two players, the *hider* (H) and the *seeker* (S), and $[n] = \{1, \ldots, n\}$ of $n \geq 2$ boxes (or cells). Each box i has a value $v_i > 0$ associated with it. The hider is endowed with $h \geq 1$ objects (*hiding resources*) that he hides in the boxes so that each box contains at most one object. Not observing the choices of the hider, the seeker chooses $s \geq 1$ of the n boxes to check (i.e. the seeker is endowed with s *seeking resources*). After the choices are made, the seeker pays the hider the value of each unchecked box with a hidden object. We are interested in characterizing and computing mixed strategy Nash equilibria (called equilibria for short) of the game. The hide and seek game is a finite zero-sum games. This implies that the set of equilibrium strategies and the set minimax strategies for each player coincide. In addition, all equilibria are payoff equivalent and the non-negative equilibrium payoff is the value of the game. We are also interested in characterizing the value the game.

The hide and seek game was defined by von Neumann in 1953 [14] and is of interests since then as it constitutes an elegant and simple model of strategic mismatch. Such a model finds natural security and military applications, but it also applies to political campaigns, where a political party needs to avoid campaigning in the areas of an incumbent, to entry games where blocking the

This work was supported by Polish National Science Centre through grant nr 2014/13/B/ST6/01807.

X. Deng (Ed.): SAGT 2018, LNCS 11059, pp. 82–86, 2018.
https://doi.org/10.1007/978-3-319-99660-8_8

entrant requires matching his design (c.f. [6]), and to auditing election results in order to prevent results manipulation (c.f. [2] and the references there). An interesting extension of this game is the hide and seek game on graphs, where the hider chooses a node and the seeker, choosing a node, checks the node and its neighbourhood. Equilibrium strategies in this game can be used to provide game theoretic interpretation for the graph theoretic notions of fractional domination and fractional packing (c.f. [8,15]).

2 The Model

There are two players, H and S (the *hider* and the *seeker*) who have a conflict over the set $[n] = \{1, \ldots, n\}$ of boxes with values $\boldsymbol{v} = (v_1, \ldots, v_n)$ such that $v_1 \geq \ldots \geq v_n$.[1] The players are endowed with fixed numbers of resources: H has $h \geq 1$ hiding resources and player S has $s \geq 1$ seeking resources. The players, simultaneously and independently allocate their resources to the boxes, so that each box receives at most one resource of a given type. A box is won by the hider (and lost by the seeker) if he is the only one to allocate a resources there. Since the cases where a player has n or more resources are trivial, we assume that each player has between 1 and $n - 1$ resources, $1 \leq s, h \leq n - 1$. The set of strategies of a player with m resources is the set of all m element subsets of $[n]$, $\binom{[n]}{m}$.

Payoff to S from a strategy profile $(X, Y) \in \binom{[n]}{s} \times \binom{[n]}{h}$ is $\Pi^{\mathsf{S}}(X, Y) = -\sum_{j \in Y \setminus X} v_j$, and payoff to the hider from the same strategy profile is $\Pi^{\mathsf{H}}(X, Y) = \sum_{j \in Y \setminus X} v_j$. We assume that both players maximise their expected payoffs.

The description above defines a zero-sum game $\Gamma(\boldsymbol{v}, s, h)$. We are interested in mixed strategy Nash equilibria (called equilibria for short) as well as the value of the game, denoted by $\mathrm{Val}(\boldsymbol{v}, s, h)$.

3 Contribution and Related Work

The contribution of our paper is threefold: firstly, we provide an analytical characterization of equilibria and the value of hide and seek games, secondly, we provide an algorithm for constructing probability distributions on fixed size sets associated with the given marginal distribution and, thirdly, we provide an efficient algorithm for computing these equilibria. We discuss the contribution and the related literature below.

Analytical Characterization of Equilibria. The main contribution of the paper is the characterization of equilibria of hide and seek games obtained in the main theorem. The equilibria for the case of the game where $h = s = 1$ were already obtained by von Neumann in [14]. In the case of more resources, the closest

[1] Throughout the paper, given a positive integer m we will follow the usual practice of using $[m]$ to denote the set $\{1, \ldots, m\}$.

results to ours were obtained by Korzhyk et al. in [13]. The authors study related, but much more complex, games called security games. They show that a subclass of these games can be reduced to zero-sum games similar to hide and seek games. In the context of hide and seek games, these games could be described as follows. Each box has two values: $v_i > 0$ and $w_i \geq 0$ associated with it. After the choices are made, apart from paying H the value v_i for each unchecked box i with a hidden object, S pays H the value w_i of each checked box with a hidden object. Clearly the hide and seek game is a special case of this game with $w_i = 0$ for all $1 \leq i \leq n$. The authors provide a complete characterization of unidimensional marginal distributions of equilibrium strategies in the case where $h = 1$. However, the formula for the marginals of equilibrium strategies of S is in terms of the value of the game (which is not characterized analytically). The authors provide also equilibria for the cases where both $s \geq 1$ and $h > 1$ and the values $(v_i)_{i \in [n]}$ and $(w_i)_{i \in [n]}$ are such that equilibrium mixed strategies of H in the game with s seeking resources and one hiding resource are such the probabilities of hiding the object in the boxes remain ≤ 1 after multiplying them by h. In the case of hide and seek games, our results cover all the numbers of hiding and seeking resources and we do not make any additional assumptions about the values of the boxes, $(v_i)_{i \in [n]}$. In addition, we show uniqueness of unidimensional marginal distributions associated with equilibrium strategies of the two players, barring non-generic cases where multiplicity is possible. To obtain this result, we reduce the calculation of the marginals to solving simple systems of linear equations.[2] To make this reduction possible, we show (under certain necessary assumptions) that any vector of unidimensional marginal distributions summing to some given integer $m \geq 1$ is associated with a probability distribution over m element subsets of n with full support.

Computation of Equilibria from Unidimensional Marginal Distributions. We provide an algorithm that, given a vector $\boldsymbol{p} = (p_i)_{i \in N}$ with $p_i \in [0, 1]$ and $\sum_{i \in [n]} p_i = m$ constructs a distribution on the set of m element subsets of $[n]$, $\binom{[n]}{m}$, with unidimensional marginal distribution $\boldsymbol{p}$ and with support of size at most n. The algorithm requires at most n iterations, each requiring $\mathcal{O}(n)$ operations. Hence it computes the output distribution in time $\mathcal{O}(n^2)$. The algorithms for computing probability distributions from marginal distributions exist in setups similar to ours. In the context of security games, Korzhyk et al. [11] consider the problem of finding the probability distribution over possible allocations of m heterogeneous resources to n targets from marginal distributions which, for each pair (resource, target) provides the probability of the resource being assigned to the target. The authors use Birkhoff-von Neumann theorem [4] and Dulmage-Halperin algorithm [7] to provide an algorithm for computing the required probability distributions in time $\mathcal{O}((n+m)^{4.5})$. The algorithm produces distributions with supports of size $\mathcal{O}((n+m)^2)$. This approach could be adopted

[2] The system of equations can be easily used to obtain full characterization in the cases with multiple equilibria. We decided to leave it out of the paper due to presentation considerations.

to compute the desired distributions in our setup (e.g. by naming the resources and considering all the permutations of them). However, it is not clear how to do it efficiently. Moreover, the computational cost of this approach is higher compared to ours. Another, related, work Ahmadinejad et al., [1], considers the problem of computing Nash equilibria in a very general class of conflicts with multiple boxes containing, in particular, the Colonel Blotto game [5]. Their approach involves computing the unidimensional marginal distributions for equilibrium strategies. To obtain the strategies from the marginals, the apply the classic result of Grötschel et al., [9], that allows for computing the strategies with support of size at most $n+1$ in polynomial time. The algorithm to compute that has to be extracted from the proof of the result in [9] and involves applying linear programming. The algorithm proposed by us is direct, simple, and has lower computational cost. Beyond the current paper, our algorithm can be applied to compute equilibrium distributions from the margins in more general security games, like those studied in [10,12].

Computation of Equilibrium Unidimensional Marginal Distributions. The analytical characterization or equilibrium marginal distributions basis for an elementary algorithm for computing these distributions in time $\mathcal{O}(nlog(n))$. The closest related algorithm is the one proposed by Korzhyk et al., [12] in the context of security games with multiple attacking resources. That algorithm computes the solution in time $\mathcal{O}(n^2)$ and is not based on an analytical result. The authors do not propose an algorithm for computing the equilibria from the marginals. The algorithm proposed in [1], for general conflicts with multiple boxes, could also be used to compute equilibria in hide and seek games. That algorithm, however, is based on the ellipsoid method for solving linear programs with exponentially many constraint. Thus its computational cost, although polynomial, is too inefficient to be used in practice (c.f. the critique of this approach in [3]).

Very recently, a paper by Behnezhad et al. [2] consider the von Neumann's hide and seek game under the name of "auditing game". The authors study computation of strategies that guarantee securing the given level of utility with given probability. The analytical results obtained in this paper could be helpful in addressing that problem as well. In particular, the complete characterisation we provide can be used for finding strategies with additional guarantees within the set of equilibrium strategies.

Full version of the paper is available at: http://www.mimuw.edu.pl/~amosild/sagt20-full.pdf.

References

1. Ahmadinejad, A., Dehghani, S., Hajiaghayi, M., Lucier, B., Mahini, H., Seddighin, S.: From duels to battlefields: computing equilibria of Blotto and other games. In: Schuurmans, D., Wellman, M.P. (eds.) Proceedings of the Thirtieth AAAI Conference on Artificial Intelligence, Phoenix, Arizona, USA, 12–17 February 2016, pp. 376–382. AAAI Press (2016)

2. Behnezhad, S., et al.: From battlefields to elections: winning strategies of Blotto and auditing games. In: Proceedings of the Twenty-Ninth Annual ACM-SIAM Symposium on Discrete Algorithms, SODA 2018, New Orleans, LA, USA, 7–10 January 2018, pp. 2291–2310 (2018)
3. Behnezhad, S., Dehghani, S., Derakhshan, M., Hajiaghayi, M.T., Seddighin, S.: Faster and simpler algorithm for optimal strategies of Blotto game. In: Proceedings of the Thirty-First AAAI Conference on Artificial Intelligence, San Francisco, California, USA, 4–9 February 2017, pp. 369–375 (2017)
4. Birkhoff, D.: Tres observaciones sobre el algebra lineal. Universidad Nacional de Tucuman Revista Serie A **5**, 147–151 (1946)
5. Borel, É.: La Théorie du Jeu et les Équations Intégrales à Noyau Symétrique. Comptes Rendus de l'Académie des Sciences **173**, 1304–1308 (1921). Translated by Savage L.J.: The theory of play and integral equations with skew symmetric kernels. Econometrica **21**, 97–100 (1953)
6. Crawford, V., Iriberri, N.: Fatal attraction: salience, naïveté, and sophistication in experimental "hide-and-seek" games. Am. Econ. Rev. **97**(5), 1731–1750 (2007)
7. Dulmage, L., Halperin, I.: On a theorem of Frobenius-König and J. von Neumann's game of hide and seek. Proc. Trans. R. Soc. Can. **49**(3), 23–29 (1955)
8. Fisher, D.: Two person zero-sum games and fractional graph parameters. Congr. Numerantium **85**, 9–14 (1991)
9. Grötschel, M., Lovász, L., Schrijver, A.: The ellipsoid method and its consequences in combinatorial optimization. Combinatorica **1**(2), 169–197 (1981)
10. Kiekintveld, C., Jain, M., Tsai, J., Pita, J., Ordóñez, F., Tambe, M.: Computing optimal randomized resource allocations for massive security games. In: Proceedings of The 8th International Conference on Autonomous Agents and Multiagent Systems-Volume 1, AAMAS 2009, IFAAMAS, Richland, SC, pp. 689–696 (2009)
11. Korzhyk, D., Conitzer, V., Parr, R.: Complexity of computing optimal Stackelberg strategies in security resource allocation games. In: Proceedings of the 24th AAAI Conference on Artificial Intelligence, AAAI 2010. AAAI Press, Menlo Park (2010)
12. Korzhyk, D., Conitzer, V., Parr, R.: Security games with multiple attacker resources. In: Proceedings of the 22nd International Joint Conference on Artificial Intelligence, IJCAI 2011, pp. 273–279. AAAI Press, Menlo Park (2011)
13. Korzhyk, D., Yin, Z., Kiekintveld, C., Conitzer, V., Tambe, M.: Stackelberg vs. Nash in security games: an extended investigation of interchangeability, equivalence, and uniqueness. J. Artif. Intell. Res. **41**, 297–327 (2011)
14. von Neumann, J.: A certain zero-sum two-person game equivalent to the optimal assignment problem. In: Contributions to the Theory of Games (AM-28), vol. II, pp. 5–12. Princeton University Press (1953)
15. Scheinerman, E., Ullman, D.: Fractional Graph Theory. Wiley, New York (1997)

An Improved Envy-Free Cake Cutting Protocol for Four Agents

Georgios Amanatidis[1], George Christodoulou[2], John Fearnley[2], Evangelos Markakis[3(✉)], Christos-Alexandros Psomas[4], and Eftychia Vakaliou[3]

[1] Centrum Wiskunde & Informatica (CWI), Amsterdam, Netherlands
[2] University of Liverpool, Liverpool, UK
[3] Athens University of Economics and Business, Athens, Greece
markakis@aueb.gr
[4] Carnegie Mellon University, Pittsburgh, USA

Abstract. We consider the classic cake-cutting problem of producing envy-free allocations, restricted to the case of four agents. The problem asks for a partition of the cake to four agents, so that every agent finds her piece at least as valuable as every other agent's piece. The problem has had an interesting history so far. Although the case of three agents is solvable with less than 15 queries, for four agents no bounded procedure was known until the recent breakthroughs of Aziz and Mackenzie [2,3]. The main drawback of these new algorithms, however, is that they are quite complicated and with a very high query complexity. With four agents, the number of queries required is close to 600. In this work we provide an improved algorithm for four agents, which reduces the current complexity by a factor of 3.4. Our algorithm builds on the approach of [3] by incorporating new insights and simplifying several steps. Overall, this yields an easier to grasp procedure with lower complexity.

1 Introduction

Producing an envy-free allocation of an infinitely divisible resource is a classic problem in fair division. As it is customary in the literature, the resource is represented by the interval $[0, 1]$, and each agent has a probability measure encoding her preferences over subsets of $[0, 1]$. The goal is to divide the entire interval among the agents so that no one envies the subset received by another agent. We note that the partition does not need to consist of contiguous pieces; the piece of an agent may be any finite collection of subintervals.

The problem has a long and intriguing history. It has been long known that envy-free allocations exist for any number of agents, via non-constructive proofs [8,16,18]. For algorithmic results, the standard approach is to assume access to the valuation functions via *evaluation* and *cut queries* (see Sect. 2). Under this model, we are interested in counting the number of queries needed for producing an envy-free allocation. For two agents, the famous cut-and-choose protocol requires only two queries. For three agents, the procedure of Selfridge and Conway [6] guarantees an envy-free allocation after at most 14 queries. For four

X. Deng (Ed.): SAGT 2018, LNCS 11059, pp. 87–99, 2018.
https://doi.org/10.1007/978-3-319-99660-8_9

agents and onwards, however, the picture changes drastically. The first finite, yet unbounded, algorithm was proposed by [5]. This was followed up by other more intuitive algorithms, which are also unbounded, e.g., [10,13]. Finding a bounded algorithm was open for decades and positive results had been known only for certain special cases, like piece-wise uniform or polynomial valuations [4,7,9]. It was only recently that a major breakthrough was achieved by Aziz and Mackenzie, presenting the first bounded algorithms, initially for four agents [3], and later for an arbitrary number of agents [2].

Despite these significant advances, the algorithms of [2,3] are still of very high complexity. For an arbitrary number of agents, n, the currently known upper bound involves a tower of exponents of n, and even for the case of four agents, the known algorithm requires close to 600 queries. On top of that, these algorithms are rather complicated and their proof of correctness requires tedious case analysis in certain steps. Hence, a clean-cut and more intuitive algorithm is still missing.

Contribution: We focus on the case of four agents and present an improved algorithm that reduces the query complexity roughly by a factor of 3.4 (requiring 61 cut queries and 110 evaluation queries). Our algorithm utilizes building blocks that are similar to the ones used by [3], but by incorporating new insights and simplifying several steps, we obtain a solution with significantly fewer queries. The main differences between our work and [3] are highlighted at the end of this section. Our algorithm works by maintaining a partial allocation along with a leftover residue. Throughout its execution, it keeps updating the allocation and reducing the residue, until certain structural properties are satisfied. These properties involve the notion of *domination*, where we say that an agent i dominates another agent j, if allocating the whole remaining residue to j will not create any envy for i. A crucial part of the algorithm is to get a partial allocation where one agent is dominated by two others. Once we establish this, we then exhibit how to produce a complete allocation of the cake without introducing any envy. Overall, this results in an algorithm with markedly lower query complexity.

Further related work: We refer the reader to the book chapter [12] for a more proper treatment of the related literature. Towards simplifying the algorithm of Aziz and Mackenzie [3], the work of Segal-Halevi et al. [15] (see their Appendix B) proposes a conceptually simpler framework, without, however, improving the query complexity. Apart from the algorithmic results mentioned above, there has also been a line of work on lower bounds. For envy-freeness, Stromquist [17] showed that there is no finite protocol for producing envy-free allocations where all the pieces are contiguous. Later on, Procaccia [11] established an $\Omega(n^2)$ lower bound for producing non-contiguous envy-free allocations. Apparently, there is still a huge gap between the known lower and upper bounds for any $n \geq 4$.

An Overview of the Algorithm. We start with a high level description of the main ideas. As with most other algorithms, our algorithm maintains throughout its execution a partial allocation of the cake, along with an unallocated residue.

The goal is to keep updating the allocation and diminishing the residue, with the invariant that the current partial allocation is always envy-free. Once the residue is eliminated, we are left with a complete envy-free allocation. As mentioned earlier, the notion of *domination* is pivotal in our approach. The algorithm creates certain domination patterns between the agents, working in phases as follows:

Phase One. We find this first phase of particular importance, as it is also the most computationally demanding one. Here the goal is to get a partial envy-free allocation in which some agent is dominated by two other agents as in Fig. 1a. In order to establish dominations among agents, we use as a subroutine the so-called CORE protocol. In the CORE protocol one agent has the role of the "cutter", and the output is a new allocation with a strictly diminished residue. The properties of CORE have several interesting and crucial consequences. First, if CORE is executed twice with the same agent as the cutter, then this cutter dominates at least one other agent in the resulting allocation. Moreover, if we run CORE two more times, we may not get any extra dominations right away but we can still make a small *correction* so that the cutter dominates one more agent. This is done by using a protocol, referred to as the CORRECTION protocol, which performs a careful redistribution. Finally, by running CORE one more time with a different cutter and the current residue, we show how further dominations arise that lead to the desired structure of one agent being dominated by two others. In total, phase one requires up to 6 calls to the CORE protocol.

Phase Two. Suppose that at the end of phase one, agent A is dominated by agents B and C. The goal in the second phase is to produce a partial envy-free allocation where both A and D dominate both B and C. To achieve this goal, we execute CORE twice on the residue with D as the cutter. Then, if we still do not have the required dominations, we use again the CORRECTION protocol to appropriately reallocate one of the last two partial allocations produced by CORE. This suffices to create the dominations shown in Fig. 1b.

Phase Three. Since both B and C are now dominated by A and D, we can simply execute the cut-and-choose protocol for B and C on the remaining residue.

Similarities and Differences with the Aziz-Mackenzie Algorithm [3]. Our algorithm uses similar building blocks as the algorithm in [3] for four agents, combined with new insights. Namely, our CORE and CORRECTION protocols on a high level serve the same purpose as the core and the permutation protocols in [3]. Conceptually, a crucial difference is the target structure of the domination graph. The initial (and most query-demanding) step of [3] is to have *every* agent dominate two other agents. Here, our goal in phase one is to *have just one agent dominated by two other agents*. Once this is accomplished, it is possible to reach a complete envy-free allocation much faster. Another important difference is the implementation of the CORE protocol itself. Our version is simpler regarding both its statement and its analysis. It also differs in the sense that it takes as input more information than in [3], such as the current allocation, and it is not required to always output a partial envy-free allocation of the current residue.

Fig. 1. Here is an illustration of the domination graphs we want to achieve at the end of the first (a) and the second (b) phase respectively. In both graphs *additional edges may be present* but are not relevant.

This extra flexibility allows us to avoid the tedious case analysis stated in the core protocol of [3] and, at the same time, further reduce the number of queries.

2 Preliminaries

Let $N = \{1, 2, 3, 4\}$ be a set of four agents. The cake is represented as the interval $[0, 1]$; a *piece* of the cake can be any finite union of disjoint intervals. Each agent $i \in N$ is associated with a valuation function v_i defined on all finite unions of intervals. We assume that the valuation functions satisfy the following standard properties for all $i \in N$:

- Normalization: $v_i([0, 1]) = 1$.
- Additivity: for all disjoint $X, X' \subseteq [0, 1]$: $v_i(X \cup X') = v_i(X) + v_i(X')$.
- Divisibility: for every $[x, y] \subseteq [0, 1]$ and every $\lambda \in [0, 1]$, there exists $z \in [x, y]$ such that $v_i([x, z]) = \lambda v_i([x, y])$. Note that this implies that $v_i([x, x]) = 0$, for all $x \in [0, 1]$.
- Nonnegativity: for every $X \subseteq [0, 1]$ it holds that $v_i(X) \geq 0$.

By $\mathcal{X} = (X_1, X_2, X_3, X_4)$ we denote the allocation where agent i is given the piece X_i.

Definition 1 (Envy-freeness). *An allocation $\mathcal{X} = (X_1, X_2, X_3, X_4)$ is* envy-free, *if $v_i(X_i) \geq v_i(X_j)$, for all $i, j \in N$, i.e., every agent prefers her piece to any other agent's piece.*

We say that $\mathcal{X}$ is a *partial* allocation, if there is some cake that has not been allocated yet, i.e., $\bigcup_{i=1}^{4} X_i \subsetneq [0, 1]$. The unallocated cake is called the *residue*. During the execution of the algorithm the residue diminishes, until eventually it becomes the empty set. As we noted, an important notion is that of *domination* or *irrevocable advantage* [6]. It will be insightful to think of a graph-theoretic representation of our goals, via the *domination graph* of the current allocation.

Definition 2 (Domination and Domination Graph). *Given a partial allocation $\mathcal{X} = (X_1, X_2, X_3, X_4)$ and a residue R, we say that an agent i* dominates *another agent j, if $v_i(X_i) \geq v_i(X_j \cup R)$. That is, i would not be envious of j even* *if j were allocated all of R. The* domination graph *with respect to $\mathcal{X}$ is a directed graph where the nodes correspond to the agents and there exists a directed edge (i, j) if and only if agent i dominates agent j.*

Achieving certain patterns in the domination graph can make the allocation of the remaining residue straightforward. For example, if there exists a node i with in-degree 3, allocating all of the residue to agent i results in an envy-free allocation. As another example, the protocol of [3] tries to get a domination graph where every node has out-degree at least 2. In our algorithm, we also enforce a certain structure on the domination graph.

The Robertson-Webb Model. The standard model in which we measure the complexity of cake cutting algorithms is the one suggested by Robertson and Webb [14] and formalized by Woeginger and Sgall [19]. In this model, two kinds of queries are allowed:

- *Cut queries:* given an agent i, a point $x \in [0, 1]$ and a value r, with $r \leq v_i([x, 1])$, the query returns the smallest $y \in [0, 1]$ such that $v_i([x, y]) = r$.
- *Evaluation queries:* given an agent i and an interval $[x, y]$, return $v_i([x, y])$.

Virtually all known discrete protocols can be analyzed within this framework. For example, the cut-and-choose protocol is implemented by making one cut query for agent 1 with $r = 1/2$, starting from $x = 0$, followed by an evaluation query on agent 2 for one of the pieces (which also reveals the value of the second piece).

Conventions on Ties, Marks, Partial Pieces, and Residues. All algorithms in this work ignore ties. However, assuming an appropriate tie-breaking scheme, this is without loss of generality (also see the discussion in [2]).

We follow some conventions—also adopted in related work—when it comes to handling trims and partial pieces. In various steps during the algorithm, one agent cuts the residue into pieces, and the other agents are asked to place marks on certain pieces. We always assume that marks are placed starting from the left endpoint of a piece, and this operation creates a partial piece, contained between the mark and the right endpoint. In particular, suppose we have a partition of the residue into four contiguous pieces. Then, an agent may be asked to place a mark on her most favorite piece so that the resulting partial piece has the same value as her second favorite piece (see Fig. 2). The types of marks that the algorithm needs are described in the following definition.

Definition 3. *Given a partition of the residue into four pieces, we say that an agent performs an x-mark, if she places a mark on each of her $x - 1$ most valuable pieces so that the resulting partial pieces all have the same value as her x-th favorite piece.*

In the description of the algorithm we use 2-marks and 3-marks. Of course, after all marks are placed, each connected piece may have multiple marks on it.

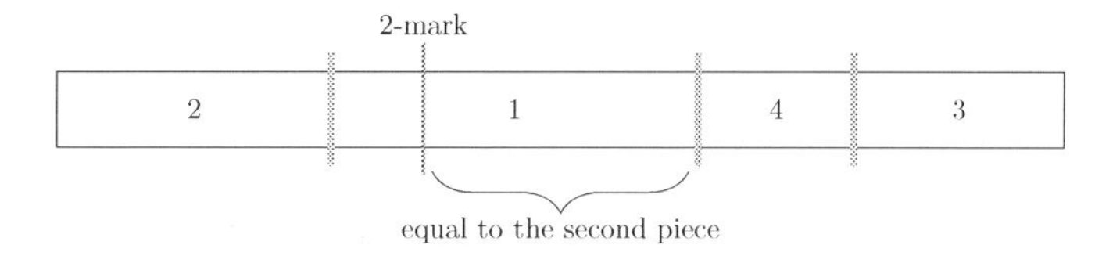

Fig. 2. The view of the residue for a non-cutter at the time she performs a 2-mark.

Whenever a connected piece p is only partially allocated, the part p' of p that is allocated is always the interval between the second rightmost mark and the right endpoint of p. While at this point it is not clear whether a *second* mark on a piece even exists, we show (see full version [1]) that marked pieces will have at least two marks. Hence, if some agent i receives a partial piece p', resulting from an initial piece p, it is not necessarily true that p' is defined by i's own mark.

Note that in the above discussion the residue is seen as a single interval, while in fact it may be a finite union of intervals. We keep this view throughout this work as it is conceptually easier and allows for a cleaner presentation.[1]

3 The Algorithm

Our main result is the following theorem. All missing proofs can be found in the full version of this work in [1].

Theorem 1. *The* MAIN PROTOCOL *returns an envy-free allocation and makes at most* 61 *cut queries, and* 110 *evaluation queries.*

We discuss first the main steps of our algorithm and provide the relevant definitions and key properties.

Phase One. This is the most important part of the protocol, and computationally the most demanding one. The goal in phase one is to get a partial envy-free allocation, where some agent is dominated by two other agents, i.e., the underlying domination graph has a node with in-degree at least 2, as depicted in Fig. 1. In order to establish dominations among agents, we use a subroutine called CORE protocol (stated in Sect. 3.1). This protocol takes as input a specified agent, called the *cutter*, the current partial allocation, and the current residue. The output of CORE is a partial (*usually* envy-free) allocation of the residue with some additional properties described below. In the initial step of CORE, the cutter divides the current residue into four equal-valued pieces according to her own valuation function. Throughout the protocol the rest of the agents—the *non-cutters*—may mark these pieces, and at the end, agents may be allocated either partial (marked) or complete pieces. Of course, if at any point CORE outputs an envy-free allocation of the whole cake, the algorithm terminates.

[1] We elaborate further on this issue in our full version [1].

For technical convenience, CORE also takes as an input a subset of agents that we choose to exclude from *competition*. This roughly means that the excluded agents will choose their piece late in the CORE protocol because they dominate the other non-cutters. In most cases, this argument is just the empty set (and when no such argument is specified we mean that it is $\emptyset$). The full description of CORE is given in Sect. 3.1 and for now, we treat it as a black box and assume that it satisfies the following properties.

CORE**Property 1.** *The cutter and at least one more agent receive complete pieces, each worth exactly* $1/4$ *of the value of the current residue according to the cutter's valuation.*

CORE**Property 2.** *The allocation output by any single execution of* CORE *when no agent is excluded from competition, is a (possibly partial) envy-free allocation.*

The above properties allow us to deduce an important fact: if CORE is executed at least twice with the same agent as the cutter, then this cutter dominates at least one agent in the resulting allocation. In fact, we can be more specific about the agent who gets dominated. The important observation here is that a second run of CORE makes the cutter dominate whoever received the so-called *insignificant piece* in the first execution.

Definition 4. *Let $\mathcal{A}$ be an allocation produced by a single run of* CORE. *Among the four pieces given to the agents, the partial piece that is least desirable to the cutter is called the* insignificant piece *of $\mathcal{A}$.*

Hence, if we run CORE twice, say with agent 1 as the cutter, we enforce one edge in the domination graph. In order to proceed further and obtain a node with in-degree two, we first attempt, as an intermediate step, to have a domination graph where one node has out-degree equal to two. One may think that the intermediate step can be achieved by running CORE more times with agent 1 as the cutter. The problem with this approach is that even if we further execute CORE *any* number of times, there is no guarantee that new dominations will appear; the same agent may receive the insignificant piece in every iteration.

To fix this issue, it suffices to run CORE 4 times with agent 1 as the cutter and then make a small *correction* to one of the 4 partial allocations produced by CORE. In particular, denote by $\mathcal{A}^k = \{p_1^k, p_2^k, p_3^k, p_4^k\}$, with $k = 1, ..., 4$, the suballocation output by the kth execution of CORE within the *for* loop of line 1 of MAIN PROTOCOL, and let R^k be the residue after the kth execution. Then clearly for each agent i, $p_i^k \subseteq R^{k-1}$, and the current allocation of the algorithm after the 4 calls to CORE is $\mathcal{X} = \{p_1, p_2, p_3, p_4\}$, with $p_i = \cup_{k=1}^4 p_i^k$. Among these 4 suballocations that $\mathcal{X}$ consists of, we identify one in which we can perform a certain redistribution without introducing any envy. To do this, we exploit the notion of *gain*, which is the difference between the value that an agent has for her own piece compared to the pieces of agents she does not dominate.

Definition 5 (Gain). *Let $\mathcal{X} = \{p_1, \ldots, p_4\}$ be the current partial allocation of the cake, and $\mathcal{A} = \{p_1', \ldots, p_4'\}$ be a suballocation of $\mathcal{X}$, i.e., $p_i' \subseteq p_i$ for $i \in N$. Further, let D_i be the set of agents that are dominated by i in $\mathcal{X}$ and*

$N_i = N \setminus (D_i \cup \{i\})$. *Then the gain of* i *with respect to* $\mathcal{A}$, $\mathrm{G}_{\mathcal{A}}(i)$, *is the difference between* $v_i(p'_i)$ *and the maximum value of* i *for a piece in* $\mathcal{A}$ *given to any agent in* N_i, *i.e.*, $\mathrm{G}_{\mathcal{A}}(i) = v_i(p'_i) - \max_{j \in N_i} v_i(p'_j)$.[2]

Using Definition 5, we identify a suballocation among $\mathcal{A}^1, \mathcal{A}^2, \mathcal{A}^3, \mathcal{A}^4$, where the gain of each agent is small compared to her combined gain from the other three suballocations (line 4 of the algorithm). The existence of such an allocation is shown in the full version. Then, the redistribution is performed via the CORRECTION protocol which takes as input an allocation $\mathcal{A}$, produced by CORE, and outputs an allocation $\mathcal{A}' = \pi(\mathcal{A})$, where π is a permutation on N. In doing so, special attention is paid to the insignificant piece of $\mathcal{A}$. For now, we treat CORRECTION as a black box and ask that it satisfies the three properties below; see Sect. 3.1 for its description.

MAIN PROTOCOL(N)

Phase One

1 **for** *count* = 1 *to* 4 **do**
2   Run CORE on the current residue with agent 1 as the cutter.
3 **if** *the same agent got the insignificant piece in all 4 executions of* CORE **then**
4   Find $\mathcal{A}^* \in \{\mathcal{A}^1, \mathcal{A}^2, \mathcal{A}^3, \mathcal{A}^4\}$ such that $\mathrm{G}_{\mathcal{A}^*}(i) \le \sum_{\mathcal{A} \ne \mathcal{A}^*} \mathrm{G}_{\mathcal{A}}(i)$ for all $i \in N \setminus \{1\}$.
5   Run CORRECTION on $\mathcal{A}^*$.
6 Run CORE on the residue with agent 1 as the cutter.
7 **if** *there is some agent* $E \in N \setminus \{1\}$ *not dominated by agent* 1 **then**
8   Run CORE on the residue with agent E as the cutter, excluding agent 1 from competition (since agent 1 dominates the other non-cutters).
9 **else**
10   Run the Selfridge-Conway Protocol on the residue for agents 2, 3, and 4, and terminate.

Now, if the algorithm has not terminated, some agent A *is dominated by two other agents* B *and* C. *Let* D *be the remaining agent.*

Phase Two

11 **for** *count* = 1 *to* 2 **do**
12   Run CORE on the current residue with agent D as the cutter, excluding from competition any one from $\{B, C\}$ who dominates two non-cutters.
13 **if** B *and* C *are not both dominated by* A *and* D **then**
14   Let $F \in \{B, C\}$ be the agent who got the insignificant piece in the last two calls of CORE.
15   Run CORRECTION on the suballocation (out of the last two) where $\mathrm{G}_{\mathcal{A}}(F)$ was smaller.

At this point both A *and* D *dominate both* B *and* C.

Phase Three

16 Run CUT AND CHOOSE on the current residue for agents B and C.

---

[2] Note that $\mathrm{G}_{\mathcal{A}}(i)$ is not defined when $N_i = \emptyset$. In fact, we never need it in such a case.

CORRECTION**Property 1** *The insignificant piece of $\mathcal{A}$ is given to a different agent in $\mathcal{A}'$. In particular, it is given to an agent that has marked it in $\mathcal{A}$.*

CORRECTION**Property 2** *If a non-cutter was allocated her favorite unmarked piece in $\mathcal{A}$, she will again be allocated a piece of the same value in $\mathcal{A}'$.*

Assume there is no agent dominating everyone else, meaning that $\mathrm{G}_{\mathcal{A}}(i)$ is defined for all $i \in N$. For a partial envy-free allocation like $\mathcal{A}$, the gain of any agent is nonnegative. However, this may not be true for $\mathcal{A}'$, as it is not necessarily envy-free. What we need is for $(\mathcal{X} \setminus \mathcal{A}) \cup \mathcal{A}'$ to be envy-free, and towards this $\mathrm{G}_{\mathcal{A}'}(i)$ should not be too small for any $i \in N$.

CORRECTION**Property 3** $\mathrm{G}_{\mathcal{A}'}(i) \geq -\mathrm{G}_{\mathcal{A}}(i)$ for all agents i.

By Correction Property 1, the insignificant piece has changed hands after line 5. This allows us to make one extra call to CORE in order to enforce one more domination (line 6). Hence, the intermediate step is completed and we know that agent 1 dominates at least 2 other agents. If she dominates all three of them, we can run any of the known procedures for 3 agents on the residue, and be done with only a few queries. The interesting remaining case is to assume that agent 1 currently dominates exactly two other agents.

At this point there are various ways to proceed. E.g., we could repeat the whole process so far, but with agents 2 and 3 as cutters, and get at least 6 edges in the domination graph. This would ensure a node with in-degree two, but it requires several calls to CORE. Instead, and quite remarkably, we show that it suffices to run CORE only one more time, with the agent who is not dominated by agent 1 as the cutter. In the full version of this paper we prove that this makes the cutter dominate one of the agents that are dominated by agent 1. Hence, phase one is now complete, as we have one agent with in-degree two.

Remark 1. The intermediate step of getting a node with out-degree two has also been utilized in [3]. The goal there however was to make *every* agent dominate two other agents, whereas we only needed this to hold for one agent.

Phase Two. Suppose phase two starts with a partial envy-free allocation where some agent, say A, is dominated by agents B and C (Fig. 1a). Our next goal is to produce a partial envy-free allocation where both A and D dominate both B and C (Fig. 1b). To achieve this goal, we execute CORE twice with D as the cutter, i.e., with the agent not involved in the dominations of phase one. Again, we need to argue about the behavior of CORE under the existing dominations, and we ask for the following property.

CORE**Property 3** *Assume we run* CORE *with* D *as the cutter, and suppose agent* A *is dominated by the other two non-cutters,* B *and* C*, neither of whom dominates the other. Then, (1)* A *gets her favorite of the four complete pieces without making any marks, (2) at least three complete pieces are allocated, and (3) if a non-cutter, say* B*, gets a partial piece, then the remaining non-cutter,* C*, is indifferent between her piece and* B*'s piece.*

Using this property, we can show that after one call to CORE (1st execution of line 12), agents A and D will both dominate either B *or* C. However, we need domination over both B *and* C. The second call to CORE (2nd execution of line 12) ensures that we can again resort to the CORRECTION protocol. If, after the two calls to CORE, only one of B and C, say B, is dominated by both A and D, then running CORRECTION on one of the two core allocations from this phase—the one where the gain of B is smaller—resolves the issue, and makes A and D dominate both B and C.

Phase Three. Since both agents B and C are dominated by A and D, we just execute the cut-and-choose protocol for B and C, where B cuts the residue in two equal pieces and C chooses her favorite piece. This completes our algorithm.

3.1 CORE and CORRECTION

Here we provide the description of the CORE and CORRECTION protocols. The main results within this subsection are the following.

Theorem 2. *The* CORE *protocol satisfies* CORE *Properties 1, 2 and 3, and makes at most* 9 *cut queries and* 15 *evaluation queries.*

Theorem 3. *The* CORRECTION *protocol satisfies* CORRECTION *Properties 1, 2 and 3, and makes no queries.*

CORE $(k, R, \mathcal{X}, \mathcal{E})$

---

1 Agent k cuts the current residue R in four equal-valued pieces (according to her).
2 Let $S = N \setminus (\{k\} \cup \mathcal{E})$ be the set of agents who may compete for pieces.
3 **if** *there exists $j \in S$ who has no competition in S for her favorite piece* **then**
4 j is allocated her favorite piece and is removed from S.
5 **if** *every agent in S has a different favorite piece* **then**
6 Everyone gets her favorite piece and the algorithm terminates.
7 **for** *every agent $i \in S$* **do**
8 **if** *(1) i has no competition for her second favorite piece p,* **or**
9 *(2) i has exactly one competitor $j \in S$ for p, j also considers p as her second favorite, and i,j each have exactly one competitor for their favorite piece* **then**
10 i makes a 2-mark.
11 **else**
12 i makes a 3-mark.
13 Allocate the pieces according to a rightmost rule:
14 **if** *an agent has the rightmost mark in two pieces* **then**
15 Out of the two partial pieces, considered until the second rightmost mark (which always exists), she is allocated the one she prefers.
16 The other partial piece is given to the agent who made the second rightmost mark on it.
17 **else**
18 Each partial piece is allocated—until the second rightmost mark—to the agent who made the rightmost mark on that piece.
19 **if** *any non-cutters were not given a piece yet* **then**
20 Giving priority to any remaining agents in S (but in an otherwise arbitrary order), they choose their favorite unallocated complete piece.
21 The cutter is given the remaining unallocated complete piece.

The CORE protocol is used for allocating part of the current residue every time it is called. CORE takes as input an agent, specified as the cutter, the current residue, and the current partial allocation. It first asks the cutter to divide the residue into four equally valued contiguous pieces. The cutter is going to be the last one to receive one of these four pieces. Regarding the remaining three agents, each of them will either be immediately allocated her favorite piece or will be asked to place a mark on certain pieces, based on the relative rankings of the non-cutters for the pieces, and on possible domination relations that have already been established. Marks essentially provide limits on how to partially allocate pieces that are desired by many agents, so that they can be given without introducing envy. There are two possible types of marks that can be placed; 2-marks and 3-marks. The type of mark that the agents will be asked to place depends mainly on the conflicts that arise for the favorite and second favorite pieces of each agent.

Definition 6. *During an execution of* CORE, *let* P *be a set of pieces and* S *be a subset of non-cutters. We say that an agent* $i \in S$ *has* competition *for a piece* $p \in P$, *if* (1) i *is not dominated by everyone in* S, *and* (2) *there exists* $j \in S$ *such that* p *is* j*'s favorite or second favorite piece in* P. *We call* j *a* competitor *of* i *for* p.

Definition 6 helps us identify whether we need to perform a 2-mark or 3-mark on the available pieces. Furthermore, in some cases where we know that certain domination patterns appear, it is convenient to prevent some agents from competing for any piece (in particular, when some agent dominates the other non-cutters).

The CORRECTION protocol takes as input an allocation $\mathcal{A}$, produced by CORE, and outputs an allocation $\mathcal{A}' = \pi(\mathcal{A})$, where π is a permutation on N, so that the envy-freeness of the overall partial allocation and certain dominations are preserved. Its description is shown below and its analysis is provided in the full version. Note that we refer to the cutter in allocation $\mathcal{A}$ as D.

CORRECTION($\mathcal{A}$)

---

1 Let A, B be the agents having the two marks on the insignificant piece, and suppose A was given this piece in allocation $\mathcal{A}$.
2 The insignificant piece is allocated to B.
3 **if** *there is no other partial piece* **then**
4   Agents choose their favorite piece in the order C, A, D.
5 **else**
6   Find the rightmost mark not made by B on the other partial piece. Let $E \in \{A, C\}$ be the agent who made it.
7   Agent E is allocated the partial piece.
8   The last non-cutter chooses her favorite among the two complete pieces.
9   The cutter is allocated the remaining (complete) piece.

References

1. Amanatidis, G., Christodoulou, G., Fearnley, J., Markakis, E., Psomas, C.A., Vakaliou, E.: An improved envy-free cake cutting protocol for four agents (2018). https://arxiv.org/pdf/1807.00317.pdf
2. Aziz, H., Mackenzie, S.: A discrete and bounded envy-free cake cutting protocol for any number of agents. In: 57th Annual IEEE Symposium on Foundations of Computer Science, FOCS 2016, pp. 416–427 (2016)
3. Aziz, H., Mackenzie, S.: A discrete and bounded envy-free cake cutting protocol for four agents. In: 48th ACM Symposium on the Theory of Computing, STOC 2016, pp. 454–464 (2016)
4. Aziz, H., Ye, C.: Cake cutting algorithms for piecewise constant and piecewise uniform valuations. In: Liu, T.-Y., Qi, Q., Ye, Y. (eds.) WINE 2014. LNCS, vol. 8877, pp. 1–14. Springer, Cham (2014). https://doi.org/10.1007/978-3-319-13129-0_1
5. Brams, S.J., Taylor, A.D.: An envy-free cake division protocol. Am. Math. Monthly **102**(1), 9–18 (1995)

6. Brams, S.J., Taylor, A.D.: Fair Division: from Cake Cutting to Dispute Resolution. Cambridge University Press, Cambridge (1996)
7. Branzei, S.: A note on envy-free cake cutting with polynomial valuations. Inf. Process. Lett. **115**(2), 93–95 (2015)
8. Dubins, L., Spanier, E.: How to cut a cake fairly. Am. Math. Monthly **68**, 1–17 (1961)
9. Kurokawa, D., Lai, J.K., Procaccia, A.D.: How to cut a cake before the party ends. In: Proceedings of the Twenty-Seventh AAAI Conference on Artificial Intelligence, AAAI 2013, pp. 555–561 (2013)
10. Pikhurko, O.: On envy-free cake division. Am. Math. Monthly **107**(8), 736–738 (2000)
11. Procaccia, A.D.: Thou shalt covet thy neighbor's cake. In: Proceedings of the 21st International Joint Conference on Artificial Intelligence, IJCAI 2009, pp. 239–244 (2009)
12. Procaccia, A.D.: Cake cutting algorithms. In: Brandt, F., Conitzer, V., Endriss, U., Lang, J., Procaccia, A.D. (eds.) Handbook of Computational Social Choice, chap. 13. Cambridge University Press (2016)
13. Robertson, J.M., Webb, W.A.: Near exact and envy-free cake division. Ars Combinatorica **45**, 97–108 (1997)
14. Robertson, J.M., Webb, W.A.: Cake Cutting Algorithms: be fair if you can. AK Peters (1998)
15. Segal-Halevi, E., Hassidim, A., Aumann, Y.: Waste makes haste: Bounded time algorithms for envy-free cake cutting with free disposal (2018). https://arxiv.org/pdf/1511.02599.pdf
16. Stromquist, W.: How to cut a cake fairly. Am. Math. Monthly **87**(8), 640–644 (1980)
17. Stromquist, W.: Envy-free cake divisions cannot be found by finite protocols. Electr. J. Comb. **15**(1) (2008)
18. Su, F.E.: Rental harmony: sperner's lemma in fair division. Am. Math. Monthly **106**(10), 930–942 (1999)
19. Woeginger, G., Sgall, J.: On the complexity of cake cutting. Discrete Optim. **4**(2), 213–220 (2007)

A Truthful Mechanism for Interval Scheduling

Jugal Garg and Peter McGlaughlin(✉)

University of Illinois at Urbana Champaign, Urbana, IL 61801, USA
{jugal,mcglghl2}@illinois.edu

Abstract. Motivated by cloud computing, we study a market-based approach for job scheduling on multiple machines where users have hard deadlines and prefer earlier completion times. In our model, completing a job provides a benefit equal to its present value, i.e., the value discounted to the time when the job finishes. Users submit job requirements to the cloud provider who non-preemptively schedules jobs to maximize the social welfare, i.e., the sum of present values of completed jobs. Using a simple and fast greedy algorithm, we obtain a $1+s/(s-1)$ approximation to the optimal schedule, where $s > 1$ is the minimum ratio of a job's deadline to processing time. Building on our approximation algorithm, we construct a pricing rule to incentivize users to truthfully report all job requirements.

1 Introduction

Cloud computing's explosive growth over the past decade is attributable to its flexible computing resources and the economy of scale provided by large data centers. This framework allows users to rent computing resources on demand, avoiding the need for costly infrastructure investment. Typically, pricing is pay-as-you-go where users pay per unit time. While simple, this pricing scheme does not reflect current market conditions, i.e., user demand versus the cloud provider's capacity, nor does it account for important job requirements such as deadlines.

We investigate an alternative market based approach for the fair allocation of reusable resources by introducing a new scheduling problem, Present Value Scheduling (PVS). Abstractly, the problem is to non-preemptively schedule jobs with hard deadlines on m identical machines. Each job $\mathcal{J}_i = (v_i, t_i, d_i)$ is defined by a processing time t_i, a deadline d_i, and a value v_i if completed immediately. Users prefer earlier completion times, leading job values to decay over time as determined by the discount factor $0 < \beta < 1$ shared by all jobs. Then, completing job $\mathcal{J}_i$ at time $\tau \in [t_i, d_i]$ provides a benefit of $v_i\beta^\tau$. Note that this is the job's present value, the standard economic model for the time value of money.

Users submit job requests to the cloud provider who determines an allocation of resources based on the jobs' requirements with the objective of maximizing

J. Garg—Supported by NSF CRII Award 1755619.

X. Deng (Ed.): SAGT 2018, LNCS 11059, pp. 100–112, 2018.
https://doi.org/10.1007/978-3-319-99660-8_10

social welfare, i.e. the sum of present values of completed jobs. The inherently difficult scheduling problem is further complicated as users may misreport any job parameter (v_i, t_i, d_i) in an effort to increase their utility, defined as present value minus payment. We aim to construct scheduling and pricing rules to incentivize truthful reporting of all job information.

Our Contribution. This paper addresses a fundamental issue in mechanism design, non-preemptive job scheduling for social welfare maximization. Our model, PVS, includes the natural preference for early completion times by discounting the value of job to the time when it finishes. In other words, we consider maximizing the present value of completed jobs.

PVS is a special case of interval scheduling with arbitrary values, i.e., a job's value is an arbitrary function of time. Theoretical work in this area centers on constant factor approximations as the allocation (scheduling) problem is NP-hard. However, most existing algorithms do not ensure truthfulness. While there is a black-box method to construct truthful mechanisms from these approximations, the conversion comes at high computational cost. Fortunately, PVS provides sufficient structure to achieve a deterministic truthful mechanism through simple and efficient allocation and pricing rules.

First we provide a $1 + s/(s-1)$ approximation to PVS for any discount factor $0 < \beta < 1$, where $s = \min_i d_i/t_i > 1$. Our algorithm greedily schedules jobs in decreasing order of weights $w_i = v_i\beta^{t_i}/(1-\beta^{t_i})$ on the machine which gives the earliest completion time, as long as jobs complete by their deadlines. Our method achieves significantly faster run time compared to other applicable approximations which do not require truthfulness, an important consideration for large scale problems encountered in practice. The $1 + s/(s-1)$ bound is essentially tight for $\beta \approx 1$ and $s \gg 1$. However, it is conservative for $\beta \ll 1$ or $s \approx 1$. Second, we show a few key properties of our greedy approximation algorithm that allow an extension of Myerson's lemma [12] to PVS. As a result, we obtain a deterministic mechanism which is truthful with respect to all parameters (v_i, t_i, d_i).

1.1 Related Work

We provide a survey of related work in interval scheduling and mechanism design, taking care to highlight competing approaches to design truthful mechanisms for PVS.

Interval Scheduling. The allocation problem's theoretical foundations lie in interval scheduling. Essentially the discrete version of machine scheduling, each job is defined by explicitly listing all available scheduling times (intervals), with each interval potentially providing a different value. In other words, jobs' values are arbitrary functions of time. Nearly all versions of the problem are NP-hard, confining theoretical work to constant factor approximations. We focus on the best known approximations applicable to PVS and note that none of the following works require truthfulness.

Bar-Noy et al. [3] presents a 2 approximation based on LP rounding. Their algorithm starts by finding the optimal fractional allocation to the natural LP relaxation of the problem. The fractional solution is rounded into a set of polynomially many integer valued (feasible) solutions using a graph coloring argument, the largest of which yields at least 1/2 the optimal schedule's value. Bar-Noy et al. [2] uses the local ratio technique to derive a combinatorial 2 approximation in a generalization of interval scheduling where each job has a width. Independently, Berman and DasGupta [4] obtained a similar algorithm which achieves better runtime by specializing to the standard interval scheduling problem.

Mechanism Design. There is extensive literature on scheduling in the context of mechanism design, starting with the seminal work of Nisan and Ronen [13]. However, the majority of existing literature focuses on makespan minimization, e.g., see [9,11]. One particularly interesting and relevant approach for PVS is the black-box method of Lavi and Swamy [10] which converts approximation algorithms for set packing problems to truthful randomized mechanisms with the same approximation ratio. The procedure starts by applying a fractional VCG mechanism to the natural LP relaxation of the problem. Using the approximation algorithm as a separation oracle, the rescaled fractional allocation is decomposed into a convex combination of integer valued allocations. These integer solutions provide a truthful in expectation randomized mechanism with the same performance guarantees as the initial approximation algorithm. Although applicable to the previously mentioned interval scheduling approximations, and therefore PVS, the technique raises practical concerns for computational efficiency. First, one must solve an LP with a large number of variables and constraints multiple times for the fractional VCG mechanism. Then, decomposing the fractional allocation requires multiple calls to the approximation algorithm.

Recent work in the related field of batch computing in cloud systems offers an alternative to the black-box method. Batch computing generalizes interval scheduling by allowing jobs to run on multiple machines in parallel, up to a given threshold. Drawing on the LP rounding approximation of [3] and the black-box method of [10], Jain et al. [6] construct a truthful in expectation mechanism which approaches a 2 approximation as the number of machines goes to infinity. Requiring only one solution to the natural LP relaxation, their mechanism addresses some of the computational efficiency issues of the black-box approach, and it also allows job values to be arbitrary non-increasing functions of time.

In the preemptive version of batch computing, Jain et al. [7] develop a deterministic truthful mechanism with near optimal performance as the number of machines goes to infinity under a slackness condition on jobs, i.e., a lower bound on the ratio of a job's deadline to its processing time d_i/t_i. From the perspective of PVS, this paper is of interest as the allocation rule is akin to our own. Their approximation greedily schedules jobs in decreasing order of value density v_i/t_i, the natural analog of our weights w_i when the discount factor $\beta = 1$. Azar et al. [1] extend this mechanism to the online setting. We note both works assume job values are constant over time.

2 Definitions and Notation

PVS Model. In PVS a set of n jobs, $\mathcal{J} = \{\mathcal{J}_1, \ldots, \mathcal{J}_n\}$, compete for processing time on m identical machines. Each job $\mathcal{J}_i$ is defined by the tuple (v_i, t_i, d_i) where: v_i is the job's valuation, t_i is the job's processing time, and d_i is the job's deadline. Note that under the identical machines assumption, a job's processing time is the same on all machines. We assume integer valued processing times and deadlines, though techniques generalize naturally to positive real values. The value of completing a job decays over time, determined by the discount factor $0 < \beta < 1$ shared by all jobs. Specifically, the value of completing job $\mathcal{J}_i$ at time $\tau \in [t_i, d_i]$ is $v_i\beta^{\tau}$.

A schedule for machine j is an ordered subset of jobs: $\mathcal{S}_j = \{\mathcal{J}_{k_1}, \ldots, \mathcal{J}_{k_a}\} \subseteq \mathcal{J}$, to be completed in the given order, i.e., $\mathcal{J}_{k_1}$ is completed first, then $\mathcal{J}_{k_2}$ and so on. The value of a schedule on machine j is the sum of values of completed jobs. That is, if $\mathcal{S}_j = \{\mathcal{J}_{k_1}, \ldots, \mathcal{J}_{k_a}\}$, then job $\mathcal{J}_{k_i}$ completes at time $\tau_{k_i} = \sum_{b=1}^{i} t_{k_b}$ and the value of the schedule $\mathcal{S}_j$ is: $V(\mathcal{S}_j) = \sum_{i=1}^{a} v_{k_i}\beta^{\tau_{k_i}}$. A full schedule consists of a schedule for each machine: $\mathcal{S} = \{S_1, \ldots, S_m\}$. We will simply say schedule when the distinction between full and machine specific schedules is clear. A schedule is feasible if each job is contained in no more than one machine schedule: $\mathcal{S}_i \cap \mathcal{S}_j = \emptyset$, $\forall i, j$. In words, each job is processed at most once across all machines. The social welfare maximizing schedule $\mathcal{S}^*$ is the feasible schedule with maximum value: $V(\mathcal{S}^*) \geq V(\mathcal{S})$ for all feasible $\mathcal{S}$. We will often refer to the social welfare maximizing schedule simply as the optimal schedule. We call the problem of finding the optimal schedule the allocation problem. We say that a schedule $\mathcal{S}$ is an α approximation (to optimal schedule) if $V(\mathcal{S}) \geq V(\mathcal{S}^*)/\alpha$.

Mechanisms. A mechanism $\mathcal{M}$ is an algorithm to produce an allocation (schedule) and a set of payments p_i. Each job in a schedule $\mathcal{J}_i \in \mathcal{S}$ is charged a payment p_i, earning utility $u_i(\mathcal{J}_i, \mathcal{J}_{-i}) = v_i\beta^{\tau_i} - p_i$, where $\tau_i > 0$ is time when $\mathcal{J}_i$ completes. Note $p_i = 0$ for all $\mathcal{J}_i \in \mathcal{J} \setminus \mathcal{S}$. Further, we assume agents receive no benefit for partially completing their job, or completing their job after their deadline. Jobs seek to maximize utility. The true parameters of a job $\mathcal{J}_i = (v_i, t_i, d_i)$ are private information and a job may misreport any or all of the values $\mathcal{J}_i' = (v_i', t_i', d_i')$ to gain higher utility. A mechanism is truthful if accurately reporting all job parameters is a dominant strategy: $u_i(\mathcal{J}_i, \mathcal{J}_{-i}) \geq u_i(\mathcal{J}_i', \mathcal{J}_{-i})$, $\forall \mathcal{J}_i'$. In words, truthfully reporting job parameters maximizes utility. We say a mechanism is social welfare maximizing if the scheduling algorithm returns the social welfare maximizing schedule $\mathcal{S}^*$, and an α approximation if it returns an α approximation.

3 Approximation for PVS

Due to space restrictions, we do not provide all proofs. Complete, detailed proofs can be found in the full version of this paper. Our first goal is solving the PVS allocation problem, i.e., maximizing the present value of completed jobs. In the

Algorithm 1. Greedy Scheduling Algorithm (GS)

Input : Job parameters (v_i, t_i, d_i) for each job $\mathcal{J}_i$
Output: Schedule $\mathcal{S}$ of jobs
1 Define: $w_i = \frac{v_i \beta^{t_i}}{1-\beta^{t_i}}$
2 Sort and relabel jobs in descending order of w_i
3 $\mathcal{S}_j \leftarrow \emptyset$, $j = 1, \ldots, m$ (schedule on machine j)
4 $\tau_j \leftarrow 0$, $j = 1, \ldots, m$ (total processing time of machine j)
5 $k = 1$ (machine offering fastest completion time)
6 **for** $i = 1$ ***to*** n **do**
7   **if** $\tau_k + t_i \leq d_i$ **then**
8         $\mathcal{S}_k \leftarrow (\mathcal{S}_k, \mathcal{J}_i)$; $\tau_k \leftarrow \tau_k + t_i$; $k \in \arg\min_j \tau_j$

full version of this paper, we show that this problem is strongly NP-hard. For this reason, we design a greedy algorithm which achieves a $1+s/(s-1)$ approximation to the optimal schedule, where $s = \min_i d_i/t_i > 1$. Before presenting our algorithm, it is instructive to consider a simpler scheduling problem on single machine without deadlines. This special case admits an exact solution.

Proposition 1. *Define the weight of job* $\mathcal{J}_i$ *as:*

$$w_i = \frac{v_i \beta^{t_i}}{1 - \beta^{t_i}}. \tag{1}$$

If there is a single machine, and there are no deadlines, then placing jobs in decreasing order of w_i *maximizes the social welfare.*

This result follows from a simple interchange argument. Proposition 1 provides the basis for a natural greedy approximation: schedule jobs in decreasing order of weights $w_i = v_i\beta^{t_i}/(1-\beta^{t_i})$ on the machine providing the earliest completion time, as long as jobs complete by their deadlines. A formal algorithm is shown in Greedy Scheduling Algorithm (GS). Despite its simplicity, GS provides performance guarantees for any discount factor $0 < \beta < 1$, under an assumption on the minimum slackness of any job $s = \min_i d_i/t_i > 1$.

Theorem 1. *Assume the minimum slackness of any job* $s = \min_i d_i/t_i > 1$*, then GS provides an approximation of* $1+s/(s-1)$ *to the PVS allocation problem.*

Remark 1. Intuitively, the assumption $s > 1$ means all agents are willing to wait at least a small amount of time proportional to the length of their job. For some applications, it is plausible that $s \approx 1$ making the performance guarantee vacuous. However, this bound is conservative for $s \approx 1$. In the full version of this paper, we show the actual approximation factor approaches $(2-\beta)/(1-\beta)$ as s goes to 1. Meaning GS gives a constant factor approximation for all $\beta < 1$. For practical application, we assume that cloud providers can use historical data to estimate s for their platform and assess our mechanism's guarantee from there.

3.1 Analysis of GS

The analysis of GS relies on dual fitting, an approach for proving approximation guarantees on greedy algorithms [16]. At a high level, we consider an LP relaxation of PVS and its dual. For any dual feasible variables λ, define $cost(\lambda)$ as the value of the dual problem evaluated at λ. Abusing notation, let GS be the value of the greedy schedule. Suppose we can show $\alpha GS \geq cost(\lambda)$ for some $\alpha \geq 1$, then weak duality implies GS is an α approximation to the optimal schedule. Under standard terminology, we say the algorithm GS is charged α to pay for the dual variables λ.

LP Relaxation of PVS and its Dual. We begin with the natural LP relaxation of PVS and its dual. Let $\mathcal{I}_i(t) = \{s : s \leq d_i, t \leq s \leq t + t_i - 1\}$ be the set of feasible finishing times for job $\mathcal{J}_i$ that overlap the time interval $[t-1, t)$, then an Integer Programming formulation of PVS is:

$$\max_x \sum_{i=1}^{n} \sum_{t \in [t_i, d_i]} v_i \beta^t x_{i,t} \tag{P}$$

$$\text{subject to: } \frac{1}{m} \sum_{i: d_i \geq t} \sum_{s \in \mathcal{I}_i(t)} x_{i,s} \leq 1 \quad t = 1, 2, \ldots, T \tag{C1}$$

$$\sum_{t \in [t_i, d_i]} x_{i,t} \leq 1 \quad i = 1, 2, \ldots, n \tag{C2}$$

$$x_{i,t} \in \{0, 1\} \quad \forall i, t,$$

where $T = \max_i d_i$, is the last deadline. Here, the variables $x_{i,t}$ indicate that job $\mathcal{J}_i$ finishes at time $t \in [t_i, d_i]$. The constraints $(C1)$ require that at most m jobs are scheduled at any point in time, and the constraints $(C2)$ require that each job is scheduled at most once. We obtain the natural LP relaxation with $x_{i,t} \geq 0$. Note that the constraints $x_{i,t} \leq 1$ are redundant due to $(C2)$.

The dual problem has a simple form. There is one dual variable γ_i for each job, and there is one dual variable λ_t for each time slot $t = 1, 2, \ldots, T$. Note that we define time slots in terms of their right endpoints, so that λ_t corresponds to the time interval $(t-1, t]$. The dual problem is:

$$\min_{\lambda, \gamma} \sum_{i=1}^{n} \gamma_i + \sum_{t=1}^{T} \lambda_t \tag{D}$$

$$\text{subject to: } \gamma_i + \frac{1}{m} \sum_{s=t-t_i+1}^{t} \lambda_s \geq v_i \beta^t \quad \forall i, \forall t \in [t_i, d_i] \tag{C3}$$

$$\gamma_i \geq 0, \forall i, \quad \lambda_t \geq 0, \forall t$$

Note that, for each job $\mathcal{J}_i$ there is exactly one constraint for each possible finishing time $t \in [t_i, d_i]$.

Approximation Guarantee. First we show how to construct dual feasible γ and λ to satisfy $(C3)$ for jobs used by GS. For each time slot $t = 1, \ldots, T$, there are at most m jobs processing in the greedy schedule, say $\mathcal{J}_{k_1}, \ldots, \mathcal{J}_{k_m}$. These jobs have weights $w_{k_1}, \ldots, w_{k_m}$. We set:

$$\lambda_t = \sum_{j=1}^{m} w_{k_j}(1-\beta)\beta^{t-1}. \tag{2}$$

Suppose GS finishes job $\mathcal{J}_i$ at time τ_i, then we set:

$$\gamma_i = v_i \beta^{\tau_i} \tag{3}$$

and $\gamma_i = 0$ otherwise.

Lemma 1. *The dual variables γ and λ ensure dual constraints $(C3)$ are satisfied for each job $\mathcal{J}_i$ used in GS.*

Proof (Sketch). We consider the three cases:

Case 1: $(C3)$ for $t \geq \tau_i$. From (3): $\gamma_i + \frac{1}{m} \sum_{l=t-t_i+1}^{t} \lambda_l \geq \gamma_i \geq v_i \beta^t$.

Case 2: $(C3)$ for $t \leq \tau_i - t_i$. GS schedules jobs in decreasing order of weight on the machine providing the earliest completion time. Since $\mathcal{J}_i$ starts processing at time slot $\tau_i - t_i \geq t$, GS must be processing m jobs with higher weight than w_i for all times $l \leq t$. By (2):

$$\gamma_i + \frac{1}{m} \sum_{l=t-t_i+1}^{t} \lambda_l \geq \frac{1}{m} \sum_{l=t-t_i+1}^{t} \lambda_l \geq \sum_{l=t-t_i+1}^{t} w_i(1-\beta)\beta^{l-1} = \beta^{t-t_i} w_i(1-\beta^{t_i}) = v_i \beta^t.$$

The last equality follows from: $w_i(1-\beta^{t_i}) = v_i \beta^{t_i}$, which is easily seen from (1).

Case 3: $(C3)$ for $\tau_i - t_i < t < \tau_i$. This case is handled with a technique similar to Case 2. □

We still need to satisfy $(C3)$ for jobs not used in GS. This is easy if GS always uses jobs with higher weight up to the unused job's deadline. That is, suppose job $\mathcal{J}_j$ is not used in GS, and for all $t \leq d_j$ GS uses a job $\mathcal{J}_i$ with $w_i \geq w_j$, then all of $\mathcal{J}_j$'s dual constraints are satisfied. The argument is essentially the same as Case 2 of Lemma 1. If this is not true, then there is some smallest time $u < d_j$ so that for all time slots $t \in [u+1, d_j]$ GS uses a job with lower weight on some machine. We call $\mathcal{J}_j$ a missed job. Covering dual constraints for missed jobs is the most challenging part of proof. For clarity, we present the remaining argument for a single machine. We show how to generalize the result to multiple machines in the full version of this paper.

Let $\mathcal{J}_j$ be a missed job. Note that $d_j - t_j$ is the last time we can start processing $\mathcal{J}_j$ and have it finish before its deadline. Let $\mathcal{J}_k$ be the job used by GS during this time slot, and τ_k be the time it completes. Since GS schedules jobs in decreasing order of weight, $w_k \geq w_j$ and $u > \tau_k \geq d_j - t_j + 1$. We say that $\mathcal{J}_j$

is missed at time τ_k. In our approach, we will go through each job $\mathcal{J}_k$ in GS and cover dual constraints for any missed jobs at τ_k by increasing λ_t's. Let $w(t)$ be the weight of job GS uses at time t. To cover all of $\mathcal{J}_j$'s dual constraints, we need to increase λ_t's by: $\hat{\lambda}_t = (w_j - w(t))(1-\beta)\beta^{t-1}$, for all times slots $t \in [u+1, d_j]$. This follows from (2), since $\lambda_t + \hat{\lambda}_t = w(t)(1-\beta)\beta^{t-1} + (w_j - w(t))(1-\beta)\beta^{t-1}$ so that:

$$\sum_{t=d_j-t_j+1}^{u} \lambda_t + \sum_{t=u+1}^{d_j} (\lambda_t + \hat{\lambda}_t) = \sum_{t=d_j-t_j+1}^{u} w_k(1-\beta)\beta^{t-1} + \sum_{t=u+1}^{d_j} w_j(1-\beta)\beta^{t-1} \geq v_j\beta^{d_j}.$$

We pay for the $\hat{\lambda}_t$'s using a portion of the value of GS up to time τ_k. In fact, we will show that an extra $1/(s-1)$ copies of the greedy schedule are enough to pay for the increased cost in dual variables for all missed jobs.

Let $Q(\tau_k)$ be the pool of resources available to cover the additional costs $\hat{\lambda}_t$'s needed to satisfy dual constraints for any missed jobs at τ_k. Formally, we set $Q(\tau_1) = v_1\beta^{\tau_1}/(s-1)$, the discounted value of the first job used by the greedy schedule scaled by $1/(s-1)$. We use $Q(\tau_1)$ to pay $C(\tau_1) = \sum_{t=u+1}^{d_j} \hat{\lambda}_t$, the cost of covering $(C3)$ for all missed jobs at τ_1. Suppose $\mathcal{J}_2$ is the second job used by GS, then the available resources to pay for missed jobs at time τ_2 is: $Q(\tau_2) = Q(\tau_1) - C(\tau_1) + v_2\beta^{\tau_2}/(s-1)$. In words, the available resources at time τ_2 are the resources remaining after covering all missed jobs up to time τ_1 plus the discounted value of the next job used in GS. Define $Q(\tau_k)$ similarly for all times τ_k when GS completes job $\mathcal{J}_k$. The following lemma provides a key result.

Lemma 2. *Assume $s = \min_i d_i/t_i > 1$, and let $\mathcal{J}_j$ be a missed job at time τ_k. If $Q(\tau_k)$ satisfies:*

$$\frac{Q(\tau_k)}{1-\beta^{\tau_k}} \geq \frac{w_j}{s-1}, \tag{4}$$

then $Q(\tau_k)$ is enough value to pay for the $\hat{\lambda}_t$'s required to cover $\mathcal{J}_j$'s dual constraints. In addition, if the next job used by the greedy schedule $\mathcal{J}_{k+1}$ completes after d_j, then:

$$\frac{Q(\tau_{k+1})}{1-\beta^{\tau_{k+1}}} \geq \frac{w_{k+1}}{s-1}. \tag{5}$$

Lemma 2 is essentially a technical result, a proof is provided in the full paper.

Proof (Theorem 1). Lemma 1 shows that setting λ and γ according to (2) and (3) respectively satisfies all dual constraints for jobs used in the greedy schedule. Clearly, this costs two copies of GS.

Lemma 2 implies that $1/(s-1)$ extra copies of GS are enough to cover dual constraints of all missed jobs. We start with $\mathcal{J}_1$ the first job of GS. Let $\mathcal{J}_j$ be the missed job at τ_1 with longest processing time. We may assume that $\mathcal{J}_j$ also has the highest weight of all missed jobs at time τ_1 since this means $\mathcal{J}_j$ is the missed job with the highest value. Therefore, satisfying $(C3)$ for $\mathcal{J}_j$ will satisfy $(C3)$ for all other missed jobs at τ_1. Since GS schedules jobs in decreasing order of weight $w_j \leq w_1$. By (1) and $Q(\tau_1) = v_1\beta^{t_1}/(s-1)$, condition (4) of Lemma 2 is satisfied. This means $Q(\tau_1)$ is enough to pay for the increased cost of dual variables needed to satisfy ($C3$) for all jobs missed at time τ_1.

We only pay for a portion of required increase in dual variables now and defer the remaining payment until time τ_2, when the second job of GS completes. Specifically, at τ_1 we only pay for the portion of w_j which exceeds w_2, i.e. $\hat{\lambda}_t = (w_j - w_2)(1-\beta)\beta^{t-1}$ for $t = \tau_1 + 1, \ldots, d_j$. Effectively, this artificially extends the deadline of $\mathcal{J}_2$ to d_j, allowing application of the second condition (5) of Lemma 2 to yield $Q(\tau_2)/(1-\beta^{\tau_2}) \geq w_2/(s-1)$. However, artificially extending the deadline of $\mathcal{J}_2$ also increases the value of GS. To account for this, we add a fictitious missed job $\hat{\mathcal{J}}_j$ at time τ_2 with weight w_2, processing time t_j, and deadline d_j. It is easily seen that this matches the value added to GS. Further, all missed jobs at τ_2, including the newly added fictitious job, have weight less than w_2 and (4) is satisfied again at τ_2. As a result, $Q(\tau_2)$ is enough value to cover the cost of all missed jobs at τ_2. Repeating the above argument for each job used in the greedy schedule we see that our $1/(s-1)$ extra copies of GS are enough to pay for the dual constraints of all missed jobs. In total we require $2 + 1/(s-1) = 1 + s/(s-1)$ copies of GS to construct dual feasible λ and γ. □

The above proof extends easily from a single machine to m identically machines. First we use $w_m(t) = m^{-1}\sum_{i=1}^{m} w(t)$ in place of $w(t)$ in Lemma 2. Then, we proceed through the jobs of GS in increasing order of completion time, covering missed jobs as we go. The full version of the paper provides all details.

Remark 2. The $1+s/(s-1)$ performance guarantee is essentially tight for $\beta \approx 1$ and $s \geq 2$. However, the bound is conservative for $\beta \ll 1$ or if $s \approx 1$. This is due to the somewhat loose analysis in Lemma 2. More careful treatment reveals $C(\beta, s) = \beta^{s-1}(1-\beta)/(1-\beta^{s-1})$ copies of GS are sufficient to cover all missed jobs, giving the performance guarantee of $1 + (1-\beta^s)/(1-\beta^{s-1})$, showing the algorithms dependence on β. We state the conservative bound since we assume most applications require β close to 1. Indeed, $\beta > 0.9$ is common in economics and finance literature.

4 Truthful Mechanism

Our $1 + s/(s-1)$ approximation to the allocation problem is only half of the mechanism design problem. As rational agents, job owners may lie about any or all of their job's parameters (v_i, t_i, d_i) to increase utility. We seek a pricing rule to ensure truthful reporting is a dominate strategy. The task is well understood in single parameter domains where the celebrated Myerson's lemma [12] provides the unique payment rule for any monotone allocation rule. Multi-parameter domains, as our own problem, present a challenge. VCG payments [14] create a truthful mechanisms when the allocation problem can be solved exactly, but many problems of interest require an approximation algorithm. It is known that generalizations of monotonicity are necessary and sufficient in these situations, see [5,8,15], but the conditions are difficult to check. Instead, we show a few simple properties of the GS allocation rule allow us to construct a pricing rule which yields a truthful mechanism. We note that this is an extension of a result first obtained by Jain et al. in [7].

Properties of GS Allocation. We begin by introducing some notation used throughout this section. In PVS, each agent i reports a bid $b_i = (v_i, r_i)$ of their job's value v_i and requirements $r_i = (t_i, d_i)$. Given the set of bids $b = (b_1, \ldots, b_n)$, the cloud provider determines a completion time τ_i for i's job. We say i receives the allocation $\mathcal{A}_i(b) = \beta^{\tau_i}$ and note that i receives a value of $v_i \mathcal{A}_i(b) = v_i \beta^{\tau_i}$.

Assume $b_i = (v_i, r_i)$ are the agent's true valuation and job requirements, but they may misreport any of these values. If the agent submits a false bid $b'_i = (v'_i, r'_i)$, the actual allocation they receive may be different from $\mathcal{A}_i(b'_i, b_{-i})$ depending on their true requirements. For example, if an agent reports $t'_i > t_i$ and is scheduled for the time slot $[0, t'_i)$, then their actual allocation is the interval $[0, t_i)$ as their job only requires t_i units of processing time. Define $\mathcal{A}_i(b|r)$ as the actual allocation received by agent i assuming the requirements r. Continuing the earlier example with $t'_i > t_i$, then $\mathcal{A}_i(b'_i, b_{-i}|r_i) = [0, t_i)$. We assume $\mathcal{A}_i(b'_i, b_{-i}|r') = \mathcal{A}_i(b'_i, b_{-i})$. Finally, we note that the true benefit received from bidding b'_i is $v_i \mathcal{A}_i(b'_i, b_{-i}|r_i)$ and the utility received is:

$$u_i(b') = v_i \mathcal{A}_i(b'|r_i) - p_i(b'). \tag{6}$$

We show how a few simple properties of the allocation $\mathcal{A}_i$ allow us to construct a pricing rule $p_i(b_i, b_{-i})$ which yields a truthful mechanism. For notational convenience, we drop the b_{-i} argument and write $\mathcal{A}_i(v, r|r')$ instead of $\mathcal{A}_i(b_i, b_{-i}|r')$ or $p_i(v, r)$ instead of $p_i(b_i, b_{-i})$.

Definition 1: An allocation rule $\mathcal{A}$ is rational if for all agents i, all bids b_{-i}, all requirements r, r': $\mathcal{A}_i(v, r'|r) > 0 \implies \mathcal{A}_i(v, r'|r') > 0$.

Definition 2: An allocation rule $\mathcal{A}$ is value monotonic if for all agents i, all bids b_{-i}, all requirements r, r', and all valuations $v < v'$:

$$\mathcal{A}_i(v, r'|r) \leq \mathcal{A}_i(v', r'|r). \tag{7}$$

Definition 3: An allocation rule $\mathcal{A}$ is requirement monotonic if for all agents i, all bids b_{-i}, all requirements r, r' the following property holds: if there exists a v such that $\mathcal{A}_i(v, r'|r) > 0$ then:

$$\mathcal{A}_i(v, r'|r') \leq \mathcal{A}_i(v, r'|r) \quad \text{and} \quad \mathcal{A}_i(v', r'|r') \leq \mathcal{A}_i(v', r|r), \forall v'. \tag{8}$$

Intuitively these definitions have the following meanings: Rationality says if an allocation satisfies an alternative set of job requirements r, then it must also satisfy the requested requirements r'. Value monotonicity asks that the allocation is non-decreasing in the valuation v after fixing a set of job requirements. Finally, requirement monotonicity says if an allocation meets an alternate set of requirements r, then these requirements must be easier to satisfy and will always receive an allocation at least as good as the original request r'. Before establishing that the GS algorithm satisfies these properties, we show how they contribute to a truthful mechanism.

Proposition 2. *Let $\mathcal{A}$ be a non-negative, rational, value monotonic and requirement monotonic allocation rule, then mechanism $\mathcal{M}(\mathcal{A}, p)$ using the pricing rule:*

$$p_i(v, r) = v\mathcal{A}_i(v, r|r) - \int_0^v \mathcal{A}_i(x, r|r)dx \quad (9)$$

is truthful and individually rational.

Note that the form of the payment rule is the same as that of Myerson's lemma, the important distinction being that jobs have additional requirements which must be satisfied, e.g., complete before their deadline. This result is an extension of [7] in which allocations are binary functions, i.e., jobs have constant value and are either completed or not. In PVS, the allocation is piece-wise constant. This means pricing rule (9) reduces to: the sum over (change in allocation) * (value where the allocation changes). The derivation is similar to the familiar single parameter case of Myerson's lemma. For more details, see the full version of the paper.

PVS Mechanism. Before showing GS satisfies the assumptions of Proposition 2, we impose a few natural constraints on what agents may misreport. It is important to note that these are not additional assumptions, rather certain types of misreporting are dominated by truthfulness. Therefore, a rational job owner would not misreport values in these ways. We assume $b_i = (v_i, t_i, d_i)$ are $\mathcal{J}_i$'s true valuation and requirements, and $b'_i = (v'_i, t'_i, d'_i)$ are alternative values. First, agents may only misreport longer processing times $t'_i \geq t_i$. This holds since agents receive no benefit from partially completed jobs. As such, a job owner reporting $t'_i < t_i$ gains no benefit from any allocation but is charged a non-negative price, implying $u_i(b'_i, b_{-i}) \leq 0$. Therefore, no rational job owner would report $t'_i < t$. Second, we assume agents may only under report their deadlines $d'_i \leq d_i$. This case is similar to the first, completing a job after the deadline provides no benefit but requires a non-negative payment creating the possibility for negative utility. We now show that GS satisfies the conditions of Proposition 2.

Proposition 3. *The GS allocation is rational, price monotonic, and requirement monotonic.*

These properties follow easily from the fact that GS greedily schedules jobs in decreasing order of weight: $w_i = v_i\beta^{t_i}/(1 - \beta^{t_i})$, which is increasing in v_i and decreasing in t_i. Full details are provided in the full paper. Proposition 3 shows

GS satisfies the assumptions of Proposition 2, providing a truthful mechanism when using pricing rule (9). Combining this with Theorem 1 we obtain:

Corollary 1. *Assuming* $s = \min_i d_i/t_i > 1$, *the mechanism consisting of the GS allocation rule and the pricing rule (9) gives a truthful* $1+s/(1-s)$ *approximation to the social welfare maximizing schedule.*

5 Conclusion

In this paper, we propose a new scheduling problem, PVS, where jobs have hard deadlines and their values decay over time. Our simple and fast greedy scheduling algorithm, GS, provides reasonable performance guarantees under the relatively mild assumption that users are willing to wait for at least a short period of time, i.e. $s > 1$. Further, we exploit the greedy nature of GS to extend the celebrated Myerson's Lemma to a special multi-parameter domains where users report processing time and deadline in addition to their job value. From this, we obtain a mechanism for PVS which truthful with respect to all job parameters.

Our model does suffer from some over simplifications. Most notably, all users must have the same discount factor β. Allowing user specific discount factor β_i is more realistic. Further, β_i should be private information so that users report bids $b_i = (v_i, t_i, d_i, \beta_i)$. This presents interesting and challenging problems in both the design of an approximation algorithm and a truthful mechanism. We leave this to future work. Another avenue for future work is the design of a truthful revenue maximizing mechanism.

References

1. Azar, Y., Kalp-Shaltiel, I., Lucier, B., Menache, I., Naor, J.S., Yaniv, J.: Truthful online scheduling with commitments. In: Proceedings of the Sixteenth ACM Conference on Economics and Computation, pp. 715–732. ACM (2015)
2. Bar-Noy, A., Bar-Yehuda, R., Freund, A., Naor, J., Schieber, B.: A unified approach to approximating resource allocation and scheduling. J. ACM (JACM) **48**(5), 1069–1090 (2001)
3. Bar-Noy, A., Guha, S., Naor, J., Schieber, B.: Approximating the throughput of multiple machines in real-time scheduling. SIAM J. Comput. **31**(2), 331–352 (2001)
4. Berman, P., DasGupta, B.: Multi-phase algorithms for throughput maximization for real-time scheduling. J. Comb. Optim. **4**(3), 307–323 (2000)
5. Bikhchandani, S., Chatterji, S., Lavi, R., Mu'alem, A., Nisan, N., Sen, A.: Weak monotonicity characterizes deterministic dominant-strategy implementation. Econometrica **74**(4), 1109–1132 (2006)
6. Jain, N., Menache, I., Naor, J.S., Yaniv, J.: A truthful mechanism for value-based scheduling in cloud computing. Theory Comput. Syst. **54**(3), 388–406 (2014)
7. Jain, N., Menache, I., Naor, J.S., Yaniv, J.: Near-optimal scheduling mechanisms for deadline-sensitive jobs in large computing clusters. ACM Trans. Parallel Comput. **2**(1), 3 (2015)
8. Kovács, A., Vidali, A.: A characterization of n-player strongly monotone scheduling mechanisms. In: IJCAI, pp. 568–574 (2015)

9. Lavi, R., Swamy, C.: Truthful mechanism design for multi-dimensional scheduling via cycle monotonicity. In: Proceedings of the 8th ACM Conference on Electronic Commerce, pp. 252–261. ACM (2007)
10. Lavi, R., Swamy, C.: Truthful and near-optimal mechanism design via linear programming. J. ACM (JACM) **58**(6), 25 (2011)
11. Mu'alem, A., Schapira, M.: Setting lower bounds on truthfulness. In: Proceedings of the Eighteenth Annual ACM-SIAM Symposium on Discrete Algorithms, pp. 1143–1152. Society for Industrial and Applied Mathematics (2007)
12. Myerson, R.B.: Optimal auction design. Math. Oper. Res. **6**(1), 58–73 (1981)
13. Nisan, N., Ronen, A.: Algorithmic mechanism design. In: Proceedings of the Thirty-First Annual ACM Symposium on Theory of Computing, pp. 129–140. ACM (1999)
14. Nisan, N., Roughgarden, T., Tardos, E., Vazirani, V.V.: Algorithmic Game Theory, vol. 1. Cambridge University Press, Cambridge (2007)
15. Rochet, J.C.: A necessary and sufficient condition for rationalizability in a quasi-linear context. J. Math. Econ. **16**(2), 191–200 (1987)
16. Vazirani, V.V.: Approximation Algorithms. Springer, Heidelberg (2013)

On Revenue-Maximizing Mechanisms Assuming Convex Costs

Amy Greenwald[1], Takehiro Oyakawa[1](✉), and Vasilis Syrgkanis[2]

[1] Brown University, Providence, RI 02912, USA
{amy_greenwald,takehiro_oyakawa}@brown.edu
[2] Microsoft Research, Cambridge, MA 02142, USA
vasy@microsoft.com

Abstract. We investigate revenue-maximizing mechanisms in settings where bidders' utility functions are characterized by convex costs. Such costs arise, for instance, in procurement auctions for energy, and when bidders borrow money at non-linear interest rates. We provide a $1/16e$ approximation guarantee for a prior-free randomized mechanism when bidders' values are drawn from MHR distributions, and their costs are polynomial. Additionally, we propose two heuristics that allocate proportionally, using either bidders' values or virtual values. Perhaps surprisingly, in the convex cost setting, it is preferable to allocate to multiple relatively high bidders, rather than only to bidders with the highest (virtual) value, as is optimal in the traditional quasi-linear utility setting.

Keywords: Mechanism design · Optimal auction · Prior-free

1 Introduction

In the field of mechanism design, a central planner attempts to implement a socially-optimal outcome for the mechanism's participants, without knowing their preferences, which are assumed to be private information. While certain problems in this space have been solved elegantly, solutions are sometimes too complex for widespread practical implementation. Consequently, over the past decade, researchers have been exploring the trade off between simplicity and optimality in mechanism design. A recent line of work has addressed whether simple mechanisms can achieve approximately optimal performance in single-dimensional environments [3,11,12,14,16]. Often, simplicity requirements take the form of prior-freeness, meaning the mechanism should not depend on any distributional knowledge about participants' private information [21,22]. Such *oblivious* designs lead to more robust guarantees, as they do not heavily depend on modeling assumptions, or on data collection to learn about the participants' private information.

All of the aforementioned work on simple, optimal auctions assumes that bidders' utilities are quasi-linear with respect to payments, i.e., $u_i = v_i x_i - p_i$, where $v_i > 0$ is i's private value, x_i is his allocation, and p_i is his payment to

X. Deng (Ed.): SAGT 2018, LNCS 11059, pp. 113–124, 2018.
https://doi.org/10.1007/978-3-319-99660-8_11

the auctioneer.[1] In this paper, we investigate the problem of simple, optimal auction design, assuming utilities of the form $u_i = v_i x_i - c_i(p_i)$. Here, $c_i(\cdot)$ can be understood to represent bidder i's *value of money*, which is naturally modelled as a convex function for risk-seeking bidders, or a concave function, for risk-averse bidders. Analogously, $-c_i(\cdot)$ can be understood to represent bidder i's *cost of money*, which is naturally modelled as a concave function for risk-seeking bidders, or a convex function, for risk-averse bidders. These costs can be hallucinated to hedge against any uncertainty a bidder might have about its private information, or about whether the auctioneer will indeed allocate as promised. These costs can also be real; they can model additional payments made to anyone other than the auctioneer: e.g., a bank, if interest is owed on the payment p_i.

Our original motivation for this study stemmed from a reverse auction design problem in the realm of renewable energy markets. Consider a procurement auction in which a government with a fixed budget is offering subsidies (in euros, say) to power companies in exchange for a supply of renewable energy (in watts, say). Suppose the power companies' utility functions take this form: $\mathfrak{u}_i = x_i - c_i(p_i)/v_i$, where x_i is some fraction of the total budget in euros, and p_i is some deliverable amount of power in watts.[2] The value v_i which is measured in watts per euro, is a private conversion (i.e., scaling) factor used to convert power from watts to euros. The assumption that $c_i(\cdot)$ is convex reflects the fact that energy production costs may not be linear; on the contrary, it may be the case that as more energy is produced, additional units become more expensive to produce due to a scarcity of raw materials.[3]

Another problem which also fits into our framework is the problem of allocating time, rather than money—for example, a media network allocating advertising time to retailers, or a cloud service provider allocating computation time. In this application, an agent's utility is calculated by converting its allocation, in time, into dollars via its private value (measured in dollars per time unit), and then subtracting the cost of production: $u_i = v_i x_i - c_i(p_i)$. Here again, production costs can be convex; for example, the agent might have to borrow funds to enable production, and in so doing, might be subject to a non-linear (and publicly disclosed) interest rate.

Whether allocating money or time, we assume the auctioneer seeks an optimal auction: i.e., one that maximizes its total expected "revenue". In the energy problem, the auctioneer's (i.e., the government's) objective is to maximize the amount of power produced, subject to its budget constraint. In the advertising and cloud service problems, the auctioneer seeks to maximize the revenue it can earn by selling access to its resources (i.e., time).

---

[1] A notable exception is [8], who study prior-free auctions for risk-averse agents, which are modelled by a very specific form of capped quasi-linear utilities.

[2] Multiplying $\mathfrak{u}_i$ by v_i yields a familiar utility function, that of the forward setting, with utility measured in units of power, rather than money: $v_i \mathfrak{u}_i = u_i = v_i x_i - c_i(p_i)$.

[3] For example, it is more expensive to convert bitumen into synthetic crude oil than it is to drill and pump conventional crude oil.

The departure from quasi-linear utilities presents both technical and qualitative differences from the standard mechanism design setting. Technically, the optimal mechanism does not appear to have an intuitive, closed-form characterization, but rather is the outcome of a mathematical program. Beyond this difficulty, the Myerson characterization [14], in which the optimal auction in a symmetric environment with regular distributions allocates to bidders with the highest values above some reserve, is no longer valid. As our next example shows, a mechanism that allocates only to the highest-value bidders cannot achieve a constant-factor approximation.

Example 1. We show that, if we insist on allocating to only bidders of the highest value, the resulting mechanism can be very suboptimal, with a suboptimality ratio that decays to zero as the number of bidders n approaches infinity, at a rate of $1/n^{1/4}$.

Let $c_i(p_i) = p_i^2$ for each bidder i, and consider the following distribution of values: the value of each bidder is either 1 with probability $\frac{\log(n)}{\sqrt{n}}$, or $1-\epsilon$ with probability $1-\frac{\log(n)}{\sqrt{n}}$.

As the number of bidders grows large, then with very high probability, there will be approximately $\sqrt{n}$ bidders with value 1 and $n-\sqrt{n}$ bidders with value $1-\epsilon$. So the allocation of a highest-bidders-win auction is approximately: $1/\sqrt{n}$ for a bidder with value 1, and 0 for a bidder with value $1-\epsilon$. It follows that the payment of a bidder with value 1 is at most $\sqrt{xv} = \sqrt{x(1)} \approx 1/n^{1/4}$ (see Sect. 2.3), while the payment of a bidder with value $1-\epsilon$ is approximately 0. Thus, the expected payment of a single bidder is approximately $\left(\frac{1}{\sqrt{n}}\right)\left(\frac{1}{n^{1/4}}\right)$, and the total expected revenue is n times this quantity, which is $n^{1/4}$.

On the other hand, if we instead allocate to all bidders uniformly at random (as long as they bid at least $1-\epsilon$), then each bidder's payment is $\sqrt{xv} = \sqrt{\frac{1}{n}(1-\epsilon)}$, leading to a total expected revenue of $\sqrt{n(1-\epsilon)}$. As $\epsilon \to 0$, the ratio of the revenue of this latter mechanism to the former's is $O(n^{1/4})$.

This finding is not specific to this pathological example. Experimentally, we find that for many value distributions, natural mechanisms that allocate only to bidders with the highest value can perform very poorly in the convex cost setting, in spite of performing well in quasi-linear settings.

Contributions. In this work, we design a prior-free mechanism that achieves a constant factor worst-case approximation ratio relative to the optimal, for the convex cost setting. We also design two heuristic mechanisms for this setting that allocate to multiple relatively high bidders, which in our experiments perform near optimally for a wide variety of distributions over values. Our main theoretical results hold when cost functions take the form $c_i(p_i) = p_i^d$, for $d \geq 2$, and when the distribution over values satisfies the monotone hazard rate condition. In this specific setting, our theoretical results can be summarized as follows:

- We characterize an upper bound on the revenue of the optimal mechanism by finding a closed-form solution to a convex program that upper bounds the optimal revenue.
- We show that a mechanism which sets a reserve price by drawing randomly from the distribution of values, and then allocating uniformly to all bidders above this reserve, is a constant-factor approximation to the optimal mechanism. This result implies a prior-free mechanism that is a constant-factor approximation of the optimal: randomly pick one bidder, and then allocate uniformly to all bidders whose values lie above that of this price-setting bidder. This idea is stylistically similar to that of Bulow and Klemperer, who remove one bidder from the Vickrey auction to achieve an approximation guarantee relative to the revenue-maximizing auction [3]. However, as we argued in Example 1, allocating only to the highest bidders can be very sub-optimal. Hence, our mechanism allocates uniformly at random to all bidders that surpass the reserve. This simple modification is crucial to obtaining a constant-factor approximation in the convex cost setting.

Related Work. [20] showed that auctions in which the highest bidder wins and pays the second-highest bid incentivize bidders to bid truthfully. [14] showed that in the single-parameter setting, with the usual quasi-linear utility function involving linear payments, total expected revenue is maximized by a Vickrey auction with reserve prices. Our setting is not captured by Myerson's classic characterization because costs in our model are not equivalent to payments.

[3], [12], and [16] study simple prior-free mechanisms. The results in [3] show that running a simple second-price auction is an $(n-1)/n$ approximation in symmetric settings with quasi-linear utilities. [12] extend this analysis to obtain similar results in asymmetric settings. Bulow and Klemperer's result can also be phrased as obtaining a constant-factor approximation by running a second-price auction with a reserve drawn from the distribution of values. Our constant-factor prior-free result stems from this intuition.

In prior work [10], we also study convex costs, albeit in a contest (i.e. indirect) setting, for which we obtain interim guarantees. Here, we use the same convex costs, but we focus on direct mechanisms and obtain ex-post guarantees (i.e., we require that constraints hold for all possible type profiles).

The technical difficulties that arise in our setting are similar in spirit to the ones faced by [15] when designing optimal auctions for budget-constrained bidders. If $c_i(p_i) = p_i^k$ for some $k \gg 0$, then $u_i = v_i x_i - p_i^k$ quickly approaches $-\infty$ if $p_i > (v_i x_i)^{1/k}$. Thus, we can interpret a utility function with convex costs as a continuous approximation of that of a budget-limited agent whose utility is $-\infty$ whenever her payment exceeds her budget.

Our model also has strong connections with the literature on optimal auctions for risk-averse buyers (see [13]), since a concave utility function can be interpreted as a form of risk aversion.

Translating a reverse auction into a direct auction by multiplying utility by the private parameter v_i was previously proposed in the literature on optimal contests (see, for example, [5,7]).

Procuring services subject to a budget constraint is also the subject of the literature on budget-feasible mechanisms initiated by [19]. However, in this literature, the service of each seller is fixed and the utility of the buyer is a combinatorial function of the set of sellers the buyer picks. In our setting, each seller can produce a different level of service by incurring a different cost, so the buyer picks not only a set of sellers, but a level of service that each seller should provide as well. This renders the two models incomparable.

Settings where bidders' utilities decrease at least as quickly as payments increase have been studied by [17], in the context of strategy-proof environments.

For a recent survey on optimal mechanism design with non-linear preferences (mostly budget constraints), we refer the reader to Chap. 8 of [11]. In principle, some formulations of our problem can be solved using Border's characterization [2] of interim feasible outcomes and an ellipsoid-style algorithm with a separation oracle. Even more generally, we can apply the algorithmic approach of [4] for computing the optimal mechanism, which again is based on an ellipsoid-style algorithm. However, such mechanisms tend to be computationally expensive and do not yield closed-form characterizations or interpretable mechanisms. Here, we seek fast allocation heuristics with potential economic justification, such as virtual-value-based maximization. Virtual-value-maximizing approximations to optimal auction design were also studied recently by [1] in the context of multi-dimensional mechanism design, and from a worst-case point of view.

2 Model and Preliminaries

There is one auctioneer/seller who would like to sell one unit-sized divisible good, and there are n bidders that would like to buy as much of it as possible. Each bidder $i \in N = \{1, \ldots, n\}$ has a private value for the good in its entirety. Each value v_i is drawn independently from an atomless distribution F, with continuous probability density f that is strictly positive on the support, which is the closed interval $T_i = [0, \bar{v}]$. We write $\mathbf{v} = (v_1, \ldots, v_n) \in T$ to denote a sample vector of values, drawn from distribution F^n.

Given a vector of reports $\mathbf{b} = (b_1, \ldots, b_n) \in \mathbb{R}^n$, with $b_i \in T_i$, for all $i \in N$, a mechanism consists of an allocation rule $\mathbf{x}(\mathbf{b}) \in [0, 1]^n$ together with a payment rule $\mathbf{p}(\mathbf{b}) \in \mathbb{R}_{\geq 0}$, where bidder i's payment to the seller is $p_i(\mathbf{b})$. For vectors such as $\mathbf{b}$, we use the notation $\mathbf{b} = (b_i, \mathbf{b}_{-i})$ to emphasize the distinction between bidder i's role in the auction, and all other bidders $N \setminus \{i\}$.

Let the utility of each bidder i be $u_i(b_i, \mathbf{b}_{-i}) = v_i x_i(b_i, \mathbf{b}_{-i}) - c_i\left(p_i(b_i, \mathbf{b}_{-i})\right)$, where $c_i : \mathbb{R}_{\geq 0} \to \mathbb{R}_{\geq 0}$ is a convex cost function such that $c_i(0) = 0$. For readability, we often write $c_i(b_i, \mathbf{b}_{-i})$ instead of $c_i\left(p_i(b_i, \mathbf{b}_{-i})\right)$. This is bidder i's *cost*, beyond his payment $p_i(\mathbf{b})$, the latter of which only includes payments made to the auctioneer. Similar to the payment rule, we refer to $\mathbf{c}(\mathbf{b}) \in \mathbb{R}^n_{\geq 0}$, comprised of variables $c_i(\mathbf{b})$, as the cost rule.

2.1 Constraints

Next, we formalize the constraints we impose on an optimal auction design. Because we restrict our attention to incentive compatible auctions, where it is optimal to bid truthfully, we write, for example, $c_i(v_i, \mathbf{v}_{-i})$ instead of $c_i(b_i, \mathbf{b}_{-i})$.

A mechanism is called **incentive compatible** (IC) if each bidder maximizes her utility by reporting truthfully (i.e., $b_i = v_i$): $\forall i \in N$, $\forall v_i, w_i \in T_i$, and $\forall \mathbf{v}_{-i} \in T_{-i}$, $v_i x_i(v_i, \mathbf{v}_{-i}) - c_i(v_i, \mathbf{v}_{-i}) \geq v_i x_i(w_i, \mathbf{v}_{-i}) - c_i(w_i, \mathbf{v}_{-i})$. **Individual rationality** (IR) ensures that bidders have non-negative utilities: $\forall i \in N$, $\forall v_i \in T_i$, and $\forall \mathbf{v}_{-i} \in T_{-i}$, $v_i x_i(v_i, \mathbf{v}_{-i}) - c_i(v_i, \mathbf{v}_{-i}) \geq 0$. We say a mechanism is **ex-post feasible** (XP) if it never overallocates: $\forall \mathbf{v} \in T$, $\sum_{i=1}^{n} x_i(v_i, \mathbf{v}_{-i}) \leq 1$. Finally, we require that $0 \leq x_i(\mathbf{v}) \leq 1$, $\forall i \in N$, $\forall \mathbf{v} \in T$.

The goal of the auctioneer is to maximize **total expected (ex-post) revenue**, which is equal to: $\mathbb{E}_{\mathbf{v}}\left[\sum_{i \in N} p_i(\mathbf{v})\right]$.

2.2 Distributions and Properties

We introduce some useful notation and terminology with respect to properties of the value distribution F. For any distribution F, let $q(v) = 1 - F(v)$ be the quantile function, and let $v(q) = q^{-1}(\cdot)$ be the inverse quantile function. The quantile of a value v is the probability that a random draw from distribution F exceeds v. Observe that quantiles are distributed uniformly on $[0, 1]$.

Define $R(q) = v(q)q = v(q)(1 - F(v(q))$. Since F is atomless with support $[0, \bar{v}]$, $R(0) = R(1) = 0$. Adopting language from the traditional quasi-linear setting, we continue to refer to the function R as the **revenue** curve, but note that the function R should not be understood as revenue in the convex cost setting.[4] Finally, let $q^* \in \arg\max_{q \in [0,1]} R(q)$ be the quantile corresponding to the optimal revenue. We will also denote by $\kappa = v(1/2)$ the median of the distribution, and by $\mu = \mathbb{E}_{v \sim F}[v]$, the mean.

We will be looking at two standard classes of distributions. The smaller class is that of monotone hazard rate (MHR) distributions, which require that $h(v) = f(v)/(1 - F(v))$ be monotone non-decreasing. The larger class is that of regular distributions, which require that $R(q)$ be a concave function, or equivalently, that $\varphi(q) \equiv R'(q) = v(q) - \frac{1-F(v(q))}{f(v(q))}$, be monotone decreasing. Since $\varphi(q) = v(q) - \frac{1}{h(v(q))}$, an MHR distribution is also regular.

We now state two lemmas describing bounds on revenue curves, depending on the assumptions made on the distributions values are drawn from.

Lemma 1 ([6]). *For any MHR distribution,* $R(q^*) \geq \frac{\mu}{e}$.

Lemma 2 ([18]). *For any regular distribution,* $R(q^*) \leq \kappa$.

---

[4] In the convex cost setting, if we interpret $v(q)$ as a posted cost, rather than a posted price (i.e., payment) then $R(q)$ can be wrongly interpreted as an expected cost function.

2.3 Allocation, Payments, and Revenue

Myerson showed that for a mechanism to satisfy IC, IR and XP, several conditions needs to hold. We restate his result below, adapted to the convex cost setting.

Theorem 1 ([14]). *Assuming a convex cost function, a mechanism is IC and IR if and only if* $\forall i \in N, \forall v_i \in T_i, \forall \mathbf{v}_{-i} \in T_{-i}$, *the following conditions hold:*

1. *The allocation rule is monotone:* $x_i(v_i, \mathbf{v}_{-i}) \geq x_i(w_i, \mathbf{v}_{-i}), \forall v_i \geq w_i \in T_i,$,
2. *Costs satisfy the following condition:* $v_i x_i(\mathbf{v}) - c_i(\mathbf{v}) = \int_0^{v_i} x_i(z, \mathbf{v}_{-i})\, \mathrm{d}z,$.

Myerson also showed that the total expected revenue of such a mechanism can be described using **virtual values**, $\varphi_i(v_i) = v_i - \frac{1-F_i(v_i)}{f_i(v_i)} = R'(q(v_i))$. We restate his findings here, adapted to the convex cost setting.

Theorem 2 ([14]). *Assuming a convex cost function, the total expected cost of an IC, IR, and XP mechanism,* $\sum_{i \in N} \mathbb{E}_{\mathbf{v}}\left[c_i(v_i, \mathbf{v}_{-i})\right]$ *is equal to* $\sum_{i \in N} \mathbb{E}_{\mathbf{v}}\left[\varphi_i(v_i)\, x_i(v_i, \mathbf{v}_{-i})\right]$.

In the traditional quasi-linear setting, when the cost function is the identity function, the fact that virtual surplus is equivalent to revenue tells us that in order to maximize revenue the good should be allocated to bidders with the highest non-negative virtual values. However, in the convex cost setting, revenue is not pinned down by Myerson's theorem; only cost is. Moreover, revenue does not have any obvious interpretation using Myerson's characterization of costs. As an early work on this convex cost setting, we restrict our attention to polynomial costs, where $c = p^d$, for some $d \geq 1$.

3 Upper Bound on Optimal Revenue

Although we cannot easily find an optimal mechanism in the convex cost setting, or even calculate the optimal revenue, we can provide an upper bound on the optimal revenue.

By IR, we have the following: $\sum_{i \in N} \mathbb{E}_{\mathbf{v}}\left[p_i(\mathbf{v})\right] \leq \sum_{i \in N} \mathbb{E}_{\mathbf{v}}\left[c_i^{-1}\left(v_i x_i(\mathbf{v})\right)\right]$. We call the quantity on the right hand side of this inequality, **pseudo-surplus**, and give a closed-form solution for maximizing it presently.

Lemma 3. *When,* $\forall i \in N$, $c_i = p_i^d$, *where* $d > 1$, *the allocations that maximize pseudo-surplus are given by*

$$x_i(v_i, \mathbf{v}_{-i}) = v_i^{\frac{1}{(d-1)}} / \sum_{j \in N} v_j^{\frac{1}{(d-1)}}, \forall i \in N, \forall \mathbf{v} \in T. \tag{1}$$

Proof. Compute the derivative of $\left(v_i x_i(v_i, \mathbf{v}_{-i})\right)^{1/d}$ with respect to $x_i(v_i, \mathbf{v}_{-i})$. By the equi-marginal principle [9], these derivatives must be equal for all bidders i. Using this fact, and the ex-post feasibility condition, we arrive at the result.

Let OPT be the maximum total expected revenue that could be generated by any IC/IR/XP mechanism. The following lemma upper bounds OPT, regardless of the distribution F.

Lemma 4. *When, $\forall i \in N$, $c_i = p_i^d$, where $d \geq 2$:*

$$OPT \leq n \left({}^{\mu}/_{n} \right)^{1/d}. \tag{2}$$

Proof (Sketch). Starting with the optimal solution to pseudo-surplus given by Lemma 3, we can upper-bound OPT: $\text{OPT} \leq \mathbb{E}_{\mathbf{v}} [(\sum_{i \in N} v_i^{1/(d-1)})^{\frac{d-1}{d}}]$. Observe that $f(x) = x^{(d-1)/d}$ is a concave function for any $d > 1$, and $g(x) = x^{1/(d-1)}$ is a concave function for any $d \geq 2$. By Jensen's inequality, for a concave function G, $\mathbb{E}_{\mathbf{v}} [G(X)] \leq G(\mathbb{E}_{\mathbf{v}} [X])$. Applying this inequality using our observations of f and g yields Eq. (2).

Finally, we upper bound OPT assuming values are drawn from an MHR distribution.

Theorem 3. *When, $\forall i \in N$, $c_i = p_i^d$, where $d \geq 2$ and the value distribution F is an MHR distribution:*

$$OPT \leq n \left({}^{e\kappa}/_{n} \right)^{1/d}. \tag{3}$$

Proof. Lemmas 1 and 2 tell us that: $\mu \leq eR(q^*) \leq e\kappa$. This, combined with Lemma 4, proves the theorem.

In the bounds given by Lemma 4 and Theorem 3, which only rely on n, d, μ, e and κ, we observe that as d tends towards infinity, the upper bound on OPT tends to n. This trend is to be expected: as the rate of growth of the cost function increases, less and less can be extracted from the bidders.

4 Reserve Price Mechanisms

We now turn to the design of simple prior-free mechanisms for the convex payment setting. We begin our analysis by looking at a *uniform-allocation reserve price mechanism*, i.e. a mechanism that allocates uniformly to all bidders whose value v_i is above some reserve price r, or equivalently, to bidders whose quantile q_i is below some quantile reserve $\hat{q}$. This mechanism charges the Z winning bidders the reserve price $c_i^{-1}(v(\hat{q})/Z)$. This payment rule makes the mechanism IC and IR.

We then describe the performance of a mechanism that selects a quantile reserve uniformly at random. Finally, we show how selecting a quantile reserve uniformly at random corresponds to a prior-free mechanism where one bidder is picked at random to be used as the reserve price setter. This leads to the main result of this section: a constant-factor approximately optimal prior-free mechanism.

We begin with the analysis of the revenue of a *uniform-allocation reserve price mechanism*, with quantile reserve $\hat{q}$.

Lemma 5. *Consider a convex cost setting with $c_i = p_i^d$, $\forall i \in N$, where $d \geq 1$. Let $APX(\hat{q})$ be the expected revenue of a mechanism that allocates uniformly across all bidders with quantile $q_i \leq \hat{q}$ and charges each of these Z bidders ex-post truthful payments $(v(\hat{q})/Z)^{1/d}$. Then:*

$$APX(\hat{q}) \geq n \left(\frac{v(\hat{q})}{1 + (n-1)\hat{q}} \right)^{1/d} \hat{q}. \tag{4}$$

Proof. The expected revenue of the mechanism with reserve price $v(\hat{q})$ is $\mathrm{APX}(\hat{q}) = \mathbb{E}_{\mathbf{v} \sim F} \left[\sum_{i=1}^{n} \left(\frac{v(\hat{q}) \mathbb{1}_{v_i \geq v(\hat{q})}}{\sum_{j=1}^{n} \mathbb{1}_{v_j \geq v(\hat{q})}} \right)^{1/d} \right]$. The probability that $v_i \geq v(\hat{q})$ is $1 - F(v(\hat{q})) = \hat{q}$, and if there exists a winner, the denominator is at least one, so

$$\mathrm{APX}(\hat{q}) = \sum_{i=1}^{n} \mathbb{E}_{\mathbf{v} \sim F} \left[\left(\frac{1}{1 + \sum_{j=1}^{n-1} \mathbb{1}_{v_j \geq v(\hat{q})}} \right)^{1/d} \right] v(\hat{q})^{1/d} \hat{q}.$$

The function $h(x) = 1/(1+x)^{1/d}$ is convex for $d \geq 1$. By Jensen's inequality, $\mathbb{E}[h(x)] \geq h(\mathbb{E}[x])$, so we have

$$\mathbb{E}_{\mathbf{v} \sim F} \left[\left(\frac{1}{1 + \sum_{j=1}^{n-1} \mathbb{1}_{v_j \geq v(\hat{q})}} \right)^{1/d} \right] \geq \left(\frac{1}{1 + (n-1)\hat{q}} \right)^{1/d}.$$

Substituting this lower bound into the sum gives Eq. (4).

Given the performance of a mechanism with quantile reserve $\hat{q}$, we now describe how well a mechanism does by selecting the quantile reserve uniformly at random.

Lemma 6 (Random Reserve Price Mechanism). *Consider a convex cost setting with $c_i = p_i^d$, $\forall i \in N$, where $d \geq 1$. Let APX be the expected revenue of a mechanism which draws a quantile reserve $\hat{q}$ uniformly at random in $[0, 1]$, and then allocates uniformly across all bidders with quantile $q_i \leq \hat{q}$, and charges each of these Z bidders ex-post truthful payments $(v(\hat{q})/Z)^{1/d}$. Then:*

$$APX \geq \frac{1}{8} n^{1-1/d} \kappa^{1/d}. \tag{5}$$

Proof. A lower bound on the total expected revenue of the mechanism can be computed by integrating $\mathrm{APX}(\hat{q})$ with respect to quantile $\hat{q}$: $\mathrm{APX} = \int_0^1 \mathrm{APX}(\hat{q})\, \mathrm{d}\hat{q}$. Invoking Lemma 5, and since $\hat{q} \in [0, 1]$, the quantity $\mathrm{APX}(\hat{q})$ can be lower-bounded as follows:

$$\mathrm{APX}(\hat{q}) \geq n^{1-1/d} v(\hat{q})^{1/d} \hat{q}.$$

Thus we get:

$$\mathrm{APX} \geq n^{1-\frac{1}{d}} \int_0^1 \frac{v(\hat{q})\hat{q}}{v(\hat{q})^{1-\frac{1}{d}}} \mathrm{d}\hat{q} \geq n^{1-\frac{1}{d}} \int_{1/2}^1 \frac{R(\hat{q})}{v(\hat{q})^{1-\frac{1}{d}}} \mathrm{d}\hat{q}.$$

Since $\kappa = v(1/2) \geq v(\hat{q})$ for $1/2 \leq \hat{q} \leq 1$, we have

$$\text{APX} \geq \frac{n^{1-\frac{1}{d}}}{\kappa^{1-\frac{1}{d}}} \int_{1/2}^{1} R(\hat{q})\,\mathrm{d}\hat{q},$$

and because $R(\hat{q})$ is concave, $\int_{1/2}^{1} R(\hat{q})\,\mathrm{d}\hat{q} \geq \frac{1}{2}\frac{1}{2}R(1/2) = \kappa/8$, where the last step follows from the proof of Lemma 2, giving us Eq. (5).

We are now ready to prove the main theorem of this section: a prior-free mechanism that is a constant-factor approximation of the optimal. Observe that to draw a random quantile reserve, we do not need to know the distribution of values, as the quantile of a randomly selected bidder is ex-ante equivalent to a randomly drawn quantile reserve. That is, we can sacrifice a randomly selected bidder by using his quantile as the reserve and run a random reserve price mechanism among the $n - 1$ bidders. Notice that this mechanism is prior-free. We show that it is approximately optimal.

Theorem 4 (Prior-Free Mechanism). *Consider a convex cost setting with $c_i = p_i^d$, $\forall i \in N$, where $d \geq 2$ and the value distribution is an MHR distribution. Let APX be the expected revenue of the random price setter mechanism. The random price setter mechanism achieves revenue APX which satisfies:*

$$\frac{APX}{OPT} > \frac{1}{8}\left(\frac{n-1}{n}\right)^{1-1/d}\frac{1}{e^{1/d}} \geq \frac{1}{16e}. \tag{6}$$

Proof. Observe that the revenue of the mechanism is equal to the revenue of the random reserve price mechanism with $n - 1$ bidders. Thus, by applying Lemma 6, we have $\text{APX} \geq \frac{1}{8}(n-1)^{1-1/d}\kappa^{1/d}$. By Theorem 3, we also have $\text{OPT} \leq n^{1-1/d}(e\kappa)^{1/d}$. Combining the two bounds, yields the theorem.

For $d \geq 2$, the approximation ratio given by Eq. (6) is $\frac{1}{8e^{1/d}} > .075$ in the limit, as n tends towards infinity, and $\frac{1}{8}$, as d tends towards infinity as well.

5 Heuristics and Experiments

Example 1 and preliminary experimental results (not reported in detail here, due to space constraints) suggest that it is desirable to allocate to multiple bidders in the convex cost setting. Thus, we propose two mechanisms, which can potentially allocate to all bidders: 1. a mechanism that allocates proportionally by positive values, and 2. a mechanism that allocates proportionally by positive virtual values. Note that the first mechanism can allocate to any bidder whose type is positive. The second mechanism is similar to that of the optimal revenue-maximizing mechanism in the usual quasi-linear setting in that it avoids bidders with negative virtual values, so certain types are never allocated.

We evaluated our mechanisms using polynomial costs $c = p^d$, where $d \in \{2, 3, 10, 20\}$. We saw that as d increases, the mechanism that allocates proportionally by value performed best. This can be explained as follows: as d increases, the amount of revenue contributed by a bidder becomes vanishingly small, so it becomes more favorable to extract payments from a larger set of bidders, including those who may not have the highest values.

6 Conclusion

We investigated the single-dimensional mechanism design problem where bidder's utility functions are not linear with respect to payments, but instead are convex—specifically, polynomial with respect to payments. We discovered that in contrast to the traditional quasi-linear setting, in our convex cost setting, it is suboptimal to allocate only to the highest bidders. We also noted that Myerson's elegant machinery does not easily extend from the quasi-linear to the convex cost setting. Nonetheless, we were able to provide an upper bound on the value of the revenue-maximizing mechanism. We additionally developed a prior-free mechanism with a constant-factor approximation guarantee. The prior-free mechanism was also empirically evaluated using values drawn from random MHR distributions, and we saw that the total expected revenue generated by this mechanism can, on average, exceed the guarantee we provide. We then proposed and evaluated two mechanisms that we provide no guarantees for, which allocate proportionally to values and virtual values, and see that they do well empirically as well, with performance exceeding that of our prior-free mechanism. In future work, we hope to construct lower bounds on these proportional allocation mechanisms.

Acknowledgments. This research was supported by NSF Grant #1217761 and Microsoft Research.

References

1. Alaei, S., Fu, H., Haghpanah, N., Hartline, J.: The simple economics of approximately optimal auctions. In: 2013 IEEE 54th Annual Symposium on Foundations of Computer Science (FOCS), pp. 628–637. IEEE (2013)
2. Border, K.C.: Implementation of reduced form auctions: a geometric approach. Econometrica **59**(4), 1175–1187 (1991). http://www.jstor.org/stable/2938181
3. Bulow, J., Klemperer, P.: Auctions versus negotiations. Am. Econ. Rev. **86**(1), 180–194 (1996)
4. Cai, Y., Daskalakis, C., Weinberg, S.M.: Understanding incentives: mechanism design becomes algorithm design. In: 2013 IEEE 54th Annual Symposium on Foundations of Computer Science (FOCS), pp. 618–627, October (2013). https://doi.org/10.1109/FOCS.2013.72
5. Chawla, S., Hartline, J.D., Sivan, B.: Optimal crowdsourcing contests. In: Proceedings of the Twenty-Third Annual ACM-SIAM Symposium on Discrete Algorithms, pp. 856–868. SIAM (2012)
6. Dhangwatnotai, P., Roughgarden, T., Yan, Q.: Revenue maximization with a single sample. Games Econ. Behav. **91**, 318–333 (2015)

7. DiPalantino, D., Vojnovic, M.: Crowdsourcing and all-pay auctions. In: Proceedings of the 10th ACM Conference on Electronic Commerce, pp. 119–128. ACM (2009)
8. Fu, H., Hartline, J., Hoy, D.: Prior-independent auctions for risk-averse agents. In: Proceedings of the Fourteenth ACM Conference on Electronic Commerce, pp. 471–488. EC '13, ACM, New York, NY, USA (2013). http://doi.acm.org/10.1145/2482540.2482551
9. Gossen, H.H.: Entwickelung der gesetze des menschlichen verkehrs, und der daraus fliessenden regeln für menschliche handeln. F. Vieweg (1854)
10. Greenwald, A., Oyakawa, T., Syrgkanis, V.: Simple vs optimal contests with convex costs. In: Champin, P., Gandon, F.L., Lalmas, M., Ipeirotis, P.G. (eds.) Proceedings of the 2018 World Wide Web Conference on World Wide Web, WWW 2018, Lyon, France, 23–27 April 2018, pp. 1429–1438. ACM (2018). http://doi.acm.org/10.1145/3178876.3186048
11. Hartline, J.D.: Mechanism design and approximation. Book draft. October (2015)
12. Hartline, J.D., Roughgarden, T.: Simple versus optimal mechanisms. In: Proceedings of the 10th ACM Conference on Electronic Commerce, pp. 225–234. EC '09, ACM, New York, NY, USA (2009). http://doi.acm.org/10.1145/1566374.1566407
13. Maskin, E., Riley, J.: Optimal auctions with risk averse buyers. Econometrica **52**(6), 1473–1518 (1984). http://www.jstor.org/stable/1913516
14. Myerson, R.B.: Optimal auction design. Math. Oper. Res. **6**(1), 58–73 (1981)
15. Pai, M.M., Vohra, R.: Optimal auctions with financially constrained buyers. J. Econ. Theory **150**, 383–425 (2014). https://doi.org/10.1016/j.jet.2013.09.015. http://www.sciencedirect.com/science/article/pii/S0022053113001701
16. Roughgarden, T., Talgam-cohen, I., Yan, Q.: Robust auctions for revenue via enhanced competition (2016)
17. Sakurai, Y., Saito, Y., Iwasaki, A., Yokoo, M.: Beyond quasi-linear utility: strategy/false-name-proof multi-unit auction protocols. In: IEEE/WIC/ACM International Conference on Web Intelligence and Intelligent Agent Technology, 2008. WI-IAT'08, vol. 2, pp. 417–423. IEEE (2008)
18. Samuel-Cahn, E.: Comparison of threshold stop rules and maximum for independent nonnegative random variables. Ann. Probab. **12**(4), 1213–1216 (1984)
19. Singer, Y.: Budget feasible mechanism design. SIGecom Exch. **12**(2), 24–31 (2014). http://doi.acm.org/10.1145/2692359.2692366
20. Vickrey, W.: Counterspeculation, auctions, and competitive sealed tenders. J. Financ. **16**(1), 8–37 (1961)
21. Wilson, R.: Incentive efficiency of double auctions. Econometrica J. Econ. Soc. **53**, 1101–1115 (1985)
22. Wilson, R.: Game-theoretic approaches to trading processes. In: Advances in Economic Theory: Fifth World Congress, pp. 33–77 (1987)

On the Price of Stability of Social Distance Games

Christos Kaklamanis, Panagiotis Kanellopoulos(✉), and Dimitris Patouchas

Computer Technology Institute and Press "Diophantus"
and Department of Computer Engineering and Informatics,
University of Patras, 26504 Rio, Greece
{kakl,kanellop,patouchas}@ceid.upatras.gr

Abstract. We consider social distance games, where a group of utility maximizing players, connected over a network representing social proximity, wish to form coalitions (or clusters) so that they are grouped together with players that are at close distance. Given a cluster, the utility of each player depends on its distance to the other players inside the cluster and on the cluster size, and a player will deviate to another cluster if this leads to higher utility. We are interested in Nash equilibria of such games, where no player has an incentive to unilaterally deviate to another cluster, and we present bounds on the price of stability both for the normal utility function and for a slightly modified one.

1 Introduction

In many social contexts, people (or companies) frequently choose to interact and collaborate with other people and economic entities, and form close personal or business ties. This process, also called *coalition formation*, captures several interesting settings, and is essentially ubiquitous, especially with the rise of social networks. The criteria for selecting with whom to coalesce (or whom to avoid) may range from personal preferences (e.g., which party to attend or whom to invite for dinner) to financial motives (e.g., which company merge should be encouraged), and give rise to complex dynamic interactions.

Consider, for example, a conference welcome reception where the attendants typically tend to form small groups (or *clusters*) and discuss recent news of personal or scientific nature. The composition of these clusters rarely stays fixed during the entire welcome reception as people may observe that another, perhaps more appealing, cluster has formed and, hence, wish to join it. For each attendant, the pleasure she derives by a particular cluster depends on how many people in this cluster she knows directly (and, actually, likes) or may have common friends with, etc., as well as on the cluster size.

We are interested in stable states of this coalition formation process, i.e., when all cluster members are satisfied with the current configuration and no one wishes to join a different cluster, and, in addition, in comparing a stable state to the optimal clustering. We consider a model where the utility obtained

X. Deng (Ed.): SAGT 2018, LNCS 11059, pp. 125–136, 2018.
https://doi.org/10.1007/978-3-319-99660-8_12

by a cluster depends on average inverse social distance to the cluster members; informally, we assume that people tend to prefer being in clusters with friends, or friends of friends, than with strangers. The class of strategic games that captures this process is that of *social distance games.*

Related Work. Social distance games were introduced by Brânzei and Larson [9] who proved that finding the optimal clustering is NP-hard and designed a 2-approximation algorithm with respect to the social welfare. In addition, they studied stability using the notion of the core. The work that is most related to ours is by Balliu et al. [4] who considered the price of stability in social distance games and presented lower bounds of 6/5 for general graphs, 169/160 for bipartite graphs, as well as an upper bound of $(\sqrt{2}+1)/2$ for graphs for girth 5. Moreover, they proved that social distance games do not admit a potential function, as best response dynamics may cycle, while computing the social welfare maximizing Nash stable clustering is also NP-hard. In another paper, Balliu et al. [5] considered the price of Pareto optimality in social distance games and presented asymptotically tight bounds.

Similar questions have been also studied for the related class of hedonic games introduced by Drèze and Greenberg [10] (see [3] for a recent survey), and, in particular, for simple, symmetric fractional hedonic games. Aziz et al. [2] introduced fractional hedonic games, where players have utility over players, and, given a clustering, the utility of each player is defined as the sum of utility it obtains from each player in its cluster divided by the cluster size. In simple symmetric fractional hedonic games, the utility obtained from a single player can be either 0 or 1 and the utility function is symmetric. Olsen [16], among other results, suggested an alternate utility function for fractional hedonic games, where the utility function of player i does not take i into account when averaging over the cluster size.

Bilò et al. [6,7] considered the price of anarchy and stability in fractional hedonic games, and, among other results, proved that the price of stability in simple symmetric fractional hedonic games is at least 2 for general graphs, at most 4 for triangle-free graphs, and presented almost tight bounds for the case of bipartite graphs. Kaklamanis et al. [15] also considered the price of stability in simple, symmetric fractional hedonic games and presented an improved lower bound of $1+\sqrt{6}/2 \approx 2.224$ for general graphs as well as upper bounds of 1 for graphs of girth 5 and for general graphs under the utility function defined by Olsen. Further notions of stability in fractional hedonic games have been investigated by Brandl et al. [8].

Apart from social distance games and simple, symmetric fractional hedonic games, Peters and Elkind [18] investigated the computational complexity of stability-related questions in hedonic games, while Peters [17] studied the computational complexity of questions related to dichotomous hedonic games, where each player either approves or disapproves a given coalition. Feldman et al. [12] considered the non-cooperative version of hedonic clustering games, where they characterize Nash equilibria and provide upper and lower bounds on the price of anarchy and price of stability, Hoefer et al. [14] studied hedonic games and characterized the structures bases on which dynamic coalition formation can stabilize quickly, while Hoefer and Jiamjitrak [13] considered proportional allocation for profit sharing in hedonic games.

Our Contribution. We consider Nash stable clusterings and focus on the price of stability. We improve upon the lower bound of 6/5 from [4] and present a lower bound of 2 for general graphs. The construction we use in the proof admits an optimal clustering consisting of clusters of diameter 1, while in the only Nash stable clustering almost all nodes are at distance 2 from each other. We also consider games played on trees and we prove that the price of stability is 1. Our final result concerns modified social distance games, where the utility is computed with respect to the cluster size minus 1, and we show that, under this utility function, the price of stability of social distance games is again 1.

Roadmap. The remainder of the paper is structured as follows. We begin, in Sect. 2, by formally introducing the class of social distance games and presenting the necessary definitions. Then, in Sect. 3, we present the results on the price of stability and we conclude with open problems in Sect. 4.

2 Preliminaries

A *social distance game* is a strategic game played on a graph $G = (V, E)$ by a set N of n utility maximizing players; each node in $V(G)$ corresponds to a strategic player. A *clustering* of the game consists of a set $\mathcal{C} = \{C_1, C_2, \dots\}$ of clusters such that $\cup_i C_i = N$, and $C_i \cap C_j = \emptyset$ for any pair $i \neq j$, i.e., each player belongs to exactly one cluster. We let $C(u)$ denote the cluster that player u belongs to.

The utility of player i is given by

$$u_i(C(i)) = \frac{1}{|C(i)|} \sum_{j \in C(i) \setminus \{i\}} \frac{1}{d_{C(i)}(i,j)},$$

where $d_{C(i)}(i,j)$ is the distance of players i and j in the subgraph defined by cluster $C(i)$. In case i and j are disconnected in cluster $C(i)$, then $d_{C(i)}(i,j) = \infty$.

The *social welfare* $\mathrm{SW}(\mathcal{C})$ of clustering $\mathcal{C}$ is defined as the sum of the player utilities, i.e., $SW(\mathcal{C}) = \sum_i u_i(C(i))$, and we denote by $\mathcal{C}^*$ the clustering that maximizes the social welfare. Similarly, we define the utility of cluster C as $u(C) = \sum_{i:C(i)=C} u_i(C)$.

Since each player is utility maximizing, given a clustering $\mathcal{C}$, player i may deviate from its current cluster $C(i)$ in $\mathcal{C}$ and join another cluster C', if it holds that $u_i(C(i)) < u_i(C' \cup i)$. A player i is *Nash stable* if there is no cluster $C' \neq C(i)$ such that i's utility improves by deviating to C', while a cluster is *Nash stable* if all players in the cluster are Nash stable. A clustering is a *Nash stable clustering* if all clusters are Nash stable.

The *price of stability* PoS (introduced in [1]) denotes the best-case performance deterioration arising from the requirement that the resulting clustering is Nash stable. Given a graph G, the corresponding social distance game Γ_G and its set of Nash stable clusterings $\mathcal{C}_s$, the price of stability for the game Γ_G is formally defined as $\text{PoS}(\Gamma_G) = \min_{\mathcal{C} \in \mathcal{C}_s} \frac{\text{SW}(\mathcal{C}^*)}{\text{SW}(\mathcal{C})}$. Similarly, the price of stability for the class of social distance games is defined as $\text{PoS} = \sup_G \text{PoS}(\Gamma_G)$.

We also consider a variant of social distance games, called *modified social distance games* in accordance to similar nomenclature for the case of fractional hedonic games (e.g., see [11] for the case of modified fractional hedonic games), where the single difference is that the utility of each player i is now defined as

$$u'_i(C(i)) = \begin{cases} \frac{1}{|C(i)|-1} \sum_{j \in C(i) \setminus \{i\}} \frac{1}{d_{C(i)}(i,j)}, & \text{if } |C(i)| > 1; \\ 0, & \text{otherwise.} \end{cases}$$

3 Price of Stability of Social Distance Games

This section contains our results on the price of stability of social distance games. We begin by presenting an improved lower bound for the general case of arbitrary graphs, and then we consider the setting where players form a tree; in this case, we show that there exists an optimal clustering that is stable, i.e., the price of stability is 1. Finally, we conclude with the case of modified social distance games where, again, we prove that the price of stability is 1, by arguing about the structure of an optimal clustering.

3.1 A Lower Bound for General Graphs

We begin with a technical lemma; in any stable clustering, any two players with the same closed neighborhood must belong to the same cluster.

Lemma 1. *For any two players x, y such that $(x, y) \in E(G)$, if $N(x) \cup x = N(y) \cup y$, then, in any stable clustering, x and y are in the same cluster, i.e., $C(x) = C(y)$.*

We are now ready to prove the main result of this section. In our construction, the only Nash stable clustering is the grand coalition, while the optimal clustering consists of cliques.

Theorem 1. *The price of stability of social distance games is at least $2 - \epsilon$ for $\epsilon > 0$.*

Proof. Let α be a positive integer. Consider the following graph G that is also presented in Fig. 1. There exist two sets S, S' of α nodes each, where each node $i \in S$ is connected only to the corresponding node $i' \in S'$. There also exist $\alpha/2$ cliques K^j, $1 \leq j \leq \alpha/2$, each of size α. Any node in clique K^1 is connected to all nodes in $S' \cup (\bigcup_{i \geq 2} K^i)$. Therefore, the total number of nodes in G is $\alpha^2/2 + 2\alpha$.

We first argue about the social welfare in the optimal clustering and then we argue that the grand coalition is the only Nash stable clustering. Consider the clustering $\mathcal{C}$ where each player $i \in S$ is paired with the neighboring player

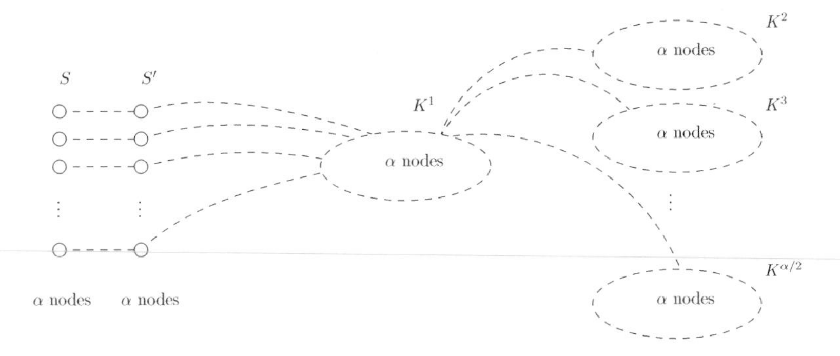

Fig. 1. The graph G used in the proof of the lower bound. Each bubble corresponds to a clique and each dashed line represents edges adjacent to all clique nodes.

$i' \in S'$, cliques K^1 and K^2 form a cluster, while each clique K^j, for $3 \leq j \leq \alpha/2$ forms a cluster. The social welfare of this clustering is

$$\begin{aligned}\mathrm{SW}(\mathcal{C}) &= \alpha + (2\alpha - 1) + (\alpha/2 - 2)(\alpha - 1) \\ &= \alpha^2/2 + \alpha/2 + 1,\end{aligned}$$

where the first term in the first equality is due to the α clusters containing the players in $S \cup S'$, the second term is due to the cluster $K^1 \cup K^2$, while the last term is due to the remaining $(\alpha/2 - 2)$ clusters, each containing a single clique of size α. Clearly, for the optimal clustering $\mathcal{C}^*$ it holds that $\mathrm{SW}(\mathcal{C}^*) \geq \mathrm{SW}(\mathcal{C})$, hence, it holds that

$$\mathrm{SW}(\mathcal{C}^*) \geq \alpha^2/2 + \alpha/2 + 1. \tag{1}$$

Consider now the clustering $\mathcal{C}'$ where all players form a single cluster, i.e., the grand coalition. Its social welfare is

$$\begin{aligned}\mathrm{SW}(\mathcal{C}') &= \frac{\alpha(1 + \frac{\alpha}{2} + \frac{\alpha^2/2-1}{3} + \frac{\alpha-1}{4}) + \alpha(\alpha + 1 + \frac{\alpha^2/2-1}{2} + \frac{\alpha-1}{3})}{\alpha^2/2 + 2\alpha} \\ &+ \frac{\alpha(\alpha^2/2 + \alpha - 1 + \frac{\alpha}{2}) + (\alpha^2/2 - \alpha)(2\alpha - 1 + \frac{\alpha^2/2-\alpha}{2} + \frac{\alpha}{3})}{\alpha^2/2 + 2\alpha} \\ &= \frac{\alpha(2\alpha^2 + 9\alpha + 5)/12 + \alpha(3\alpha^2 + 16\alpha + 2)/12 + \alpha(\alpha^2 + 3\alpha - 2)/2}{\alpha^2/2 + 2\alpha} \\ &+ \frac{(\alpha^2/2 - \alpha)(3\alpha^2 + 22\alpha - 12)/12}{\alpha^2/2 + 2\alpha} \\ &= \frac{\alpha(3\alpha^3 + 38\alpha^2 + 30\alpha + 14)/24}{\alpha^2/2 + 2\alpha} \\ &= \frac{3\alpha^3 + 38\alpha^2 + 30\alpha + 14}{12\alpha + 48},\end{aligned} \tag{2}$$

where, in the first equality, the first term is due to the α players in S, the second term is due to the α players in S', the third term is due to the α players in clique K^1, while the last term is due to the $(\alpha^2/2-\alpha)$ players in the remaining cliques.

In the following, we show that $\mathcal{C}'$ is the only Nash stable clustering. First, observe that in any stable clustering, any player $i \in S$ must be in the same cluster as its neighbor $i' \in S'$. If this is not the case, then $u_i(C(i)) = 0$ as there is no path inside $C(i)$ connecting i to any other player in $C(i)$, while by joining cluster $C(i')$ the utility of player i would be strictly positive. Then, due to Lemma 1, observe that, in any stable clustering, all players in a clique K^j, where $1 \le j \le \alpha/2$, must belong in the same cluster, as by the construction of graph G, any pair of players in K^j has the same closed neighborhood.

We continue by showing that any player in S' belongs to the same cluster as the players in clique K^1. Assume otherwise and consider such a player $i' \in S'$. By the discussion above, i' is in the same cluster as its neighboring player $i \in S$ and, by our assumption that i' is not in the same cluster as clique K^1, i' has no path inside $C(i')$ connecting it to any other player in $C(i')$, hence $u_{i'}(C(i')) \le 1/2$. Consider the cluster C that contains the players of clique K^1 and let us assume that C contains also κ players of S' (but different than i'), where $0 \le \kappa < \alpha$ and the players of λ additional cliques $K^{j>1}$, where $0 \le \lambda < \alpha/2$; by the discussion above, C also contains κ players of S. Then, the utility of player i' when joining cluster C is

$$u_i(C \cup i') = \frac{\alpha + (\kappa + \lambda\alpha)/2 + \kappa/3}{(\lambda+1)\alpha + 2\kappa + 1},$$

which is strictly greater than $1/2$ for any $\kappa < \alpha$.

So far we have established that, in any Nash stable clustering, the players of S, S', and K^1 are necessarily in the same cluster, while the players of any clique $K^{j>1}$ are also together in a cluster; note that we have not yet ruled out the possibility that different cliques belong to different clusters. We conclude the argument that the grand coalition is the only Nash stable clustering by showing that if there exists a clique $K^{j'>1}$ that is not in the same cluster as the clique K^1, then any player in K^1 has an incentive to deviate and join the same cluster as the players of $K^{j'}$. Indeed, in this case, the utility of any player k in K^1 is maximized whenever $C(k)$ contains, apart from players in $S \cup S' \cup K^1$, all players except those in $K^{j'}$ and, hence, it holds that

$$\begin{aligned} u_k(C(k)) &\le \frac{(\alpha/2-2)\alpha + 2\alpha - 1 + \frac{\alpha}{2}}{(\alpha/2-2)\alpha + 3\alpha} \\ &= \frac{\alpha^2 + \alpha - 2}{\alpha^2 + 2\alpha} \\ &= \frac{\alpha - 1}{\alpha}, \end{aligned}$$

where the first term in the inequality is due to the distance to the other cliques, the last term is due to the distance to players in S while the remaining terms are due to the distance to players in S' and the remaining players of K^1. By deviating to cluster $C_{j'}$ containing at least the players of clique $K^{j'}$, player

k would obtain utility $u_k(C_{j'} \cup k) \geq \frac{\alpha}{\alpha+1}$,[1] i.e., strictly greater utility than $u_k(C(k))$. This concludes the argument that the grand coalition is the only Nash stable clustering.

By combining inequalities (1) and (2), we conclude that the price of stability for graph G is

$$\begin{aligned} \text{PoS}(G) &= \frac{\text{SW}(\mathcal{C}^*)}{\text{SW}(\mathcal{C}')} \\ &\geq \frac{\alpha^2/2 + \alpha/2 + 1}{(3\alpha^3 + 38\alpha^2 + 30\alpha + 14)/(12\alpha + 48)} \\ &= \frac{6\alpha^3 + 30\alpha^2 + 36\alpha + 48}{3\alpha^3 + 38\alpha^2 + 30\alpha + 14} \\ &\geq 2 - \epsilon, \end{aligned}$$

as α tends to infinity, where ϵ is an arbitrarily small positive number. □

3.2 The Case of Trees

We consider the case where the graph G is a tree and prove that an optimal clustering is stable. We call a clustering *compact* if it consists only of stars and paths of 4 nodes and we first argue that there exists a compact optimal clustering. We begin by showing that at least one optimal clustering is compact.

Lemma 2. *In social distance games on trees, there exists an optimal clustering where each cluster is either a path of* 4 *nodes or a star.*

We now argue that, starting from an unstable compact clustering, we can obtain another compact clustering with at least the same social welfare and fewer paths of 4 nodes.

Lemma 3. *A compact clustering $\mathcal{C}$ is either stable or it can be turned into another compact clustering $\mathcal{C}'$, where $SW(\mathcal{C}) \leq SW(\mathcal{C}')$ and the number of paths of* 4 *nodes is strictly less in $\mathcal{C}'$ than in $\mathcal{C}$.*

Proof. Consider a compact clustering $\mathcal{C}$. We argue about the stability of $\mathcal{C}$ and we will show that when $\mathcal{C}$ is unstable, then we can obtain another compact clustering $\mathcal{C}'$ satisfying the lemma.

In star clusters, each player that is a leaf has utility 1/2 and by joining a path of 4 nodes would obtain utility 5/12, if its neighbor is a path endpoint, or utility 7/15 if its neighbor is not a path endpoint. Similarly, by joining a star, it would obtain utility of 1/2, so no leaf player has any incentive to deviate. Each player that is a center in a star cluster has utility at least 1/2 and, as in the case of leaf nodes, has no incentive to deviate.

In paths of 4 nodes, each player that is not a path endpoint has utility 5/8 and would obtain strictly less utility by any deviation. A player that is a path

[1] $u_k(C_{j'} \cup k)$ obtains its minimum value when $C_{j'}$ contains a single clique, i.e., $K^{j'}$.

endpoint obtains utility 11/24 and has no incentive to deviate by connecting to an endpoint of another path of 4 nodes, or to a leaf in a star cluster of at least 3 players. When such a player has an incentive to connect as a leaf to a center of a star cluster, then by allowing this deviation we obtain another compact clustering with strictly greater social welfare and a reduced number of paths of 4 nodes, as the original path has now become a star with 3 players. When a player that is a path endpoint wishes to deviate and connect to a non-endpoint node in a path of 4 nodes, then we can rearrange these two clusters, that are both paths of 4 nodes, into three star clusters of size 2, 3, and 3 with the same social welfare; again, we obtain a compact clustering that satisfies the lemma. □

By combining Lemmas 2 and 3 we obtain the following theorem, as there exists an optimal clustering $\mathcal{C}^*$ that is compact and, in case it is unstable, we can obtain another compact optimal clustering with strictly less paths of nodes 4; clearly, this process will eventually halt at a stable compact optimal clustering.

Theorem 2. *The price of stability of social distance games on trees is* 1.

3.3 Modified Social Distance Games

This section contains our results for the setting of modified social distance games. In our proofs, we exploit the following technical lemma that argues about the social welfare in a clustering that consists only of stars.

Lemma 4. *Any clustering where* m *star clusters span a set of* n *nodes has total utility* $m + n/2$.

Proof. Consider the m star clusters $C_1, C_2, \ldots, C_m$ and let n_i, $1 \leq i \leq m$, denote the number of nodes in cluster C_i. For a given cluster C_i, the cluster center has utility 1 while each of the remaining $n_i - 1$ nodes has utility $(1 + (n_i - 2)/2)/(n_i - 1) = n_i/(2n_i - 2)$. By summing over all players in C_i, we obtain that $u(C_i) = 1 + n_i/2$. The lemma follows by summing over all m star clusters and since $\sum_i n_i = n$. □

We first show that there exists an optimal clustering where each cluster is a triangle, a star or a single disconnected node; we begin by an optimal clustering that may not exhibit these properties and we show how to transform it into another optimal clustering that satisfies them.

Lemma 5. *There exists an optimal clustering* $\mathcal{C}^*$ *where each cluster* C *with* $|C| \geq 4$ *is a star.*

Proof. Consider an optimal clustering $\mathcal{C}^*$ and a cluster C with $m \geq 4$ nodes that is not a star. Let $\mathcal{D}$ be a decomposition of cluster C into a collection of disjoint sets $\{T, P, I\}$ where T contains x disjoint cliques of size 3, i.e., triangles, P contains y disjoint cliques of size 2, i.e., pairs, and I contains z isolated nodes, so that z is the minimum among all such decompositions. Clearly, $m = 3x+2y+z$.

We now argue about the structure inside cluster C. First, observe that there cannot be an edge connecting a node belonging to set T in $\mathcal{D}$ to a node belonging to set I in $\mathcal{D}$, or two nodes belonging to set I in $\mathcal{D}$, as then another decomposition $\mathcal{D}'$ exists with strictly less isolated nodes than $\mathcal{D}$. Furthermore, each isolated node in $\mathcal{D}$ may be directly connected to at most one node of each pair in $\mathcal{D}$ as otherwise we could form a triangle and reduce the number of isolated nodes. In addition, for each pair in $\mathcal{D}$, at most one node may be directly connected to any isolated node, as otherwise we could "break" the pair, connect its endpoints to the isolated nodes and again reduce the number of isolated nodes. Let P' be the set of the $y' \leq y$ pairs in $\mathcal{D}$ where exactly one endpoint has at least one isolated node as neighbor, while, for the remaining $y - y'$ pairs, no endpoint has an isolated node as neighbor. Furthermore, for any pair where one endpoint is directly connected to an isolated node, the other endpoint cannot be connected to a triangle node, as then we could again reduce the number of isolated nodes by rearranging the triangle, the pair, and the isolated node. Finally, any isolated node i is directly connected to at least one node (which, by the discussion above, must be a pair node), as otherwise the cluster $C \setminus \{i\}$ has strictly greater utility than C, contradicting the optimality of $\mathcal{C}^*$.

By the last observation about isolated nodes, it follows that we can decompose cluster C into smaller clusters such that each triangle in T forms a cluster, while there are also y star clusters that contain the nodes in P as well as the isolated nodes, i.e., $2y + z$ nodes in total. By Lemma 4, we have that the total utility $u(\mathcal{D})$ of the new set of clusters is

$$u(\mathcal{D}) = 3x + 2y + z/2. \tag{3}$$

Let us now argue about the utility of cluster C. We upperbound the utility of each player i in C according to its type in the decomposition $\mathcal{D}$. If i is in a triangle,

$$\begin{aligned} u_i(C) &\leq \frac{3x + y' + 2(y - y') - 1 + (y' + z)/2}{3x + 2y + z - 1} \\ &= \frac{3x + 2y - y'/2 + z/2 - 1}{3x + 2y + z - 1}, \end{aligned} \tag{4}$$

since, by the discussion above, i can be directly connected to at most the nodes in triangles, one endpoint for each of y' pairs, both endpoints for the remaining $(y - y')$ pairs, while it can have distance at least 2 from the remaining $y' + z$ nodes.

If i is an isolated node, then

$$\begin{aligned} u_i(C) &\leq \frac{y' + (y' + 2(y - y') + 3x + z - 1)/2}{3x + 2y + z - 1} \\ &= \frac{3x/2 + y + y'/2 + z/2 - 1/2}{3x + 2y + z - 1}, \end{aligned} \tag{5}$$

since it can be directly connected to at most y' nodes and has distance at least 2 to all remaining nodes.

If i is a node in the y' pairs and has isolated nodes as neighbors, then the utility is

$$u_i(C) \leq \frac{3x + 2y + z - 1}{3x + 2y + z - 1}, \tag{6}$$

as it may be directly connected to all remaining nodes, while if i is an endpoint in one of the y' pairs and does not have isolated nodes as neighbors, then the utility is

$$u_i(C) \leq \frac{2y - 1 + (3x + z)/2}{3x + 2y + z - 1}, \tag{7}$$

since i can be directly connected only to the remaining nodes belonging to pairs in $\mathcal{D}$.

Finally, if i is part of the $y - y'$ pairs where no endpoint has an isolated node as neighbor, then the utility is

$$u_i(C) \leq \frac{3x + 2y - 1 + z/2}{3x + 2y + z - 1}, \tag{8}$$

as it may be directly connected to all remaining nodes apart from those that are isolated in $\mathcal{D}$.

The total utility of cluster C can be bounded from above by using inequalities (4)–(8) as

$$\begin{aligned} u(C) &= \sum_{i \in C} u_i(C) \\ &= \sum_{i \in T} u_i(C) + \sum_{i \in P'} u_i(C) + \sum_{i \in P \setminus P'} u_i(C) + \sum_{i \in I} u_i(C) \\ &\leq \frac{3x(3x + 2y - y'/2 + z/2 - 1) + y'(3x + 2y + z - 1 + 2y - 1 + (3x + z)/2)}{3x + 2y + z - 1} \\ &+ \frac{2(y - y')(3x + 2y - 1 + z/2) + z(3x/2 + y + y'/2 + z/2 - 1/2)}{3x + 2y + z - 1} \\ &= \frac{9x^2 + 4y^2 + z^2/2 + 12xy - 3xy' + 3xz + 2yz + y'z - 3x - 2y - z/2}{3x + 2y + z - 1}. \end{aligned} \tag{9}$$

The proof follows as, by combining (3) and (9), we obtain $u(\mathcal{D}) \geq u(C)$, since $(3x+2y+z/2)(3x+2y+z-1) \geq 9x^2+4y^2+z^2/2+12xy-3xy'+3xz+2yz+y'z-3x-2y-z/2$ for any $y' \leq y$, i.e., we can decompose cluster C into triangles and stars without losing social welfare; clearly, we can repeatedly perform this process until all clusters are as desired. □

We are now ready to prove the main result of this section, i.e., that there exists a stable optimal clustering.

Theorem 3. *The price of stability of modified social distance games is* 1.

Proof. Consider an optimal clustering $\mathcal{C}^*$ that exhibits the properties of Lemma 5. Clearly, if $\mathcal{C}^*$ is stable, the theorem follows, so we assume that $\mathcal{C}^*$ is unstable and we show how to modify it in order to obtain a stable optimal clustering.

First, observe that any player that is in a triangle cluster and any player that is a root in a star cluster is satisfied since its utility is 1. Therefore, the players that may wish to deviate from $\mathcal{C}^*$ are those that are either in singleton clusters, or leaves in a star. We now argue that players in singleton clusters are disconnected in graph G and, hence, obtain utility equal to 0 in any clustering; so, they have no incentive to deviate and can be ignored, without loss of generality. Indeed, if i is connected by an edge $(i, j) \in E(G)$ to another player j in a singleton cluster, then we can merge the two singleton clusters $C^*(i)$ and $C^*(j)$ to a single cluster and strictly increase the social welfare; a contradiction since $\mathcal{C}^*$ is an optimal clustering. A similar reasoning applies if player i is connected by an edge $(i, k) \in E(G)$ to a player k that is in a triangle cluster or that is a leaf in a star cluster, as then we can replace $C^*(i)$ and $C^*(k)$ by $C^*(i) \cup \{k\}$ and $C^*(k) \setminus \{k\}$ and obtain strictly greater social welfare. Finally, if i is connected to a player ℓ that is root in a star cluster, we can merge clusters $C^*(i)$ and $C^*(\ell)$ and obtain strictly greater social welfare as $u(C^*(i)) = 0$, $u(C^*(\ell)) = 1 + |C^*(\ell)|/2$, while $u(C^*(\ell) \cup \{i\}) = 1 + (|C^*(\ell)| + 1)/2$.

We now consider the players that are leaves in star clusters. As in the case of singleton clusters, a leaf player i cannot be connected by an edge to a triangle cluster or to another leaf player, as again we would obtain strictly greater welfare by letting player i and its neighbor form a new cluster. Hence, i can only deviate by becoming a leaf to another star cluster. By Lemma 4, the social welfare of any set of disjoint star clusters spanning a given set of nodes does not depend on how these stars are actually formed, and, hence, any such deviation does not change the social welfare. Observe that such a deviating move of player i from the star $C^*(i)$ to another star C, requires that $|C^*(i)| > |C| + 1$. Therefore, by considering the lexicographic order π of all star clusters in the clustering $\mathcal{C}^*$ based on the number of nodes, from the minimum to the maximum, we observe that any deviating move that leads to a clustering $\mathcal{C}'$ satisfies $\pi(\mathcal{C}^*) < \pi(\mathcal{C}')$ and, hence, this process is guaranteed to end as we can apply this reasoning to any subsequent deviation. This concludes the proof of the theorem. □

4 Conclusions

We have presented new bounds on the price of stability of social distance games. The most important open question concerns the upper bound for general graphs, where no better bound than the trivial $O(n/\log n)$ is known. Similarly, no better bounds are known even for the case of bipartite graphs; we remark that for the class of simple, symmetric fractional hedonic games, the price of stability for bipartite graphs is a very small constant [7]. We also conjecture that the price of stability is 1 even for graphs of girth at least 5, i.e., the upper bound in [4] is not tight.

References

1. Anshelevich, E., Dasgupta, A., Kleinberg, J., Tardos, E., Wexler, T., Roughgarden, T.: The price of stability for network design with fair cost allocation. SIAM J. Comput. **38**(4), 1602–1623 (2008)
2. Aziz, H., Brandt, F., Harrenstein, P.: Fractional hedonic games. In: 13th International Conference on Autonomous Agents and Multiagent Systems (AAMAS), pp. 5–12 (2014)
3. Aziz, H., Savani, R.: Hedonic games. In: Brandt, F., Conitzer, V., Endriss, U., Lang, J., and Procaccia, A. (eds.) Handbook of Computational Social Choice. Cambridge University Press (2016)
4. Balliu, A., Flammini, M., Melideo, G., Olivetti, D.: Nash stability in social distance games. In: 31st AAAI Conference on Artificial Intelligence (AAAI), pp. 342–348 (2017)
5. Balliu, A., Flammini, M., Olivetti, D.: On Pareto optimality in social distance games. In: 31st AAAI Conference on Artificial Intelligence (AAAI), pp. 349–355 (2017)
6. Bilò, V., Fanelli, A., Flammini, M., Monaco, G., Moscardelli, L.: Nash stability in fractional hedonic games. In: Liu, T.-Y., Qi, Q., Ye, Y. (eds.) WINE 2014. LNCS, vol. 8877, pp. 486–491. Springer, Cham (2014). https://doi.org/10.1007/978-3-319-13129-0_44
7. Bilò, V., Fanelli, A., Flammini, M., Monaco, G., Moscardelli, L.: On the price of stability of fractional hedonic games. In: 14th International Joint Conference on Autonomous Agents and Multiagent Systems (AAMAS), pp. 1239–1247 (2015)
8. Brandl, F., Brandt, F., Strobel, M.: Fractional hedonic games: individual and group stability. In: 14th International Conference on Autonomous Agents and Multi-Agent Systems (AAMAS), pp. 1219–1227 (2015)
9. Brânzei, S., Larson, K.: Social distance games. In: 22nd International Joint Conference on Artificial Intelligence (IJCAI), pp. 91–96 (2011)
10. Drèze, J.H., Greenberg, J.: Hedonic coalitions: optimality and stability. Econometrica **48**(4), 987–1003 (1980)
11. Elkind, E., Fanelli, A., Flammini, M.: Price of Pareto optimality in hedonic games. In: 30th AAAI Conference on Artificial Intelligence (AAAI), pp. 475–481 (2016)
12. Feldman, M, Lewin-Eytan, L., Naor, J.: Hedonic clustering games. ACM Trans. Parallel Comput. **2**(1), Article 4 (2015)
13. Hoefer, M., Jiamjitrak, W.: On proportional allocation in hedonic games. In: Bilò, V., Flammini, M. (eds.) SAGT 2017. LNCS, vol. 10504, pp. 307–319. Springer, Cham (2017). https://doi.org/10.1007/978-3-319-66700-3_24
14. Hoefer, M., Vaz, D., Wagner, L.: Hedonic coalition formation in networks. In: 29th AAAI Conference on Artificial Intelligence (AAAI), pp. 929–935 (2015)
15. Kaklamanis C., Kanellopoulos P., Papaioannou K.: The price of stability of simple symmetric fractional hedonic games. In: 9th Symposium on Algorithmic Game Theory (SAGT), pp. 220–232 (2016)
16. Olsen, M.: On defining and computing communities. In: 18th Computing: Australasian Theory Symposium (CATS), pp. 97–102 (2012)
17. Peters, D.: Complexity of hedonic games with dichotomous preferences. In: 30th AAAI Conference on Artificial Intelligence (AAAI), pp. 579–585 (2016)
18. Peters, D., and Elkind, E.: Simple causes of complexity in hedonic games. In: 24th International Joint Conference on Artificial Intelligence (IJCAI), pp. 617–623 (2015)

Schelling Segregation with Strategic Agents

Ankit Chauhan, Pascal Lenzner(✉), and Louise Molitor

Algorithm Engineering Group, Hasso Plattner Institute,
University of Potsdam, Potsdam, Germany
{ankit.chauhan,pascal.lenzner,louise.molitor}@hpi.de

Abstract. Schelling's segregation model is a landmark model in sociology. It shows the counter-intuitive phenomenon that residential segregation between individuals of different groups can emerge even when all involved individuals are tolerant. Although the model is widely studied, no pure game-theoretic version where rational agents strategically choose their location exists. We close this gap by introducing and analyzing generalized game-theoretic models of Schelling segregation, where the agents can also have individual location preferences.

For our models we investigate the convergence behavior and the efficiency of their equilibria. In particular, we prove guaranteed convergence to an equilibrium in the version which is closest to Schelling's original model. Moreover, we provide tight bounds on the Price of Anarchy.

1 Introduction

Segregation is a well-known sociological phenomenon which is intensely monitored and investigated by sociologists and economists. It essentially means that a community of people which is mixed along e.g. ethical, racial, linguistic or religious dimensions tends to segregate over time such that almost homogeneous sub-communities emerge. The most famous example of this phenomenon is residential segregation along racial lines in many urban areas in the US.[1]

To explain the emergence of residential segregation Schelling [19] proposed in a seminal paper a very simple and elegant agent-based model. In Schelling's model two types of agents, say type A and type B agents, are placed on a line or a grid which models some residential area. Each agent is aware of its neighboring agents and is content with her current residential position if at least a τ fraction of agents in her neighborhood is of the same type, for some $0 \leq \tau \leq 1$. If this condition is not met, then the agent becomes discontent with her current position and exchanges positions with a randomly chosen discontent agent of the other type or the agent jumps to a randomly chosen empty spot.[2] Schelling showed with simple experiments using coins, graph paper and random numbers that even with $\tau \leq \frac{1}{2}$, i.e., with tolerant agents, the society of agents

[1] See the racial dot map [8] for an impressive visualization.
[2] A playful interactive demonstration can be found in [13].

X. Deng (Ed.): SAGT 2018, LNCS 11059, pp. 137–149, 2018.
https://doi.org/10.1007/978-3-319-99660-8_13

will eventually segregate into almost homogeneous communities. This surprising observation caught the attention of many economists, physicists, demographers and computer scientists who studied related random models and verified experimentally that tolerant local neighborhood preferences can nonetheless induce global segregation in social and residential networks, see e.g. [1,4,11,14].

To the best of our knowledge, all agent-based models of segregation are essentially random processes where discontent agents choose their new location at random. In this paper we depart from this assumption by introducing and analyzing a game-theoretic version of Schelling's model where agents strategically choose their location. Empirically, our model yields outcomes which are very similar to Schelling's original model - see Fig. 1 for an example.

random inital placement

stable placement in the u-SSG

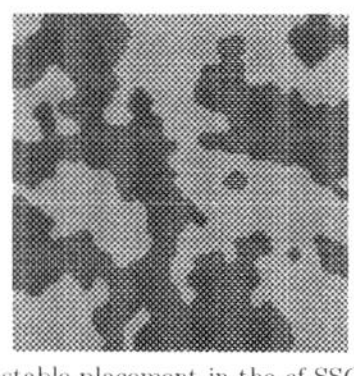

stable placement in the cf-SSG (common favorite node: center)

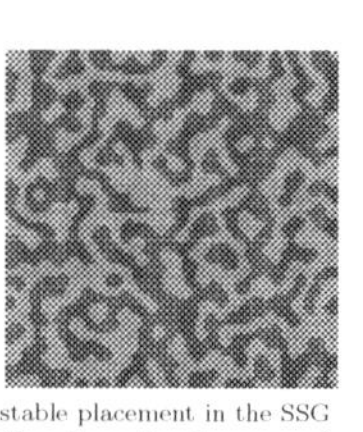

stable placement in the SSG (favorite nodes chosen u.a.r)

Fig. 1. Sample equilibria of our model showing significant segregation. Here $|A| = |B| = 5000$, nodes of type A are blue, type B nodes are green and $\tau = \frac{1}{2}$. (Color figure online)

Moreover, our model generalizes Schelling's model since we allow agents to have preferences over the available locations. Hence, we introduce and explore the influence of such individual location preferences.

1.1 Related Work

There is a huge body of work on Schelling's model and variations thereof, see e.g. [4,5,10,17,18,20–23]. Most related work is purely empirical and provides simulation results. We focus here on the surprisingly small amount of related work, which rigorously proves properties of (variants of) Schelling's model.

Young [23] was the first to rigorously analyze a variant of the one-dimensional segregation model by using techniques from evolutionary game theory. He considered the specific dynamics where a pair of agents is chosen at random and they swap places with a suitably chosen probability. Then he analyzes the induced Markov chain and proves that under certain conditions total segregation will be with high probability a stochastically stable state. Later Zhang [24,25] proved similar results in 2-dimensional models.

The first rigorous analysis of the original Schelling model was achieved by Brandt et al. [7] for the case where agents with tolerance parameter $\tau = \frac{1}{2}$ are located on a ring and agents can only swap positions. They prove that the process converges with high probability to a state where the average size of monochromatic neighborhoods is polynomial in w, where w is the window-size for determining the neighborhood. Interestingly, Barmpalias et al. [2] have proven a drastically different behavior for $0.3531 < \tau < \frac{1}{2}$ where the size of monochromatic neighborhoods is exponential in w. Later, Barmpalias et al. [3] analyzed a 2-dimensional variant where both agent types have different tolerance parameters and agents may change their type if they are discontent. Finally, Immorlica et al. [15] considered the random Schelling dynamics on a 2-dimensional toroidal grid with $\tau = \frac{1}{2} - \epsilon$, for some small $\epsilon > 0$. Their main result is a proof that the average size of monochromatic neighborhoods is exponential in w.

Not much work has been done on the game theory side. To the best of our knowledge only the model by Zhang [25] is game-theoretic and closely related to Schelling's model. In this game agents are placed on a toroidal grid and are endowed with a noisy single peaked utility function which depends on the ratio of the numbers of the two agent types in any local neighborhood. The highest utility is attained in perfectly balanced neighborhoods and agents slightly prefer being in the majority over being in the minority. In contrast to our model, Zhang's model [25] assumes transferable utilities and it can happen that after a randomly chosen swap one or both agents are worse off. Moreover Zhang's model does not incorporate the threshold behavior at τ. However, despite the different model Zhang [25] uses a similar potential function as we do in this paper.

We note that hedonic games [6,12] are also remotely related to Schelling's model, but there the utility of an agent only depends on her chosen coalition. In Schelling's model the neighborhood of an agent could be considered as her coalition, but then not all agents in a coalition derive the same utility from it.

1.2 Model and Notation

We consider a network $G = (V, E)$, where V is the set of nodes and E is the set of edges, which is connected, unweighted and undirected. If in G every node has the same degree, i.e., the same number of incident edges, then we call G *regular*. The distance $d_G(u, v)$ between two nodes $u, v \in V$ in network G is the number of edges on a shortest path between u and v. The diameter of G is the length of the longest shortest path between any pair of nodes and is denoted by $D(G)$. For a given node $u \in V$ let $N_w(u)$ be the set of nodes $v \neq u$ which are in distance at most w from node u. We call $N_w(u)$ the w-neighborhood of u and $w \geq 1$ is the window size. We will omit w whenever a statement holds for all $w \geq 1$.

Agents of two different types are located on the nodes of network G. There are two disjoint sets of agents A and B, with $|A|, |B| \geq 2$ and we say that all agents $a \in A$ are of type A and agents $b \in B$ are of type B. In each state of our game, there is an injective mapping $p_G : \{A \cup B\} \rightarrow V$ between agents and nodes which we call a *placement*. In any placement p_G a node of G can be occupied by exactly one agent either from A or from B or the node can be empty. Let p_G be any placement and let x, y, with $x \in A$ and $y \in B$, be agents which are neighbors under placement p_G. In this case, we call x, y a *colored pair*.

For any agent $x \in A \cup B$, $p_G(x) = u$, we define $N_w^+(p_G(x)) \subseteq N_w(p_G(x))$, as the set of other nodes v in the w-neighborhood of node u, with $u \neq v$, which are occupied by the same type of agents as agent x and $N_w^-(p_G(u)) \subseteq N_w(p_G(u))$ is the corresponding set of other nodes which are occupied by agents of the other type. Note that $p_G(x) \notin N_w^+(p_G(x))$. If $|N_w^+(p_G(x))| + |N_w^-(p_G(x))| = 0$, then agent x has no neighboring agents and we say that agent x is *isolated*.

Let $\tau \in [0,1]$ be the *tolerance parameter*. Similar to Schelling's model we say that an un-isolated agent x is *happy* or *content* with placement p_G if at least a τ-fraction of the agents which occupy the nodes in her w-neighborhood under p_G are of the same type as her. I.e., an un-isolated agent x is happy if $\frac{|N_w^+(p_G(x))|}{|N_w^+(p_G(x))|+|N_w^-(p_G(x))|} \geq \tau$, otherwise x is *unhappy* or *discontent* with placement p_G. Moreover, we will assume that isolated agents are always unhappy. We call the ratio $\frac{|N_w^+(p_G(x))|}{|N_w^+(p_G(x))|+|N_w^-(p_G(x))|}$ the *local happiness ratio* of agent x. Besides having preferences about the neighborhood structure, every agent may have a *favorite node* $fav_x \in V$ in the network G.

The cost function of our agents is based on two main assumptions:

(1) An agent's high priority goal is to find a location where she is happy.
(2) An agent's low priority goal is to find a location which is as close as possible to her favorite location.

Thus, a happy agent x strives for locations where she is happy, but as close as possible to fav_x. If an agent x is unhappy, she will try to improve her local happiness ratio. If this is not possible then she will select a location which has maximum possible local happiness ratio and which is closest to fav_x.

We incorporate these assumptions as follows in our cost function: The cost of an un-isolated agent x with placement p_G in network G is the vector

$$cost_x(p_G) = \left(\max\left(0, \tau - \frac{|N_w^+(p_G(x))|}{|N_w^+(p_G(x))| + |N_w^-(p_G(x))|}\right), d_G(fav_x, p_G(x)) + 1\right)$$

and for an isolated agent x the cost is $cost_x(p_G) = (\tau,\ d_G(fav_x, p_G(x)) + 1)$.

Thus, an agent x is happy with placement p_G, if and only if $cost_x(p_G) = (0, \cdot)$. Note that we use $d_G()+1$ instead of $d_G()$ as second component of the cost vector for technical reasons. This has no influence on the behavior of the agents.

We choose the lexicographic order $\leq_{lex}$ [3] for comparing cost vectors. Agents want to minimize their cost vector lexicographically, i.e., it is more important for an agent to be happy than to be close to her favorite node.

The *social cost* $cost(p_G)$ of a placement p_G in network G is the vector consisting of the number of unhappy players and the sum of all distance terms:

$$cost(p_G) = \left(|\{x \in A \cup B \mid cost_x(p_G) = (\alpha, \cdot), \alpha \neq 0\}|, \sum_{x \in A \cup B} (d_G(fav_x, p_G(x)) + 1) \right).$$

The *strategy space* of an agent is the set of all nodes of G. A strategy vector is *feasible* if all of its entries are pairwise disjoint. Clearly, there is a bijection

[3] $(\alpha, \beta) <_{lex} (\gamma, \delta)$, if $\alpha < \gamma$ or $\alpha = \gamma$ and $\beta < \delta$. $(\alpha, \beta) =_{lex} (\gamma, \delta)$, if $\alpha = \gamma$ and $\beta = \delta$. $(\alpha, \beta) >_{lex} (\gamma, \delta)$, if $\alpha > \gamma$ or $\alpha = \gamma$ and $\beta > \delta$.

between feasible strategy vectors and placements p_G and we will use them interchangeably. For the possible strategy changes of an agent there are two versions, which yield the *Swap Schelling Game* and the *Jump Schelling Game*.

The Swap Schelling Game: In the *Swap Schelling Game (SSG)* only pairs of agents can jointly change their strategies by swapping locations. Two agents x and y agree to swap their nodes if both agents strictly decrease their cost by swapping. A placement p_G is *stable* if no pair of agents can both improve their cost via swapping. Hence, stable placements correspond to 2-coalitional pure Nash equilibria. Since locations can only be swapped, we will assume throughout the paper that there are no empty nodes in G, that is, p_G is also surjective.

The Jump Schelling Game: In the *Jump Schelling game (JSG)* an agent can change her strategy to any currently empty node, which constitutes a "jump" to that node. An agent will jump to another empty node, if this strictly decreases her cost. Here a *stable* placement p_G corresponds to a pure Nash equilibrium.

Different Variants: Besides assuming that every agents has some individual favorite position, we will consider two additional variants of the SSG and the JSG, depending on the favorite nodes of the agents. If the agents do not have a favorite node, then we call these versions *uniform* (*u-SSG* or *u-JSG*) and we simply ignore the second entry in the cost vector. Note that the uniform versions are very close to Schelling's original model. If all agents have the same favorite node, then we call these games *common favorites* (*cf-SSG* or *cf-JSG*). Observe that this variant is especially interesting since it models the case where some particular location is intrinsically more attractive than others to all agents, e.g. it could be the most popular location in a city.

Dynamic Properties: We will use *ordinal potential functions.* Such a function Φ maps placements to real numbers with the property that $\Phi(p'_G) < \Phi(p_G)$ if and only if p'_G is the placement which results from an improving move by a (pair of) agent(s) under placement p_G. If an ordinal potential function for some special case of the game exists, then this implies that this special case has the *finite improvement property (FIP)*, which states that any sequence of improving moves must be finite. Having the FIP is equivalent to the game being a potential game [16]. Such games have many attractive properties like guaranteed existence of pure equilibria and often a fast convergence to such a stable state. Moreover, a potential function is useful for analyzing the quality of equilibria. In contrast, if an infinite sequence of improving moves, usually called *improving response cycle (IRC)*, exists then there cannot exist an ordinal potential function.

1.3 Our Contribution

We introduce the first agent-based model for Schelling segregation where the agents strategically choose their locations. For this, we consider a generalization of Schelling's model where agents besides having preferences over their local neighborhood structure also have preferences of the possible locations.

Table 1. Convergence results. "✓": potential game for any w and G. "reg.": potential game for any w and regular networks G. "ring": potential game on a ring. IRC: an improving response cycle exists, i.e., not a potential game.

	u-SSG	cf-SSG	SSG	u-JSG	cf-JSG & JSG
$\tau < \frac{1}{2}$	✓ (Theorem 1)	reg. (Theorem 3)	ring, $w = 1$ (Theorem 5)	ring, $w = 1$ (Theorem 6)	IRC: $\frac{1}{3} < \tau < \frac{1}{2}$ (Theorem 7)
$\tau = \frac{1}{2}$	✓ (Theorem 1)	reg. (Theorem 3)	reg. (Theorem 4)	ring, $w = 1$ (Theorem 6)	IRC (Theorem 7)
$\tau > \frac{1}{2}$	reg. (Theorem 2)	reg. (Theorem 3)	reg. (Theorem 4)	ring, $w = 1$ (Theorem 6)	IRC: $\frac{1}{2} < \tau \leq \frac{2}{3}$(Theorem 7)

This introduces the important aspect of individual location differentiation which has a significant influence on residential decisions in real life.

Our main contribution is a thorough investigation of the convergence properties of many variants of our model. See Table 1 for details. In particular, we prove guaranteed convergence to an equilibrium for u-SSG, which essentially is Schelling's model, if tolerant agents are restricted to location swaps or if the underlying network is regular. In contrast, previous work [2,3,7,15] has established, that the process converges with high probability. Moreover, also the (cf)-SSG behaves nicely on regular networks. In contrast to this, we show that location preferences have a severe impact in the (cf-)JSG, since improving response cycles exist, which imply that there cannot exist a potential function.

Furthermore, we investigate basic properties of stable placements and their efficiency in the (u-)SSG. In particular, we prove tight bounds on a variant of the Price of Anarchy for the (u-)SSG.

Due to space constraints all omitted details can be found in [9].

2 Dynamic Properties

We analyze the convergence behavior of the Schelling game. Our main goal is to investigate under which conditions an ordinal potential function Φ exists.

2.1 Dynamic Properties of the Swap Schelling Game

We prove for various special cases of the SSG that they are actually potential games. For this we analyze the change in the potential function value for a suitably chosen potential function Φ for an arbitrary location swap of two agents x and y. Such a swap changes the current placement p_G only in the locations of agents x and y and yields a new placement p'_G.

Theorem 1. *If $\tau \leq \frac{1}{2}$ then the u-SSG is a potential game for any $w \geq 1$.*

Proof. We prove the theorem by showing that $\Phi(p_G) = \frac{1}{2}\sum_{x\in A\cup B}|N^-(p_G(x))|$ is an ordinal potential function. Note that Φ is the number of colored pairs. First of all, notice that a swap between two agents $x \in A$ and $y \in B$ will only executed when both agents are unhappy and of different types, since a swap between agents of the same type cannot be an improvement for at least one of the involved agents. Furthermore a happy agent has no possibility to improve, so there is no incentive to. An agent will decrease her cost if and only if she is unhappy and reduces the ratio of neighbors with different type by swapping. It holds that

$$\frac{|N^+(p_G(x))|}{|N^+(p_G(x))| + |N^-(p_G(x))|} < \tau \text{ and } \frac{|N^+(p_G(y))|}{|N^+(p_G(y))| + |N^-(p_G(y))|} < \tau.$$

Hence,

$$\frac{|N^-(p_G(x))|}{|N^+(p_G(x))| + |N^-(p_G(x))|} > 1-\tau \text{ and } \frac{|N^-(p_G(y))|}{|N^+(p_G(y))| + |N^-(p_G(y))|} > 1-\tau.$$

Since $\tau \leq \frac{1}{2}$ it follows that $|N^+(p_G(x))| < \tau \cdot (|N^+(p_G(x))| + |N^-(p_G(x))|) \leq (1-\tau)\cdot(|N^+(p_G(x))|+|N^-(p_G(x))|) < |N^-(p_G(x))|$ and analogously we get for agent y that $|N^+(p_G(y))| < |N^-(p_G(y))|$. Thus,

$$\begin{aligned}&|N^+(p_G(x))| + |N^+(p_G(y))| < |N^-(p_G(x))| + |N^-(p_G(y))|\\ &= |N^-(p'_G(y))| + |N^-(p'_G(x))| < |N^-(p_G(x))| + |N^-(p_G(y))|.\end{aligned}$$

This implies that for the change in the potential function value that $\Phi(p_G) - \Phi(p'_G) = (|N^-(p_G(x))| + |N^-(p_G(y))|) - (|N^-(p'_G(y))| + |N^-(p'_G(x))|) > 0$. □

Remark 1. The function $\Phi(p_G) = \frac{1}{2}\sum_{x\in A\cup B}|N^-(p_G(x))|$ is not a potential function for the (cf-)SSG. See Fig. 2 in [9].

Theorem 2. *For any $w \geq 1$ the u-SSG on regular networks is a potential game.*

Proof. We prove the theorem by showing that $\Phi(p_G) = \frac{1}{2}\sum_{x\in A\cup B}|N^-(p_G(x))|$ is an ordinal potential function. Analogously to Theorem 1 there is no incentive for an agent $x \in A$ to swap with another agent $y \in A$ who has the same type or to swap if x is happy. Since we consider the u-SSG on regular networks, we have

$$\forall x \in A \cup B\colon |N(p_G(x))| = |N^+(p_G(x))| + |N^-(p_G(x))| = k.$$

So an agent $x \in A$ will only swap when she is unhappy with another agent $y \in B$ of different type. Since the swap is an improvement, it is valid that $\frac{|N^+(p_G(x))|}{k} < \frac{|N^+(p'_G(x))|}{k}$ and $\frac{|N^+(p_G(y))|}{k} < \frac{|N^+(p'_G(y))|}{k}$. Observe that all agents x', y' who were in $N^+(p_G(x'))$, $N^+(p_G(y'))$ before the swap are after the swap in $N^-(p'_G(y'))$, $N^-(p'_G(x'))$, respectively. The same holds the other way around. Hence, we have that $|N^-(p'_G(y))| = |N^+(p_G(x))| < |N^+(p'_G(x))| = |N^-(p_G(y))|$ and $|N^-(p'_G(x))| = |N^+(p_G(y))| < |N^+(p'_G(y))| = |N^-(p_G(x))|$.

Since a swap between two agents x and y just affects colored pairs where x or y are involved, we have that $\Phi(p_G) - \Phi(p'_G) = |N^-(p_G(x))| + |N^-(p_G(y))| - (|N^-(p'_G(x))| + |N^-(p'_G(y))|) > 0$. □

Remark 2. The function $\Phi(p_G) = \frac{1}{2}\sum_{x \in A \cup B} |N^-(p_G(x))|$ is not a potential function for the u-SSG on non-regular networks. See Fig. 3 in [9].

Theorem 3. *For any $w \geq 1$ the cf-SSG is a potential game on regular networks.*

Proof (Sketch). Like in Theorem 2 we prove the statement by showing that $\Phi(p_G) = \frac{1}{2}\sum_{x \in A \cup B} |N^-(p_G(x))|$ is an ordinal potential function. When the involved agents are both happy or both unhappy the proof is very similar to the proof of Theorem 2. If one agent x is happy and the other one y unhappy, it holds that $\frac{|N^+(p_G(y))|}{k} < \frac{|N^+(p'_G(y))|}{k}$, which leads to $|N^-(p'_G(x))| < |N^-(p_G(x))|$ and $|N^-(p'_G(y))| < |N^-(p_G(y))|$. □

Theorem 4. *For any $w \geq 1$ and any $\tau \geq \frac{1}{2}$ the SSG is a potential game on regular networks.*

Proof (Sketch). We show that a function Φ decreases lexicographically for every improving swap, which implies that Φ is an ordinal potential function. For this let $\Phi(p_G) = \left(\frac{1}{2}\sum_{x \in A \cup B} |N^-(p_G(x))|, \sum_{x \in A \cup B} d_G(fav_x, p_G(x))\right)$.

Analogous to the proof of Theorem 2, the first entry of Φ decreases if the swap improves the local happiness ratio of the involved agents.

Hence we just have to consider the case when two agents x and y swap only to decrease their distances from fav_x and fav_y, respectively. It holds that in this case the first entry of Φ doesn't change and the agents just swap when both decreases their distance cost, which reduces the second entry of Φ. □

Theorem 5. *If $\tau < \frac{1}{2}$ and $w = 1$ then the SSG on a ring is a potential game.*

Proof (Sketch). We use an argument similar to the one in the proof of Theorem 4 with $\Phi(p_G) = \left(\frac{1}{2}\sum_{x \in A \cup B} |N^-(p_G(x))|, \sum_{x \in A \cup B} d_G(fav_x, p_G(x))\right)$.

Beyond that we need to consider two happy agents who swap to get closer to their favorite node. We show that after such a swap, the number of colored pairs stays the same. This implies that Φ decreases lexicographically by such a swap, since the first entry stays the same but the second entry decreases. □

2.2 Dynamic Properties of the Jump Schelling Game

Now we consider the JSG. Remember that in the JSG we assume that agents can only decrease their cost by jumping to empty nodes. Such a jump of an agent x changes the current placement p_G only in the location of agent x. We prove for the ring network that the u-JSG is a potential game. Furthermore we show that the cf-JSG and JSG are not potential games for different ranges of τ.

Theorem 6. *If $w = 1$ and the underlying graph is ring network then, the u-JSG is a potential game.*

Proof (Sketch). For any ring network $G = (V, E)$ we define the weight w_e of any edge $e = (u, v) \in E$ as follows:

$$w_e = \begin{cases} 1, & \text{if } u \text{ and } v \text{ have agents of different type} \\ \frac{1}{3}, & \text{if either } u \text{ or } v \text{ is empty} \\ 0, & \text{otherwise.} \end{cases}$$

We use the function $\Phi(p_G) = \sum_{e \in E} w_e$ and prove that if any agent makes an improving jump to some other node in the ring then Φ decreases. □

Theorem 7. *There cannot exist an ordinal potential function for the cf-JSG and the JSG for* $\frac{1}{3} < \tau \leq \frac{2}{3}$.

Proof (Sketch). We prove this statement by giving two examples of an improving response cycle for the cf-JSG on a grid. Since the cf-JSG is a special case of the JSG the statement holds for both variants. The improving response cycle for $\frac{1}{3} < \tau \leq \frac{1}{2}$ can be found in Fig. 2. The improving response cycle for $\frac{1}{2} < \tau \leq \frac{2}{3}$ can be found in Fig. 3. □

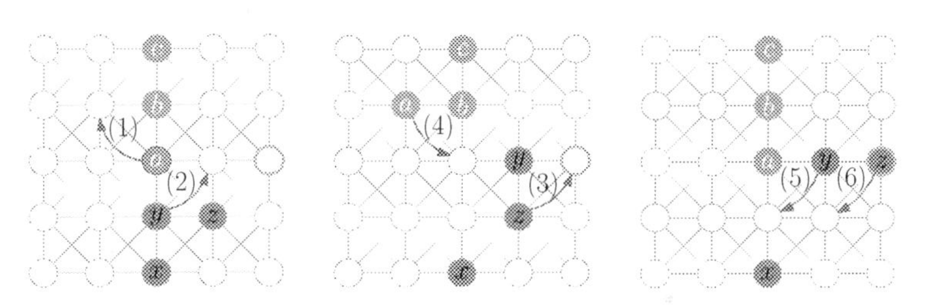

Fig. 2. An improving response cycle for the cf-JSG for $\tau \in (\frac{1}{3}, \frac{1}{2}]$. Agents of type A are blue, type B agents are green. The common favorite node is purple. (Color figure online)

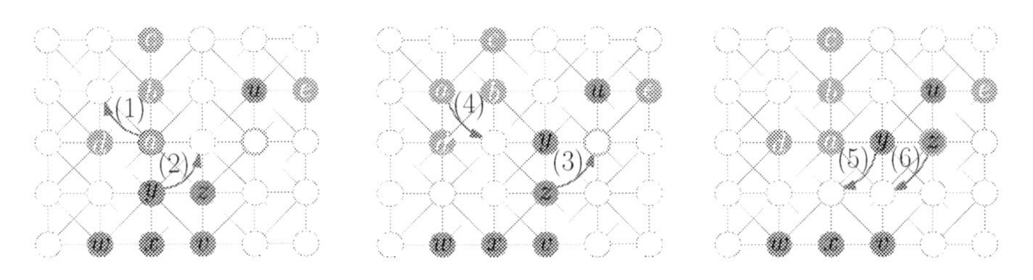

Fig. 3. An improving response cycle for the cf-JSG for $\tau \in (\frac{1}{2}, \frac{2}{3}]$. Agents of type A are blue, type B agents are green. The common favorite node is purple. (Color figure online)

3 Efficiency of Stable Placements

In this section we investigate the properties of stable placements. In particular, we investigate their (in-)efficiency.

We start with proving that stable placements exist for many of our versions.

Theorem 8. *Stable placements exist for the u-SSG, the cf-SSG and the u-JSG.*

Now we move on to proving basic properties of stable placements.

Theorem 9. *Let p_G^{NE} be a stable placement for the SSG on some graph G. The following statements hold:*

(a) If $D(G) > 1$ and $\tau \leq \frac{1}{2}$ then for any p_G^{NE}, at most one type of agents can be unhappy. Moreover, there exists a stable placement with unhappy agent(s).
(b) If $\tau > \frac{1}{2}$ there exist stable placements p_G^{NE} with unhappy agents of both types.
(c) For the SSG there is a graph G such that p_G^{NE} has a better total distance cost than the socially optimal placement.
(d) For the SSG with $\tau > \frac{1}{2} - \frac{1}{2n}$, for $n = |A| = |B|$, there exists a graph G such that there is no placement where at least one agent is happy.

Remember that the social cost $cost(p_G)$ of a placement p_G is a vector. Hence we cannot use the state-of-the-art notions for the Price of Anarchy or the Price of Stability for investigating the efficiency of stable placements. For this, we first introduce suitable measures.

Definition 1. *The* ratio of happiness *RoH and the* ratio of distance *RoD of two arbitrary placements p_G and $\tilde{p_G}$ with $Dist(p''_G) = \sum_{u \in A \cup B} d_G(fav_u, p''_G(u)) + 1$ is*

$$RoH(p_G, \tilde{p_G}) = \frac{\#of\ unhappy\ players\ in\ p_G + 1}{\#of\ unhappy\ players\ in\ \tilde{p_G} + 1} \text{ and } RoD(p_G, \tilde{p_G}) = \frac{Dist(p_G)}{Dist(\tilde{p_G})}.$$

Note that the RoH has the additional "+1" terms to handle the case where all agents in placement $\tilde{p_G}$ are happy. Essentially the RoH compares the social cost vectors of p_G and $\tilde{p_G}$ by their first entry, and the RoD by their second entry. The $\text{RoD}(p_G, \tilde{p_G}) = 1$ in the versions of u-SSG and u-JSG.

We now define our notion of PoA and PoS.

Definition 2. *For a given underlying network G, let NE be the set of all possible stable placements in G and p_G^{opt} be the socially optimal placement. Then,*

$$PoA = \max_{p_G \in \mathsf{NE}} \{(RoH(p_G, p_G^{opt}), RoD(p_G, p_G^{opt}))\},$$
$$PoS = \min_{p_G \in \mathsf{NE}} \{(RoH(p_G, p_G^{opt}), RoD(p_G, p_G^{opt}))\}.$$

Observation 1. *For any G we have $PoA \leq (n+1, D(G)-1)$ and $PoS \geq (1, 1)$.*

Theorem 10. *For the u-SSG with $\tau \leq \frac{1}{2}$, there exists a network G and a stable placement p_G^{NE} such that in the worst case $RoH = \max\{|A|, |B|\} + 1$ and in the best case $RoH = 1$.*

Proof. To prove the upper bound we already know from Theorem 9 that at most one type of agents can be unhappy. Thus the maximum number of unhappy agents in a placement is equal to the number of agents of the bigger group. When the optimal placement p_G^{opt} is stable the lower bound follows straightforwardly.

We show that the bounds are tight by an example. Consider Fig. 4a. The green agents are happy since at least half of their neighbors are of their type. The placement is stable, since they have no incentive to swap. All blue agents are discontent. Notice that there are more blue agents than green ones. However, there is a stable placement, which is shown in Fig. 4b where all agents are content. The example can be easily extended to larger networks. □

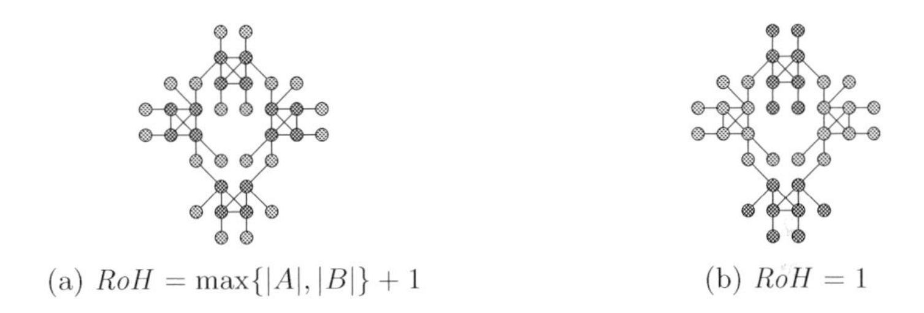

(a) $RoH = \max\{|A|, |B|\} + 1$ (b) $RoH = 1$

Fig. 4. Two different stable placements in which in (a) all blue agents are unhappy whereas in (b) all agents are happy. (Color figure online)

From Theorem 10 the corollary below follows.

Corollary 1. *In the u-SSG there exists a network G such that the PoA $= (max\{|A|, |B|\} + 1, 1)$ and the PoS $= (1, 1)$.*

Now we give a tight bound for the Price of Anarchy.

Theorem 11. *The SSG has PoA $\in (\Theta(n), \Theta(D(G)))$ for some network G.*

Proof (Sketch). Consider the network G and the placement in Fig. 5(top) for $\frac{1}{2w} < \tau < 1 - \frac{1}{2w}$. We assume that $w \in \Theta(1)$ and the number of agents of type A is $|A| = |B| + w + (w-1)\left\lceil \frac{|B|}{2} \right\rceil + w \left\lfloor \frac{|B|}{2} \right\rfloor$. Let $n = |A| + |B|$ be the total number of agents. The favorite node of agent a_i is $p_G(b_i)$ and of agent b_i it is $p_G(a_i)$, respectively, $i = 1, \ldots, |B|$. All other agents are placed at their favorite node. This placement is stable and has high cost. □

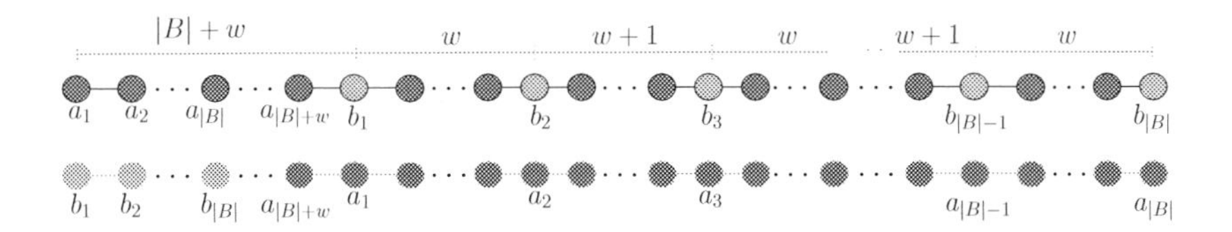

Fig. 5. A stable placement (top) and the social optimum placement (bottom).

Remark 3. The construction in the proof of Theorem 11 can be easily extended to the JSG. Thus, also for the JSG $PoA \in (\Theta(n), \Theta(D(G)))$ for some network G.

4 Conclusion and Open Questions

In this work we have introduced the first truly game-theoretic version of the well-known Schelling segregation model. The selfish agents in our model strategically choose their locations and take the structure of their local neighborhood as well as their individual location preferences into account.

We have established that many variants of our model actually are potential games, which implies the existence of pure (2-coalitional) Nash equilibria and guaranteed convergence in the sequential version. However, we have also identified cases, e.g. the (cf)-JSG, which have improving response cycles. Moreover, we have investigated the efficiency of stable placements in the (u-)SSG and proved high tight bounds on the Price of Anarchy. This implies that the outcomes of our game vary significantly in their social cost.

The most intriguing open problem is to settle the convergence behavior for the SSG for $\tau < \frac{1}{2}$ and the u-JSG. We conjecture, that a potential function exists. Moreover, it is open whether stable placements exist for the SSG and the (cf-)JSG. We conjecture that a stable placement exists for all variants. Another ambitious endeavor would be to prove bounds on the size of the monochromatic regions similar to the works [2,3,7,15]. In particular, it would be interesting to explore the impact of the location preferences on the induced stable placements.

References

1. Alba, R.D., Logan, J.R.: Minority proximity to whites in suburbs: an individual-level analysis of segregation. Am. J. Sociol. **98**(6), 1388–1427 (1993)
2. Barmpalias, G., Elwes, R., Lewis-Pye, A.: Digital morphogenesis via schelling segregation. In: FOCS 2014, pp. 156–165 (2014)
3. Barmpalias, G., Elwes, R., Lewis-Pye, A.: Unperturbed schelling segregation in two or three dimensions. J. Stat. Phys. **164**(6), 1460–1487 (2016)
4. Benard, S., Willer, R.: A wealth and status-based model of residential segregation. J. Math. Sociol. **31**(2), 149–174 (2007)
5. Benenson, I., Hatna, E., Or, E.: From schelling to spatially explicit modeling of urban ethnic and economic residential dynamics. Sociol. Methods Res. **37**(4), 463–497 (2009)
6. Bogomolnaia, A., Jackson, M.O.: The stability of hedonic coalition structures. Games Econ. Behav. **38**(2), 201–230 (2002)
7. Brandt, C., Immorlica, N., Kamath, G., Kleinberg, R.: An analysis of one-dimensional schelling segregation. In: STOC 2012, pp. 789–804 (2012)
8. Cable, D.: The racial dot map. Weldon Cooper Center for Public Service, University of Virginia (2013). https://demographics.coopercenter.org/Racial-Dot-Map/
9. Chauhan, A., Lenzner, P., Molitor, L.: Schelling segregation with strategic agents. arXiv:1806.08713 (2018)
10. Clark, W.A.V.: Residential segregation in American cities: a review and interpretation. Popul. Res. Policy Rev. **5**(2), 95–127 (1986)

11. Clark, W.A.V.: Geography, space, and science: perspectives from studies of migration and geographical sorting. Geogr. Anal. **40**(3), 258–275 (2008)
12. Dreze, J.H., Greenberg, J.: Hedonic coalitions: optimality and stability. Econom. J. Econom. Soc. **48**(4), 987–1003 (1980)
13. Hart, V., Case, N.: Parable of the polygons (2016). http://ncase.me/polygons
14. Henry, A.D., Prałat, P., Zhang, C.-Q.: Emergence of segregation in evolving social networks. PNAS **108**(21), 8605–8610 (2011)
15. Immorlica, N., Kleinberg, R., Lucier, B., Zadomighaddam, M.: Exponential segregation in a two-dimensional schelling model with tolerant individuals. In: SODA 2017, pp. 984–993 (2017)
16. Monderer, D., Shapley, L.S.: Potential games. Games Econ. Behav. **14**(1), 124–143 (1996)
17. Pancs, R., Vriend, N.J.: Schelling's spatial proximity model of segregation revisited. J. Public Econ. **91**(1), 1–24 (2007)
18. Schelling, T.C.: Models of segregation. Am. Econ. Rev. **59**(2), 488–493 (1969)
19. Schelling, T.C.: Dynamic models of segregation. J. Math. Sociol. **1**(2), 143–186 (1971)
20. Schelling, T.C.: Micromotives and Macrobehavior. WW Norton & Company, New York (2006)
21. Vinković, D., Kirman, A.: A physical analogue of the schelling model. PNAS **103**(51), 19261–19265 (2006)
22. White, M.J.: Segregation and diversity measures in population distribution. Popul. Index **52**(2), 198–221 (1986)
23. Young, H.P.: Individual Strategy and Social Structure: An Evolutionary Theory of Institutions. Princeton University Press, Princeton (1998)
24. Zhang, J.: A dynamic model of residential segregation. J. Math. Sociol. **28**(3), 147–170 (2004)
25. Zhang, J.: Residential segregation in an all-integrationist world. J. Econ. Behav. Organ. **54**(4), 533–550 (2004)

Efficient Rational Proofs with Strong Utility-Gap Guarantees

Jing Chen[1], Samuel McCauley[2], and Shikha Singh[2](✉)

[1] Stony Brook University, Stony Brook, USA
jingchen@cs.stonybrook.edu
[2] Wellesley College, Wellesley, USA
{samuel.mccauley,shikha.singh}@wellesley.edu

Abstract. As modern computing moves towards smaller devices and powerful cloud platforms, more and more computation is being delegated to powerful service providers. Interactive proofs are a widely-used model to design efficient protocols for verifiable computation delegation.

Rational proofs are payment-based interactive proofs. The payments are designed to incentivize the provers to give correct answers. If the provers misreport the answer then they incur a payment loss of at least $1/u$, where u is the *utility gap* of the protocol.

In this work, we tightly characterize the power of rational proofs that are super efficient, that is, require only logarithmic time and communication for verification. We also characterize the power of single-round rational protocols that require only logarithmic space and randomness for verification. Our protocols have strong (that is, polynomial, logarithmic, and even constant) utility gap. Finally, we show when and how rational protocols can be converted to give the completeness and soundness guarantees of classical interactive proofs.

1 Introduction

Most computation today is not done locally by a client, but instead is outsourced to third-party service providers in exchange for money. Trading computation for money brings up two problems—(a) how the client can guarantee correctness of the outsourced computation (without redoing the computation), and (b) how to design the payment scheme. The two problems are closely related: ideally, we want the payment scheme to be such that it incentivizes service providers to perform the computation correctly.

This work has been partially supported by NSF CAREER Award CCF 1553385, CNS 1408695, CCF 1439084, IIS 1247726, IIS 1251137, CCF 1217708, by Sandia National Laboratories, and by the European Union's 7th Framework Programme (FP7/2007-2013) / ERC grant agreement no. 614331. BARC, Basic Algorithms Research Copenhagen, is supported by the VILLUM Foundation grant 16582.

X. Deng (Ed.): SAGT 2018, LNCS 11059, pp. 150–162, 2018.
https://doi.org/10.1007/978-3-319-99660-8_14

Interactive proofs (IP) are the most well-studied and widely-used theoretical framework to verify correctness of outsourced computation [7,10,11,17,18,22,28,33]. In an IP, a weak client (or *verifier*) interacts with powerful service providers (or *provers*) to determine the correctness of their claim. At the end, the verifier probabilistically accepts or rejects the claim.[1] Interactive proofs guarantee that, roughly speaking, the verifier accepts a truthful claim with probability at least 2/3 (*completeness*) and no strategy of the provers can make the verifier accept a false claim with probability more than 1/3 (*soundness*).[2].

Rational proofs are payment-based interactive proofs for computation outsourcing which leverage the incentives of the service providers. In rational proofs, the provers act *rationally* in the game-theoretic sense, that is, they want to maximize their payment. The payment is designed such that when the provers maximize their payment, they also end up giving the correct answer. The model of rational proofs (RIP) was introduced by Azar and Micali in [2]. Since then, many simple and efficient rational protocols have been designed [3,9,13,23,24,26,35].

While rational proofs provide strong theoretical guarantees, there are two main barriers that separate them from what is often desired in practice. First, many rational protocols require a polynomial-time verifier—but a "weak" client is unlikely to be able to spend (say) quadratic time or linear extra space on verification. Second, many of these protocols strongly rely on the rationality of the provers. An honest prover may receive only a fraction of a cent more than a dishonest prover, yet a rational prover is assumed to be incentivized by that small amount. However, service providers may not always be perfectly rational.

The goal of this paper is to give protocols that overcome these barriers.

Utility Gap. The strength of the guarantee provided by rational proofs is captured by the notion of *utility gap*. The high level idea behind utility gap is that provers who are not perfectly rational may not care about small losses in payments and may lazily give the incorrect answer. If a rational protocol has a utility gap of u, then the provers who mislead the verifier to an incorrect answer are guaranteed to lose at least $1/u$. (This is under a normalized budget of 1; if the budget is scaled up to B, such provers can be made to lose at least B/u.) Thus, protocols with small utility gap are sound even against provers with *bounded rationality*; that is, provers who are only sensitive to large losses.

In this paper, we design efficient rational protocols with strong utility gap—that is, polynomial, logarithmic, and even constant utility gap. In Section 5, we show when and how a noticeable utility gap of a rational protocol can be utilized to achieve the strong completeness and soundness guarantees of a classical proof.

Efficient Protocols. In this paper, we focus on designing rational protocols with very small overheads in terms of verification time, space, communication cost and number of rounds. In particular, we design constant-round rational

---

[1] In classical interactive proofs there is no payment—simply acceptance or rejection.

[2] More formally, given an input x and a language L, if $x \in L$, the verifier accepts with probability at least 2/3 (*completeness*); if $x \notin L$, then no strategy of the provers can make the verifier accept with probability more than 1/3 (*soundness*).

protocols where the verification time and communication cost are logarithmic in the input size n. We also design single-round rational protocols that have only logarithmic overhead on the verifier's use of space and randomness.

1.1 Results and Contributions

In this section, we summarize our results and contributions.

Time-Efficient Rational Proofs. We study the effect of different communication costs and an additional prover on the power of rational proofs with a highly time-efficient verifier. The utility gap of these protocols is polynomial.

- **Constant Communication.** We show that multiple provers do not add any power when the communication complexity of the protocol is restricted to be extremely small—a constant number of bits. That is, we show that the class of languages that admit a multi-prover rational proof with a $O(\log n)$-time verifier and $O(1)$ communication is exactly $\mathsf{UniformTC}_0$, which is the same as the power of single-prover version under the same costs [3,23]. $\mathsf{UniformTC}_0$ is the class of constant depth, polynomial size uniform threshold circuits, that includes problems such as integer division and iterated multiplication [1,25].
- **Logarithmic Communication.** We show that any rational proof with polynomial communication can be simulated by a rational proof with logarithmic communication that uses an additional prover. Using this property, we improve the communication complexity of Azar and Micali's [3] single-prover rational protocol and show that the class of languages that admit a two-prover rational proof with logarithmic communication is exactly the class of languages decidable by a polynomial time machine that can make polynomially many queries in parallel to an NP oracle, denoted $\mathsf{P}_{||}^{\mathsf{NP}}$.[3] This is an important class (e.g., see [8,30,34]) and includes optimization problems such as maximum clique, longest paths, and variants of the traveling salesman problem.

Space-Efficient Rational Proofs. We achieve even better utility gap guarantees when the verifier's use of space and randomness is extremely small—logarithmic, but its running time may be polynomial. In particular, we exactly characterize the class of single-round rational proofs with $\gamma(n)$ utility gap and logarithmic space and randomness as the class of languages decidable by a polynomial-time machine that makes $O(\gamma(n))$ queries to an NP oracle, denoted $\mathsf{P}_{||}^{\mathsf{NP}[\gamma(n)]}$. Even when $\gamma(n) = O(1)$ this bounded-query class is still sufficiently powerful and contains many of the optimization problems mentioned above.

Rational Proofs with Completeness and Soundness Guarantees. Finally, we closely compare the two proof systems—rational and classical. We construct a condition on the expected payments of rational proofs which, if satisfied, turns them into a classical interactive proof with completeness and

[3] For parallel oracle queries, both notations $\mathsf{P}_{||}^{\mathsf{NP}}$ [34] and $\mathsf{P}^{||\mathsf{NP}}$ [3] are used in literature.

soundness guarantees. We first show how to convert a payment-based protocol for a language L to an accept-reject protocol (without payments) for L such that the expected payment of the former is exactly the probability with which the verifier accepts in the latter. We use this to prove that if the expected payments of all inputs $x \in L$ are noticeably far away from that of all inputs $x \notin L$, the rational protocol can be converted to a classical interactive protocol.

1.2 Additional Related Work

Azar and Micali [3] also characterize the classes $\mathsf{UniformTC}_0$ and $\mathsf{P}_{||}^{\mathsf{NP}}$. Their characterization of $\mathsf{P}_{||}^{\mathsf{NP}}$ requires polynomial communication, which we improve to logarithmic using a second prover. We also note that all protocols in [3] have a polynomial utility gap (under a constant budget).

Rational arguments, super-efficient rational proofs where the prover is restricted to be polynomial time, were introduced by Guo et al. [23]. Rational arguments for all languages in P were given in [24]. Campanelli and Rosario [9] study sequentially composable rational proofs. Zhang and Blanton [35] design protocols to outsource matrix multiplications to a rational cloud.

The model of multi-prover rational interactive proofs was introduced by Chen et al. [13], where they study the power of the model in its full generality (that is, polynomial-time verifier and polynomial communication). In this paper, we restrict our focus on proofs with log-time verifiers and log-size communication.

Different variants of the rational-proof models have also been studied. Chen et al. [14] consider rational proofs where the rational provers are *non-cooperative* [14]. Inasawa and Kenji [27] consider rational proofs where the verifier is also rational and wants to minimize the payment to the provers.

Interestingly, the log-space verifier studied in this paper also happens to be a *streaming algorithm*, that is, the verifier does not need to look again at any input or message bits out of order. Thus, our space-efficient rational proofs are closely related to the work on streaming interactive proofs [11,17,18].

Refereed games is another multi-prover interactive-proof model that leads to game-theoretic characterizations of various complexity classes (e.g. [12,20,21, 29,32]). The model of refereed games requires at least one honest prover.

2 Preliminaries

We begin by reviewing the model of rational proofs [2,13].

Let L be a language, x be an input string and $n = |x|$. An *interactive protocol* is a pair $(V, \vec{P})$, where V is the *verifier* and $\vec{P} = (P_1, \ldots, P_{p(n)})$ is the vector of *provers*, and $p(n)$ a polynomial in n. The goal of the verifier is to determine if $x \in L$. In general, the verifier runs in time polynomial in n and uses polynomial space as well. In Sect. 3, the verifier's running time is $O(\log n)$. In Sect. 4, the verifiers may use polynomial time but are restricted to use $O(\log n)$ space and randomness. The provers are computationally unbounded.[4]

The verifier can communicate with each prover privately, but no two provers can communicate with each other. In a *round*, either each prover sends a message to the verifier, or the verifier sends a message to each prover, and these two cases alternate. Without loss of generality, provers send the first round of messages. The first bit of the first round is the *answer bit*, denoted by c, and indicates whether $x \in L$; that is, $x \in L$ iff $c = 1$. We define the *communication* of the protocol to be the maximum number of total bits transmitted (summed over all provers and all rounds) during the protocol.

Let r be the random string used by V. Let $\vec{m}$ be the vector of all messages exchanged. At the end, the verifier computes the total payment to the provers, given by a payment function $R(x, r, \vec{m})$. We restrict the verifier's budget to be constant, that is, $R \in [0, 1]$ for convenience. We may use negative payments to emphasize penalties but they can shifted to be non-negative. The protocol (including the payment function R) is public knowledge.

The verifier outputs the answer bit c at the end of the protocol—thus the verifier always agrees with the provers.

Each prover P_i can choose a *strategy* $s_{ij} : \{0,1\}^* \to \{0,1\}^*$ for each round j, which maps the transcript he has seen up until the beginning of round j to the message he sends in round j. Note that P_i does not send any message when j is even; in this case s_{ij} can be treated as a constant function. Let $s_i = (s_{i1}, \ldots, s_{ik})$ be the strategy vector of P_i and $s = (s_1, \ldots, s_{p(n)})$ be the strategy profile of the provers. Given any input x, randomness r and strategy profile s, we may write the vector $\vec{m}$ of messages exchanged in the protocol more explicitly as $(V, \vec{P})(x, r, s)$.

The provers are *cooperative* and jointly act to maximize the total expected payment. Thus, before the protocol starts, the provers pre-agree on a strategy profile s that maximizes $u_{(V,\vec{P})}(s; x) = \mathbf{E}_r[R(x, r, (V, \vec{P})(x, r, s))]$. When $(V, \vec{P})$ and x are clear from the context, we write $u(s)$ for $u_{(V,\vec{P})}(s; x)$.

Definition 1 ([13]). *For any language L, an interactive protocol $(V, \vec{P})$ is a rational interactive proof protocol for L if, for any $x \in \{0,1\}^*$ and any strategy profile s of the prover(s) such that $u(s) = \max_{s'} u(s')$, $c = 1$ if and only if $x \in L$.*

Similar to classical proofs, single-prover rational interactive protocols, that is, when $\vec{P} = P$, are denoted by RIP. Multi-prover interactive protocols, where $\vec{P} = (P_1, \ldots, P_{p(n)})$ are denoted by MRIP. In this paper we study both single-prover and multi-prover rational proof protocols.

We use poly(n) as a shorthand for a polynomial n^k, for some constant k.

2.1 Utility Gap and Budget in Rational Proofs

In the above definitions, we assume that a prover is fully rational, and will give the correct answer for *any* increase in expected payment, no matter how small.

[4] While the model allows for extremely powerful provers, those considered in this paper essentially only need to be powerful enough to determine if $x \in L$ or $x \notin L$.

However, a prover may be lazy, and unwilling to give the correct answer unless the correct answer increases its payment by some minimum amount.

The notion of utility gap captures the payment loss incurred by provers who misreport the answer bit. We recall the formal definition below.

Definition 2 ([13]). *Let L be a language with a rational proof protocol $(V, \vec{P})$ and let $\gamma(n) \geq 0$. We say that $(V, \vec{P})$ has an $\gamma(n)$-*utility gap *if for any input x with $|x| = n$, any strategy profile s of $\vec{P}$ that maximizes the expected payment, and any other strategy profile s', where the answer bit c' under s' does not match the answer bit c under s, i.e., $c' \neq c$, then $u(s) - u(s') > 1/\gamma(n)$.*

Relationship Between Utility Gap and Budget. The *budget* is the total expected payment that a verifier can give in a protocol.

Utility gap and budget are closely related. To study utility gaps consistently, we maintain a fixed $O(1)$ budget.[5] This is because utility gap scales naturally with the payment—a polynomial utility gap under a constant budget is the same as a constant utility gap under a sufficiently-large polynomial budget.

2.2 Analyzing Computational Costs of Rational Proofs

Our primary focus in this paper is analyzing the various computational costs of rational interactive proofs. The different parameters fall into two categories.

Verification Costs. A verifier has three main resources: running time, space usage and its randomness.

In Sect. 3, we focus on time-efficient $O(\log n)$ time verifiers. Thus, their space and randomness is also $O(\log n)$. We denote the class of languages that have time-efficient RIP protocols, that is, protocols with a $O(\log n)$ time verifier as RIP^t. Multi-prover notation MRIP^t is analogous. Similar to the literature on "probabilistically checkable proofs of proximity" (PCPPs) [5,6,33], we assume that the verifier has random access to the input string and the proof tape. Thus, if the messages sent by the provers are of size $C(n)$ bits, the verifier needs at least $O(\log C(n))$ time to index a random location of the transcript.

To achieve better utility gap, in Sect. 4, we restrict the verifier's space usage and randomness, instead of its running time and consider verifiers that use $O(\log n)$ space and $O(\log n)$ randomness. We denote the class of languages that have an RIP protocol with space- and randomness-efficient verifiers, that is, verifiers with $O(\log n)$ space and $O(\log n)$ randomness as $\mathsf{RIP}^{s,r}$.

Protocol Costs. A rational interactive proof protocol has three main ingredients: communication cost, number of rounds of interaction and utility gap.[6]

[5] In contrast, Azar and Micali [3] maintain a polynomial-size budget.

[6] The number of provers is an additional parameter in MRIP protocols, but we ignore this so as not to overload notation. All the MRIP protocols in this paper have two provers and all the upper bounds work even with polynomially many provers.

In Section 3, we study the effect of varying the communication complexity of a protocol on its power when we have a logarithmic time verifier. The number of rounds in all the protocols in the paper is $O(1)$.

We denote the class of languages that have an RIP protocol with communication cost $C(n)$, number of rounds $k(n)$ and utility gap $\gamma(n)$ as $\mathsf{RIP}[C(n), k(n), \gamma(n)]$. The multi-prover version is defined similarly.

3 Verification in Logarithmic Time

In this section we consider *time-efficient* verifiers that run in time logarithmic in the input size. We show that for time-efficient verifiers, access to multiple provers is fundamentally linked to the communication cost of the protocol: any single-prover protocol with high communication costs can be reduced to a communication-efficient multi-prover protocol. On the other hand, multiple provers give no extra power for communication-efficient protocols.

Since the utility gap of all the protocols in this section is polynomial in n, we drop it from the notation for simplicity. Thus, an RIP protocol with a $O(\log n)$-time verifier that has communication complexity $C(n)$ and round complexity $k(n)$ is denoted as $\mathsf{RIP}^t[C(n), k(n)]$.

We omit the proofs, which can be found in the full version [15].

Constant Communication. We first show that multiple provers do not increase the power of a rational proof system when the communication complexity of the protocol is very small, that is, only $O(1)$ bits. Recall that with a single prover, $\mathsf{RIP}^t[O(\log n), O(\log n)] = \mathsf{RIP}^t[O(\log n), O(1)] = \mathsf{UniformTC}_0$ [3,23].

Theorem 1. $\mathsf{MRIP}^t[O(1), O(1)] = \mathsf{UniformTC}_0$.

Logarithmic and Polynomial Communication. We characterize the power of MRIP protocols with $O(\log n)$-time verification, when the communication complexity of the protocol is logarithmic and polynomial in n.

Theorem 2. $\mathsf{MRIP}^t[\mathrm{poly}(n), \mathrm{poly}(n)] = \mathsf{MRIP}^t[O(\log(n)), O(1)] = \mathsf{P}_{||}{}^{\mathsf{NP}}$.

Azar and Micali [3] characterized the class $\mathsf{P}_{||}^{\mathsf{NP}}$ in terms of single-prover rational proofs with $O(\log n)$ verification and $O(\mathrm{poly}(n))$ communication. In particular, they proved that $\mathsf{RIP}[O(\mathrm{poly}(n)), O(1)] = \mathsf{P}_{||}^{\mathsf{NP}}$.

To prove Theorem 2, we first show that using two provers reduces the communication complexity of the RIP protocol for $\mathsf{P}_{||}^{\mathsf{NP}}$ exponentially. In fact, we show prove a more general statement—any MRIP protocol (thus any RIP protocol as well) with a logarithmic time verifier and polynomial communication can be simulated using two provers, five rounds and logarithmic communication.

Lemma 1. *A MRIP protocol with $p(n)$ provers, $k(n)$ rounds, verification complexity $T(n)$, and communication complexity of $C(n)$ can be simulated by an MRIP protocol with 2 provers, 5 rounds, verification complexity $O(T(n) + \log C(n))$ and communication complexity $O(T(n) + \log C(n))$.*

The main idea behind the proof of Lemma 1 is to use the first prover to obtain the entire "effective transcript" of the original protocol. An effective transcript is all the bits that, for a given randomness r, a log-time time verifier ever accesses in the original protocol. The size of the effective transcript is at most $T(n)$. Then, the second prover is used to verify the correctness of this transcript.

Lemma 1 demonstrates the importance of two provers over one to save on communication cost in rational proofs.

Corollary 1. $\mathsf{RIP}^t[O(\text{poly}(n)), O(1)] = \mathsf{P}_{||}^{\mathsf{NP}} \subseteq \mathsf{MRIP}^t[O(\text{poly}(n)), O(\text{poly}(n)] \subseteq \mathsf{MRIP}^t[O(\log n), O(1)]$.

To complete the proof Theorem 2, we show the following upper bound.

Lemma 2. $\mathsf{MRIP}^t[O(\log(n)), O(1)] \subseteq \mathsf{P}_{||}^{\mathsf{NP}}$.

4 Verification in Logarithmic Space

The protocols in Section 3 have a polynomial utility gap. For a constant budget this means that the provers who mislead the verifier to an incorrect answer lose at least $1/\text{poly}(n)$ of their expected payment.

As utility gap is analogous to the soundness gap in classical proofs, which is constant (independent of n), it is desirable to have rational protocols with constant utility gap as well.

Constant utility gap is difficult to achieve when the verifier is $O(\log n)$ time and cannot even read the entire input. This is true even for classical proofs with a $O(\log n)$-time verifier where the soundness conditioned is weakened to design PCPPs [5,6,33]. In particular, the soundness guarantees of such proofs depend on how far (usually in terms of hamming distance) the input string x is from the language L. We note that all existing $O(\log n)$-time rational proofs [3,23,24] have polynomial utility gap (under a constant budget).

To design protocols with a strong utility gap such as logarithmic or constant, in this section we consider verifier's that use only $O(\log n)$ space and randomness.

Let $\gamma(n)$ be a polynomial-time computable and polynomially bounded function, e.g., $O(1)$, $\log n$, or $\sqrt{n}$. We prove the characterization for utility gap $\gamma(n)$.

Theorem 3. *Let* $\mathsf{P}_{||}^{\mathsf{NP}[\gamma(n)]}$ *be a polynomial-time Turing machine that can make* $O(\gamma(n))$ *non-adaptive queries to an* NP *oracle. This class is equivalent to the class of languages that have a one-round RIP protocol with a logspace verifier, polynomial communication and* $\gamma(n)$*-utility gap. That is,*

$$\mathsf{RIP}^{r,s}[\text{poly}(n), 1, \gamma(n)] = \mathsf{P}_{||}^{\mathsf{NP}[\gamma(n)]}.$$

First, we give a space-efficient RIP for the class NP using the log-space interactive proof for the language given by Condon and Ladner [16] as a blackbox.

Lemma 3. $\mathsf{NP} \in \mathsf{RIP}^{r,s}[\text{poly}(n), 1, \gamma(n)]$.

For the lower bound, we use a different but equivalent complexity class. Let $\mathbf{L}_{||}^{\mathsf{NP}[\gamma(n)]}$ be a logarithmic space machine that can make $O(\gamma(n))$ non-adaptive queries to an NP oracle. Wagner [34] showed that $\mathbf{L}_{||}^{\mathsf{NP}[\gamma(n)]} = \mathsf{P}_{||}^{\mathsf{NP}[\gamma(n)]}$.

Lemma 4. $\mathsf{P}_{||}^{\mathsf{NP}[\gamma(n)]} = \mathbf{L}_{||}^{\mathsf{NP}[\gamma(n)]} \subseteq \mathsf{RIP}^{r,s}[\text{poly}(n), 1, \gamma(n)]$.

The main idea of the proof is that the prover sends all messages (the overall answer bit, the answer bit of all NP queries and their proofs) in one round. The verifier checks all oracle queries simultaneously using the blackbox protocol [16] and scales the payment appropriately; see full version [15] for the proof.

To complete the proof of Theorem 3 we prove the following upper bound.

Lemma 5. $\mathsf{RIP}^{r,s}[\text{poly}(n), 1, \gamma(n)] \subseteq \mathsf{P}_{||}^{\mathsf{NP}[\gamma(n)]}$.

5 Relationship Between Classical and Rational Proofs

In this section, we show under what conditions does a rational interactive proof reduces to a classical interactive proof. The results in this section are stated in terms of the multi-prover model (that is, MRIP and MIP) which is more general, and thus they also hold for the single prover model (that is, RIP and IP).

To compare the two proof models, we explore their differences. In rational interactive proofs, the provers are allowed to claim $c = 1$ (that is, $x \in L$) or $c = 0$ (that is, $x \notin L$) based on their incentives.[7] Furthermore, for a particular input x of size n, if the provers' claim c about x is incorrect, they lose at least a $1/\gamma(n)$, where $\gamma(n)$ is the utility gap.

On the other hand, in classical proofs, the provers are only allowed to prove $x \in L$. Furthermore, given completeness and soundness parameters c and s respectively, where $0 \leq s < c \leq 1$, for any $x \in L$, there exists a strategy such that V accepts with probability $\geq c$ and for any $x \notin L$, for any strategy V rejects with probability $\leq s$. Thus, given L, the guarantees are independent of x.

In this section, we show when a rational proof reduces to a classical proof. Intuitively, this happens when the utility gap guarantee of a rational protocol is made to hold for all x and in particular, it is enforced to be the gap between the expected payments for all $x \in L$ and all $x \notin L$.

We first show that without loss of generality we can restrict the payments of the provers in a rational proof protocol to be either 1 or 0, where 1 corresponds to "accept" and 0 to "reject" respectively.

Lemma 6. *Any MRIP protocol $(V, \vec{P})$ with payment $R \in [0, 1]$ and utility gap $\gamma(n)$ can be simulated by a MRIP protocol $(V', \vec{P})$ with payment $R' \in \{0, 1\}$ and utility gap $\gamma(n)/2$. In particular, for any strategy s and any input x,*

$$u_{(V,\vec{P})}(x; s) \leq u_{(V',\vec{P})}(x; s) \leq u_{(V,\vec{P})}(x; s) + \gamma(n)/2.$$

V' uses $1 + \lceil \log_2 \gamma(n) \rceil$ more random bits than V.

[7] Thus it is not surprising that rational proofs are closed under complement.

In the proof of Lemma 6, V' simulates V, but instead of giving a payment $R \in [0, 1]$, it gives a payment of 1 with probability R, and 0 otherwise. This preserves the expected reward for each transcript (and thus for each strategy).

Given any rational protocol with zero-one payments, we note that it immediately gives us an accept-reject protocol such that for a given x, the probability that the verifier accepts is exactly the expected payment of the original protocol. More formally let $(V, \vec{P})$ be a rational protocol with $R \in \{0, 1\}$ and utility gap $\gamma(n)$. Let $(V', \vec{P'})$ be defined as follows: V' simulates V, ignores the answer bit c, and if the payment in $(V, \vec{P})$ is $R = 1$ then accept, else reject.

Thus, for a given input string x, the expected payment in $(V, \vec{P})$ is equal to the probability that V' accepts in $(V', \vec{P'})$. That is,

$$\begin{aligned} u_{(V,\vec{P})}(x; s) &= \mathbf{E}_r[R(x, r, (V, \vec{P})(x, r, s))] = \sum_r \Pr(r \mid R(x, r, (V, \vec{P})(x, r, s)) = 1) \\ &= \sum_r \Pr(r \mid V'\text{accepts } (V', \vec{P'})) = \Pr(V' \text{ accepts } (V', \vec{P'})). \end{aligned} \tag{1}$$

Furthermore, $(V', \vec{P'})$ satisfies the following: for any $x \in L$, let s^* denote the optimal strategy of the provers $\vec{P}$, that is, s^* maximizes their expected payment. Then for $\vec{P'}$ following s^*, V' accepts with probability exactly $c(x, n) = u_{(V,\vec{P})}(x; s^*)$. Furthermore, we know from the utility gap condition that for any $x \notin L$, for any strategy s', the probability that V' accepts is at most $u_{(V,\vec{P})}(x; s') < u_{(V,\vec{P})}(x; s^*) - 1/\gamma(n)$, that is, the probability that V' accepts is at most $s(x, n) < c(x, n) - 1/\gamma(n)$. Similar guarantees hold for any $x \notin L$.

However, if we want $(V', \vec{P'})$ to be an interactive proof protocol in the classical sense, that is, with completeness and soundness guarantees that hold for all $x \in L$ and for all $x \notin L$ respectively, we need to impose restrictions on the expected payment function of the rational protocol.

Theorem 4. *Let $(V, \vec{P})$ be an MRIP protocol for a language L such that*

$$\min_{x \in L} u_{(V,\vec{P})}(x; s^*) > \max_{x \notin L} u_{(V,\vec{P})}(x; s^*) + \frac{1}{\gamma(n)} \tag{2}$$

where x is any input of length n, s^* is the strategy of the provers that maximizes their expected payment in $(V, \vec{P})$ and $\gamma(n)$ is any function such that $\gamma(n) > 1$ and $\gamma = O(\text{poly}(n))$. Then, $(V, \vec{P})$ can be simulated by a MIP protocol for L.

We prove this theorem in two parts. First, we show prove the following lemma which proves Theorem 4 with weak completeness and soundness guarantees.

Lemma 7. *Let $(V, \vec{P})$ be an MRIP protocol for a language L that satisfies the condition 2 in Theorem 4. Then, $(V, \vec{P})$ can be simulated by MIP protocol with completeness and soundness parameters $c(n)$ and $s(n)$ respectively such that $c(n) > s(n) + 1/2\gamma(n)$ and $c(n), s(n) \geq 0$.*

We amplify the "gap" of an MIP by repeating the protocol sufficiently many times and then using Chernoff bounds. The techniques are mostly standard, although the parameters must be set carefully to deal with the case $s(n) = 0$.

Lemma 8. *Given an MIP protocol for a language L, with completeness $c(n) > 0$ and soundness $s(n) \geq 0$ such that $c(n) > s(n) + 1/\gamma'(n)$ for some $\gamma'(n) > 1$ and $\gamma' = O(\text{poly}(n))$, can be converted to an MIP protocol for L with completeness at least $1 - 1/\text{poly}(n)$ and soundness at most $1/\text{poly}(n)$.*

Remark 1. The repetition of the MIP protocol to amplify its completeness and soundness guarantee used in Lemma 8 is not efficient as it blows up the number of rounds. There exist more efficient techniques to amplify IP guarantees by parallel repetition that can be used instead; for example, see [4,19,31].

References

1. Allender, E., Hertrampf, U.: On the power of uniform families of constant depth threshold circuits. In: Symposium on Mathematical Foundations of Computer Science, pp. 158–164 (1990)
2. Azar, P.D., Micali, S.: Rational proofs. In: Proceedings of 44th Symposium on Theory of Computing, pp. 1017–1028 (2012)
3. Azar, P.D., Micali, S.: Super-efficient rational proofs. In: Proceedings of 14th Conference on Electronic Commerce, pp. 29–30 (2013)
4. Bellare, M., Goldreich, O., Goldwasser, S.: Randomness in interactive proofs. Comput. Complex. **3**(4), 319–354 (1993)
5. Ben-Sasson, E., Goldreich, O., Harsha, P., Sudan, M., Vadhan, S.: Short PCPs verifiable in polylogarithmic time. In: Proceedings of Conference on Computational Complexity, pp. 120–134 (2005)
6. Ben-Sasson, E., Goldreich, O., Harsha, P., Sudan, M., Vadhan, S.: Robust PCPs of proximity, shorter PCPs, and applications to coding. SIAM J. Comput. **36**(4), 889–974 (2006)
7. Bitansky, N., Chiesa, A.: Succinct arguments from multi-prover interactive proofs and their efficiency benefits. In: Safavi-Naini, R., Canetti, R. (eds.) CRYPTO 2012. LNCS, vol. 7417, pp. 255–272. Springer, Heidelberg (2012). https://doi.org/10.1007/978-3-642-32009-5_16
8. Buhrman, H., Kadin, J., Thierauf, T.: On functions computable with nonadaptive queries to NP. In: Proceedings of 9th Structure in Complexity Theory Conference, pp. 43–52 (1994)
9. Campanelli, M., Gennaro, R.: Sequentially composable rational proofs. In: Proceedings of Decision and Game Theory for Security, pp. 270–288 (2015)
10. Canetti, R., Riva, B., Rothblum, G.N.: Refereed delegation of computation. Inf. Comput. **226**, 16–36 (2013)
11. Chakrabarti, A., Cormode, G., McGregor, A., Thaler, J., Venkatasubramanian, S.: Verifiable stream computation and Arthur-Merlin communication. In: Proceedings of Conference on Computational Complexity, pp. 217–243 (2015)
12. Chandra, A.K., Stockmeyer, L.J.: Alternation. In: Proceedings of 17th Symposium on Foundations of Computer Science, pp. 98–108 (1976)

13. Chen, J., McCauley, S., Singh, S.: Rational proofs with multiple provers. In: Proceedings of 7th Innovations in Theoretical Computer Science Conference, pp. 237–248 (2016)
14. Chen, J., McCauley, S., Singh, S.: Rational proofs with non-cooperative provers. arXiv preprint arXiv:1708.00521 (2017)
15. Chen, J., McCauley, S., Singh, S.: Efficient Rational Proofs with Strong Utility-Gap Guarantees http://arxiv.org/abs/1807.01389 (2018)
16. Condon, A., Ladner, R.: Interactive proof systems with polynomially bounded strategies. J. Comput. Syst. Sci. **50**(3), 506–518 (1995)
17. Cormode, G., Thaler, J., Yi, K.: Verifying computations with streaming interactive proofs. Proc. VLDB Endow. **5**(1), 25–36 (2011)
18. Daruki, S., Thaler, J., Venkatasubramanian, S.: Streaming verification in data analysis. In: Elbassioni, K., Makino, K. (eds.) ISAAC 2015. LNCS, vol. 9472, pp. 715–726. Springer, Heidelberg (2015). https://doi.org/10.1007/978-3-662-48971-0_60
19. Feige, U., Kilian, J.: Two prover protocols: low error at affordable rates. In: Proceedings of 26th Symposium on Theory of Computing, pp. 172–183 (1994)
20. Feige, U., Kilian, J.: Making games short. In: Proceedings of 29th Symposium On Theory of Computing, pp. 506–516 (1997)
21. Feigenbaum, J., Koller, D., Shor, P.: A game-theoretic classification of interactive complexity classes. In: Proceedings of 10th Structure in Complexity Theory Conference, pp. 227–237 (1995)
22. Goldwasser, S., Kalai, Y.T., Rothblum, G.N.: Delegating computation: interactive proofs for muggles. In: Proceedings of 40th Symposium on Theory of Computing, pp. 113–122 (2008)
23. Guo, S., Hubáček, P., Rosen, A., Vald, M.: Rational arguments: single round delegation with sublinear verification. In: Proceedings of 5th Innovations in Theoretical Computer Science, pp. 523–540 (2014)
24. Guo, S., Hubáček, P., Rosen, A., Vald, M.: Rational sumchecks. In: Kushilevitz, E., Malkin, T. (eds.) TCC 2016. LNCS, vol. 9563, pp. 319–351. Springer, Heidelberg (2016). https://doi.org/10.1007/978-3-662-49099-0_12
25. Hesse, W., Allender, E., Barrington, D.A.M.: Uniform constant-depth threshold circuits for division and iterated multiplication. J. Comput. Syst. Sci. **65**(4), 695–716 (2002)
26. Hubáček, P.: Rationality in the Cryptographic Model. Ph.D thesis, Department Office Computer Science, Aarhus University (2014)
27. Inasawa, K., Yasunaga, K.: Rational proofs against rational verifiers. Fundam. Electron. Commun. Comput. Sci. **100**(11), 2392–2397 (2017)
28. Kalai, Y.T., Rothblum, R.D.: Arguments of proximity. In: Gennaro, R., Robshaw, M. (eds.) CRYPTO 2015. LNCS, vol. 9216, pp. 422–442. Springer, Heidelberg (2015). https://doi.org/10.1007/978-3-662-48000-7_21
29. Koller, D., Megiddo, N.: The complexity of two-person zero-sum games in extensive form. Games Econ. Behav. **4**(4), 528–552 (1992)
30. Krentel, M.W.: The complexity of optimization problems. J. Comput. Syst. Sci. **36**(3), 490–509 (1988)
31. Raz, R.: A parallel repetition theorem. SIAM J. Comput. **27**(3), 763–803 (1998)
32. Reif, J.H.: The complexity of two-player games of incomplete information. J. Comput. Syst. Sci. **29**(2), 274–301 (1984)
33. Rothblum, G.N., Vadhan, S., Wigderson, A.: Interactive proofs of proximity: delegating computation in sublinear time. In: Proceedings of 45th Symposium on Theory of Computing, pp. 793–802 (2013)

34. Wagner, K.W.: Bounded query classes. SIAM J. Comput. **19**(5), 833–846 (1990)
35. Zhang, Y., Blanton, M.: Efficient secure and verifiable outsourcing of matrix multiplications. In: Chow, S.S.M., Camenisch, J., Hui, L.C.K., Yiu, S.M. (eds.) ISC 2014. LNCS, vol. 8783, pp. 158–178. Springer, Cham (2014). https://doi.org/10.1007/978-3-319-13257-0_10

Removal and Threshold Pricing: Truthful Two-Sided Markets with Multi-dimensional Participants

Moran Feldman[1(✉)] and Rica Gonen[2(✉)]

[1] Department of Mathematics and Computer Science, The Open University of Israel, The Dorothy de Rothschild Campus, 1 University Road, 43107 Raanana, Israel
moranfe@openu.ac.il

[2] Department of Management and Economics, The Open University of Israel, The Dorothy de Rothschild Campus, 1 University Road, 43107 Raanana, Israel
ricagonen@gmail.com

Abstract. We consider mechanisms for markets that are two-sided and have agents with multi-dimensional strategic spaces on at least one side. The agents of the market are strategic and act to optimize their own utilities, while the mechanism designer aims to optimize a social goal, *i.e.*, the gain from trade. We focus on one example of this setting motivated by a foreseeable privacy-aware future form of online advertising.

Recently, it has been suggested that markets of user information built around information brokers could be introduced to the online advertising ecosystem to overcome online privacy concerns. Such markets give users control over which data gets shared in online advertising exchanges. We describe a model for the above form of online advertising and design two mechanisms for this model. The first is a deterministic mechanism which is related to the vast literature on mechanism design through trade reduction and allows agents with a multi-dimensional strategic space. The second is a randomized mechanism that can handle a more general version of the model. We provide theoretical analyses of our mechanisms and study their performance using simulations based on real-world data.

Keywords: Mechanism design · Double-sided market
Multi-dimensional players · Online advertising market

1 Introduction

Billions of transactions are carried out via exchanges at every given day, and the numbers continue to grow. The design of one-sided incentive compatible mechanisms for exchanges is relatively well understood. However, incentive compatible multi-sided mechanisms present a much more significant challenge as they introduce more sophisticated requirements such as budget balance.

This work was supported by the Horizon 2020 funded project TYPES (Project number: 653449. Call Identifier H2020-DS-2014-1).

X. Deng (Ed.): SAGT 2018, LNCS 11059, pp. 163–175, 2018.
https://doi.org/10.1007/978-3-319-99660-8_15

We are interested in designing exchanges (mechanisms) for multi-sided markets with strategic agents. The agents of the market act to optimize their own utilities, while the mechanism designer aims to optimize a social goal, *i.e.*, the gain from trade (the difference between the total value of the sold goods for the buyers and the total costs of these goods for the sellers). The design of such mechanisms raises a few interesting questions. For instance, whether the mechanism can simultaneously maintain different desirable economic properties such as: individual rationality (IR)—participants do not lose by participation, incentive compatibility (IC)—agents are incentivized to report their true information to the mechanism and weak budget balance (WBB)—the mechanism does not end up with a loss or strong budget balance (SBB)—the mechanism should not gain or lose any money. Moreover, can the mechanism maintain such properties while only suffering a bounded loss compared to the optimal gain from trade? Finally, can this be done when **all** the agents have a **multi-dimensional strategic space**?[1]

The above questions can be studied in the context of many multi-sided markets. We focus on one such market motivated by recent privacy concerns in the online advertising world. Online advertising currently supports some of the most important Internet services, including: search, social media and user generated content sites. However, the amount of information that companies collect about users increasingly creates privacy concerns in society. Such concerns were actively raised by EU regulators in recent years in efforts to find solutions to guarantee users' privacy. Recently privacy concerns have also reached the U.S. Senate and Congress as a response to Facebook's information leak to Cambridge Analytica. It was evident in Facebook's hearing before the U.S. Senate, particularly in Senator Schatz's line of questioning [13], that Facebook is expected to develop tools to enable end users to configure their privacy settings and that the notion of a data fiduciary was put forward to apply pressure to Facebook in this area.

The market we study is motivated by a solution we suggest for the above privacy issue. In this solution mediators serve as the interface between end-users and the other agents in the online advertising market. Each user informs her mediator of the attributes she is willing to reveal, and her cost, *i.e.*, the compensation she requires for every ad she views. The mediator then tries to "sell" access to the user on the advertising market based *only* on the attributes she agreed to reveal. If successful, the mediator pays her the appropriate cost out of the amount he got from the sale.

As revealing more personal attributes is likely to result in a more profitable sale, our solution incentivizes users to share their information with the advertising market while allowing them to retain control of the amount of information they share. Notice that since our solution motivates users to participate in the advertising market and to provide more precise information for targeting campaigns, the efficiency of the advertising system and the digital economy as a whole improves (in addition to answering the privacy concerns discussed above).

[1] We often refer to agents with a multi-dimensional strategic space as multi-dimensional agents.

This benefit is in sharp contrast to other natural approaches for dealing with privacy issues, such as cryptography based approaches, which reduce the amount of information available to the advertisers but give them nothing in return.

The advertising market induced by the above solution has mediators on one side and advertisers on the other side. Each mediator has a set of users associated with him, and he strives to assign these users to advertisers using the market. Each one of the users has a non-negative cost which she must be paid if she is assigned to an advertiser. The mediators themselves have no cost of their own; however, each of them has to pay his users their cost if they are assigned to advertisers. Thus, a mediator's utility is the amount paid to him minus the total cost of his users who are assigned. Finally, each advertiser has a positive capacity that determines the number of users she is interested in targeting. The advertiser gains a non-negative value for each of the users assigned to her, as long as her capacity is not exhausted. Thus, the advertiser's utility is her value multiplied by the number of users assigned to her (as long as this number does not exceed her capacity) minus her total payment.

A mechanism for the above market knows the mediators and the advertisers, but has no knowledge about their parameters or about the users. The mechanism's objective is to assign users to advertisers in a gain from trade maximizing way. In addition, the mechanism also decides how much to charge (pay) each advertiser (mediator). In order to achieve these goals, the mechanism receives reports from the advertisers and mediators. Each advertiser reports her capacity and value, and each mediator reports the number of his users and their costs. The mediators and advertisers are strategic, and thus, free to send incorrect reports. For example, a mediator may report any subset of his users and associate an arbitrary cost with each user. We say that an advertiser is *truthful* if she correctly reports her capacity and value. A mediator is considered *truthful* if he reports his true number of users and the true costs of these users.

In this paper we present a simplified variant of our model, where the costs of the users are known to their corresponding mediators, *i.e.*, the users are non-strategic. This captures the practical situation where users are individual people (unlike the advertisers and mediators), and thus, their mode of interaction with their mediator is likely to be too simple to allow them to create an automated agent that dynamically updates their strategy, which is necessary if one wants to use strategy in the super-fast online advertising ecosystem.

To better understand the design challenge raised by this market, we observe that even if our setting is reduced to a single buyer-single seller exchange, it is well known from [20] that maximizing gain from trade while maintaining IR and IC perforce runs a deficit (is not WBB). A well known circumvention of [20]'s impossibility is [19]'s trade reduction for a simple setting of double sided auctions. In [19]'s setting, trade is conducted between multiple strategic sellers offering identical goods to multiple strategic buyers, where each seller is selling a single good and each buyer is interested in buying a single good. [19]'s result relaxes the requirement for optimal trade by means of a *trade reduction*. The trade reduction leads to an IR, IC and budget balance mechanism. Following [19]'s work, several other mechanisms were designed using the trade reduction technique, but all these known mechanisms only allow agents with single-dimensional strategic spaces (even in settings where agents can hold multiple items).[2]

1.1 Our Contribution

Given that existing trade reduction solutions do not apply in our setting, we describe new double-sided mechanisms able to handle mediators and advertisers with multi-dimensional strategic spaces. Our mechanisms guarantee desirable economic properties and at the same time yield a gain from trade approximating the optimal gain from trade. If truthfulness is a dominant strategy[3] for each advertiser and each mediator, regardless of other agents' strategies, then the mechanism is *incentive compatible* (IC). If no advertiser and no mediator can have a negative utility by participating truthfully in the mechanism then it is *individually rational* (IR). The mechanisms we construct are IC, IR and WBB.

We first study a special case of our setting where the advertisers' capacities are publicly known (but need not be all equal). The set of users of each mediator, in contrast, remains unknown to the mechanism (*i.e.*, the mechanism only learns about it through the mediator's report). For this case we present a deterministic mechanism named "Price by Removal Mechanism" (PRM) that works as follows: for every mediator find a threshold cost and remove users of the mediator whose cost exceeds this threshold. Add a dummy advertiser with value that is the maximum threshold cost computed for the mediators and a capacity equal to the total number of users remaining. Assign the non-removed users to the advertisers using a VCG auction [9,18,23] in which the users are the goods and the bidders are the advertisers. Price the mediators according to their threshold cost and price the advertisers according to the prices of the VCG auction.

The method used to calculate the threshold costs of the above mechanism induces its properties. We prove that for appropriately chosen threshold costs the above mechanism is IC, IR, WBB and provides a non-trivial approximation for the optimal gain from trade. More formally, if τ is the size of the optimal trade and γ is an upper bound known to the mechanism on the maximum capacity of any agent (mediator or advertiser), then

Theorem 1. *PRM is WBB, IR, IC and* $\left(1 - \frac{5\gamma}{\tau}\right)$*-competitive.*

An online advertising system constructed based on PRM and beta tested with real users and real advertising campaigns allowed us to collect real-world data to study PRM's practical performance empirically. Interestingly, our simulation shows that although the practical performance of PRM is significantly better than its theoretical one, both performances exhibit a similar dependence on γ/τ.

[2] The sole exception for this is the work of Segal-Halevi et al. [21], which was done independently in parallel to our work. However, [21] does not offer a solution for the deterministic multi-dimensional strategic spaces case.

[3] Here and throughout the paper, a reference to domination of strategies should be understood as a reference to weak domination. We never refer to strong domination.

PRM generalizes the trade reduction ideas used so far in the literature for single-dimensional strategic agents, but is much more involved. Intuitively, PRM differs from previous ideas by the following observation. A trading set is the smallest set of agents that is required for trade to occur. In the existing literature for single-dimensional strategic agents, a trade reduction mechanism makes a binary decision regarding every trading set of the optimal trade, *i.e.*, either the trading set is removed as a whole, or it is kept. On the other hand, dealing with multi-dimensional agents requires PRM to remove only parts of some trading sets.

Our deterministic mechanism PRM handles one type of multi-dimensional agent (the mediators) and one type of single dimensional strategic space agent (the advertisers). To further enrich our strategic space and allow advertisers to also have multi-dimensional strategic spaces, we present next a randomized mechanism termed "Threshold by Partition Mechanism" (TPM). TPM applies to our general setting, *i.e.*, we no longer assume that any capacity is known to the mechanism. It works as follows: divide the set of mediators uniformly at random into two sets (M_1 and M_2) and divide the set of advertisers uniformly at random, as well, into two sets (A_1 and A_2). Then use the optimal trade for M_2 and A_2 to produce threshold cost and threshold value that allow WBB pricing and the needed reduction in trade for M_1 and A_1. Analogously, use the optimal trade for M_1 and A_1 to produce appropriate threshold cost and value for M_2 and A_2.

The above description of TPM is not complete since the use of threshold cost and value from one pair (M_i, A_i) to reduce the trade in the other pair might create an unbalanced reduction. To overcome this issue we create two random low priority sets: one of advertisers and the other of mediators. Then, whenever the reduction in trade is unbalanced, we restore balance by removing additional low priority mediators or advertisers, which can be done with high probability. The following theorem shows that the above mechanism is IC, IR, WBB and provides a non-trivial approximation for the optimal gain from trade. The parameter α is an upper bound, known to the mechanism, on the ratio between the maximum capacity of any agent (mediator or advertiser) and the size of the optimal trade.[4]

Theorem 2. *TPM is WBB, IR, IC and* $(1 - 28\sqrt[3]{\alpha} - 20e^{-2/\sqrt[3]{\alpha}})$*-competitive.*

We note that TPM is universally truthful, *i.e.*, its IC property holds for every given choice of the random coins of the mechanism. Observe also that the competitive ratio of TPM approaches 1 when α approaches 0, *i.e.*, when the market is large enough to make the market power of all agents very low. Unfortunately, when the market is not large enough to make α very small, the theoretical competitive ratio guaranteed by Theorem 2 is not so good, and might even be meaningless. Nevertheless, our simulations suggest that in practice the performance of TPM is quite good even for moderate size markets.

Both our mechanisms have a common drawback, namely their need to have access to a good bound on the maximum market power of any agent (which is

[4] The parameters γ and α both bound the maximum capacity of the agents, and they are related by $\alpha = \gamma/\tau$. We chose to formulate Theorems 1 and 2 in terms of the parameter that the mechanism corresponding to each theorem assumes access to.

captured by the parameters γ and α). From a practical point of view we believe this is a minor issue, as the mechanism can usually use the large quantity of historical data available to it to estimate the necessary bound very well.

2 Related Work

From a motivational point of view, the most closely related literature to our work consists of works that involve mediators and online advertising markets, such as [1,14,22]. These works differ from ours in two crucial points. First, despite being motivated by exchanges, the models studied by these works are actually auctions (*i.e.*, one-sided mechanisms). Second, our focus is on maximization of the gain from trade, unlike the above works which focus on revenue maximization.

We now move our attention to the above mentioned literature on trade reduction. Mechanisms using trade reduction were described for various settings with single dimensional agents [2,3,7,8,16,19]. Later [16] developed a single trade reduction procedure applicable to a class of problems generalizing all the above settings. Essentially this procedure works when the agents can be partitioned into few equivalence classes. In our model each mediator might require its own equivalence class (because a mediator with many users can always replace a mediator with few users within a trading set, but the reverse is often not true). Thus, [16]'s procedure does not yield a non-trivial guarantee for our setting.

Recent related research on maximizing gain from trade in two-sided markets was published by [5,6,10,21]. [21] is the work most relevant to ours. As noted in Sect. 1, our work was developed independently in parallel to [21]. Though both works are based on some similar ideas, their results are incomparable. First, [21] only presents a randomize mechanism, while we present also a deterministic one; and second, in settings in which both our randomized mechanism and [21]'s mechanism apply, each mechanism achieves a superior competitive ratio for a different range of the parameters.

Last but not least, we note that any result for our objective function applies also to social welfare maximization. Hence, our work is also related to works on the maximization of this objective in multi-sided markets [4,11,12,17].

3 Notation and Basic Observations

We begin this section with a more formal presentation of our model. Our model consists of a set P of users, a set M of mediators and a set A of advertisers. Each user $p \in P$ has a non-negative cost $c(p)$ which she has to be paid if she is assigned to an advertiser. The users are partitioned among the mediators, and we denote by $P(m) \subseteq P$ the set of users associated with mediator $m \in M$ (*i.e.*, the sets $\{P(m) \mid m \in M\}$ form a disjoint partition of P). The utility of a mediator $m \in M$ is the amount he is paid minus the total cost he has to forward to his assigned users; hence, if $x(p) \in \{0, 1\}$ is an indicator for the event that user $p \in P(m)$ is assigned and t is the payment received by m, then the utility of m is $t - \sum_{p \in P(m)} x(p) \cdot c(p)$. Finally, each advertiser $a \in A$ has a positive capacity $u(a)$, and she gains a non-negative value $v(a)$ from every one of the first $u(a)$ users assigned to her; thus, if advertiser a is assigned $n \leq u(a)$ users and has to pay t then her utility is $n \cdot v(a) - t$.

A mechanism for our model accepts reports from the advertisers and mediators, and based on these reports outputs an assignment of users to advertisers (recall that the report of an advertiser a consists of her capacity $u(a)$ and value $v(a)$, and the report of a mediator m consists of the number of his users $|P(m)|$ and the costs of these users). In addition, the mechanism also decides how much to charge (pay) each advertiser (mediator). The objective of the mechanism is to output an assignment of users to advertisers maximizing the gain from trade.

For ease of the presentation, it is useful to associate a set $B(a)$ of $u(a)$ slots with each advertiser $a \in A$. We then think of the users as assigned to slots instead of directly to advertisers. Formally, let B be the set of all slots (*i.e.*, $B = \bigcup_{a \in A} B(a)$), then an assignment is a set $S \subseteq P \times B$ in which no user or slot appears in more than one ordered pair. We say that an assignment S assigns a user p to slot b if $(p, b) \in S$. Similarly, we say that an assignment S assigns user p to advertiser a if there exists a slot $b \in B(a)$ such that $(p, b) \in S$. It is also useful to define values for the slots. For every slot b of advertiser a, we define the value $v(b)$ of b as equal to the value $v(a)$ of a. Using this notation, the gain from trade of an assignment S can be stated as $\texttt{GfT}(S) = \sum_{(p,b) \in S}[v(b) - c(p)]$.

In addition to the above notation, we need also the following shorthands. Given a set $A' \subseteq A$ of advertisers, we denote by $B(A') = \bigcup_{a \in A'} B(a)$ the set of slots belonging to advertisers of A'. Similarly, given a set $M' \subseteq M$ of mediators, $P(M') = \bigcup_{m \in M'} P(m)$ is the set of users associated with mediators of M'.

To make the presentation of our mechanisms simpler, we assume that the values of slots and the costs of users are all unique. Clearly, this is unrealistic, but it can be emulated using an appropriate tie-breaking rule.

3.1 Canonical Assignment

Given a set $B' \subseteq B$ of slots and a set $P' \subseteq P$ of users, the canonical assignment $S_c(P', B')$ is the assignment constructed as follows. First, we order the slots of B' in a decreasing value order $b_1, b_2, \ldots, b_{|B'|}$ and the users of P' in an increasing cost order $p_1, p_2, \ldots, p_{|P'|}$. Then, for every $1 \leq i \leq \min\{|B'|, |P'|\}$, $S_c(B', P')$ assigns user p_i to slot b_i if and only if $v(b_i) > c(p_i)$. The canonical assignment is an important tool used frequently by the mechanisms we describe in the next section. In some places we refer to the user or slot at location i of a canonical solution $S_c(P', B')$, by which we mean user p_i or slot b_i, respectively.

The following lemma, whose proof is deferred to the full version of this paper [15], shows that $|S_c(P, B)|$ is the size of the optimal trade, and thus, equal to the notion τ used in Sect. 1.1.

Lemma 1. *Among all the possible assignments of users of P' to slots of B', the canonical assignment $S_c(P', B')$ maximizes* $\texttt{GfT}(S_c(P', B'))$.

4 Mechanisms

In this section we formally present our new mechanisms. Unfortunately, due to space constraints, we defer the analysis of these mechanisms to the full version of this paper [15]. Let us begin by presenting our deterministic mechanism "Price by Removal Mechanism" (PRM). Recall that PRM assumes public knowledge of the advertisers' capacities (or that the advertisers are not strategic about them). We also assume that PRM has access to a value $\gamma \geq 1$ such that $u(a) \leq \gamma$ and $|P(m)| \leq \gamma$ for every advertiser $a \in A$ and mediator $m \in M$, respectively. In other words, γ is an upper bound on how large can the capacity of an advertiser or the number of users of a mediator be. Informally, γ can be understood as a bound on the importance every single advertiser or mediator can have.

A description of PRM is given as Mechanism 1.1. Notice that both Mechanism 1.1 and the other mechanism that we present in this paper often refer to parameters of the model that are not known to the mechanism (such as values of advertiser or the number of users of mediators). Whenever this happens, this should be understood as referring to the reported values of these parameters.

Mechanism 1.1. Price by Removal Mechanism (PRM)

---

1. For every mediator $m \in M$, if the canonical assignment $S_c(P \setminus P(m), B)$ is of size more than 4γ, denote by p_m the user at location $|S_c(P \setminus P(m), B)| - 4\gamma$ of the canonical assignment $S_c(P \setminus P(m), B)$, and let c_m be the cost of p_m. Otherwise, set c_m to $-\infty$.
2. For every mediator $m \in M$, let $\hat{P}(m)$ be the set of users of mediator m whose cost is less than c_m.
3. Assign the users of $\bigcup_{m \in M} \hat{P}(m)$ to the advertisers using a VCG auction. More specifically, the users of $\bigcup_{m \in M} \hat{P}(m)$ are the items sold in the auction, and the bidders are the advertisers of A plus a dummy advertiser a_d whose value and capacity are $v(a_d) = \max_{m \in M} c_m$ and $u(a_d) = \sum_{m \in M} |\hat{P}(m)|$, respectively. It is important that in case of a tie between $v(a_d)$ and the value of a non-dummy advertiser, the VCG auction breaks the tie in favor of the non-dummy advertiser.
4. Charge every non-dummy advertiser by the same amount she is charged (as a bidder) by the VCG auction.
5. For every user p assigned by the auction, if m is p's mediator, pay c_m to m.[a]

---

[a]Note that m is WBB as he forwards to each of his assigned users her cost—which is less than c_m.

Let us now move to the formal presentation of our randomized mechanism "Threshold by Partition Mechanism" (TPM). Unlike the deterministic mechanism PRM, TPM need not assume public knowledge about the advertisers' capacities, *i.e.*, the advertisers now have multi-dimensional strategy spaces. On the other hand, TPM assumes access to a value $\alpha \in [|S_c(P, B)|^{-1}, 1]$ such that we are guaranteed that $u(a) \leq \alpha \cdot |S_c(P, B)|$ and $|P(m)| \leq \alpha \cdot |S_c(P, B)|$ for every advertiser a and mediator m, respectively. In other words, α is an upper bound on how large can the capacity of an advertiser or the number of users of a mediator be compared to the size of the optimal assignment $S_c(P, B)$. We remind the reader that α is related to the value γ by the equation $\alpha = \gamma/|S_c(P, B)|$, and thus, α, like γ, can be informally understood as a bound on the importance of every single agent. It is important to note also that α is well-defined only when $|S_c(P, B)| > 0$, and thus, we assume this inequality is true throughout the rest of the section.

Mechanism 1.2. Threshold by Partition Mechanism (`TPM`)

---

1. Let M_L be a set of mediators containing each mediator $m \in M$ with probability $\min\{17\sqrt[3]{\alpha}, 1\}$, independently. Similarly, A_L is a set of advertisers containing each advertiser $a \in A$ with probability $\min\{17\sqrt[3]{\alpha}, 1\}$, independently. Intuitively, the subscript L in M_L and A_L stands for "low priority".

2. Let σ_A be an arbitrary order over the advertisers that places the advertisers of A_L after all the other advertisers and is independent of the reports received by the mechanism. Similarly, σ_M is an arbitrary order over the mediators that places the mediators of M_L after all the other mediators and is independent of the reports received by the mechanism.

3. Partition the mediators of M into two sets M_1 and M_2 by adding each mediator $m \in M$ with probability 1/2, independently, to M_1 and otherwise to M_2. Similarly, partition the advertisers of A into two sets A_1 and A_2 by adding each advertiser $a \in A$ with probability 1/2, independently, to A_1 and otherwise to A_2. The rest of the algorithm explains how to assign users of mediators from M_1 to slots of advertisers from A_1, and how to charge advertisers of A_1 and pay mediators of M_1. Analogous steps, which we omit, should be added for handling the advertisers of A_2 and the mediators of M_2.

4. Let $\hat{p}$ and $\hat{b}$ be the user and slot at location $\lceil(1-4\sqrt[3]{\alpha})\cdot|S_c(P(M_2), B(A_2))|\rceil$ of the canonical solution $S_c(P(M_2), B(A_2))$. If $(1-4\sqrt[3]{\alpha})\cdot|S_c(P(M_2), B(A_2))| \leq 0$, then the previous definition of $\hat{p}$ and $\hat{b}$ cannot be used. Instead define $\hat{p}$ as a dummy user of cost $-\infty$ and $\hat{b}$ as a dummy slot of value ∞. Using $\hat{p}$ and $\hat{b}$ define now two sets $\hat{P} = \{p \in P(M_1) \mid c(p) < c(\hat{p})\}$ and $\hat{B} = \{b \in B(A_1) \mid v(b) > v(\hat{b})\}$. It is important to note that $\hat{P}$ and $\hat{B}$ are empty whenever $\hat{p}$ and $\hat{b}$ are dummy user and slot, respectively.

5. While there are unassigned users in $\hat{P}$ and unassigned slots in $\hat{B}$ do:
- Let m be the earliest mediator in σ_M having unassigned users in $\hat{P}$.
- Let a be the earliest advertiser in σ_A having unassigned slots in $\hat{B}$.
- Assign the unassigned user of $\hat{P} \cap P(m)$ with the lowest cost to an arbitrary unassigned slot of $\hat{B} \cap B(a)$, charge a payment of $v(\hat{b})$ to advertiser a and transfer a payment of $c(\hat{p})$ to mediator m.[a]

---

[b]Note that m is paid $c(\hat{p})$ for the assignment of each one of his users. Hence, m is always WBB since the membership of p in $\hat{P}$ implies $c(p) < c(\hat{p})$.

Intuitively, TPM's analysis exploits concentration results showing that the canonical assignments $S_c(P(M_1), B(A_1))$ and $S_c(P(M_2), B(A_2))$ are quite similar. This similarity allows us to use information from $S_c(P(M_2), B(A_2))$ to set the payments charged to advertisers of $B(A_1)$ and payed to mediators of $P(M_1)$, and vice versa, while keeping a reasonable competitive ratio. The advantage of setting the payments this way is that it reduces the control agents have on the payments they have to pay or are paid, which helps the mechanism be IC.

5 Experiments

We have used simulations with real-world data to study the empirical performance of our mechanisms. The data was collected by an online advertising system constructed based on PRM and beta tested with real users and real advertising campaigns as part of the above mentioned Horizon 2020 project.

We begin this section by describing the simulations we used to study the performance of our deterministic mechanism (PRM). These simulations included 30 advertisers. Each one of these advertisers was associated with a different real world campaign, and we used the cost-per-click data of the campaign for choosing the value of its corresponding advertiser. The capacity of the advertisers was chosen by a different method. Specifically, in each execution of the simulation we picked a random upper bound between 1 and 65, and then picked for every advertiser a uniformly random capacity between 1 and this upper bound. This capacities generation method was chosen because it results in simulations with diverse γ values. Our simulations also included 328 users based on the data fed by 328 real users of the above mentioned online advertising system, which were assigned at random to 30 mediators.

The results of 500 executions of our simulations are summarized in Fig. 1. Interestingly, the empirical competitive ratio of PRM turned out to be larger than its theoretical guarantee by roughly 0.2 for every given γ/τ ratio.

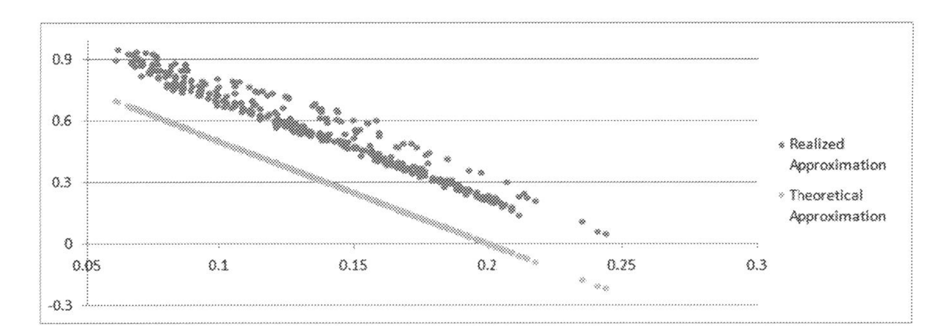

Fig. 1. Simulation results for PRM. The orange dots represent the theoretical performance guarantee for the inputs on which the simulations were run: the y-axis value of the dot is the competitive ratio and the x-axis value is the γ/τ ratio. Similarly, the blue dots represent the performance of the mechanism in reality.

The theoretical guarantee of TPM (Theorem 2) is meaningful only for very small values of α, which usually arise only in very large markets. Nevertheless, we conjectured that TPM should work well in practice for moderate size markets, despite the failure of the theoretical analysis to show that. To test this conjecture, we generated inputs of moderate size for TPM. Each one of these inputs consisted of 20000 users, each based on a random one of the above mentioned 328 real users. The generated users were than grouped into equal size groups (the size was varied between simulations in order to produce different α values), and each group of users was assigned to a different mediator. Similarly, each one of the simulation advertisers was based on a random one out of the above mentioned 30 campaigns. The number of advertisers was also varied between simulations, but the total capacity of all the advertisers was always made 20000.

The results of the simulations of TPM on the inputs generated by the above technique are summarized in Fig. 2. As is evident from this figure, TPM achieves in these simulations roughly 30% of the optimal trade even when α is as large as roughly 10^{-2}. Moreover, the empirical competitive ratio of TPM improves rapidly for smaller values of α, reaching roughly 80% for $\alpha \approx 10^{-3}$. Thus, our simulations support our conjecture that TPM works well for moderate size markets, despite the lack of a meaningful theoretical guarantee for this range of α values.

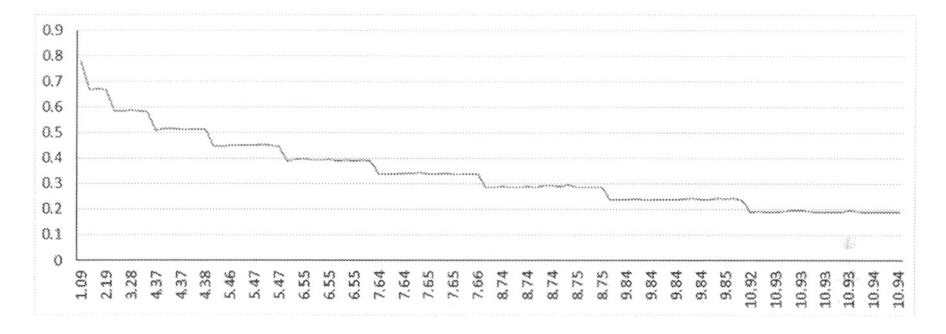

Fig. 2. The competitive ratio obtained in our TPM simulations as a function of α. The values on the x-axis are α times 10^3, and the values on the y-axis are the competitive ratios (obtained by averaging 500 independent executions of TPM).

To better understand the gap between the theoretical and empirical performance of TPM, we ran experiments also for values of α for which the theoretical guarantee of TPM is meaningful. Due to space constraints, we defer the presentation of these simulations and their results to the journal version of this paper. Nevertheless, we note that these simulations suggest the following crude rule of thumb: the empirical performance of TPM for a given value of α is similar to its theoretical performance for α values that are smaller by two orders of magnitude.

References

1. Ashlagi, I., Monderer, D., Tennenholtz, M.: Mediators in position auctions. Games Econ. Behav. **67**, 2–21 (2009)
2. Babaioff, M., Nisan, N., Pavlov, E.: Mechanisms for a spatially distributed market. Games Econ. Behav. **66**(2), 660–684 (2009)
3. Babaioff, M., Walsh, W.E.: Incentive-compatible, budget-balanced, yet highly efficient auctions for supply chain formation. Decis. Support Syst. **39**(1), 123–149 (2005)
4. Blumrosen, L., Dobzinski, S.: Reallocation mechanisms. In: EC, pp. 617–640. ACM, New York (2014)
5. Blumrosen, L., Mizrahi, Y.: Approximating gains-from-trade in bilateral trading. In: WINE, pp. 400–413 (2016)
6. Brustle, J., Cai, Y., Wu, F., Zhao, M.: Approximating gains from trade in two-sided markets via simple mechanisms. In: EC, pp. 589–590 (2017)
7. Chaib-draa, B., Müller, J. (eds.): Multiagent Based Supply Chain Management. SCI, vol. 28. Springer, Heidelberg (2006). https://doi.org/10.1007/978-3-540-33876-5
8. Chen, R.R., Roundy, R.O., Zhang, R.Q., Janakiraman, G.: Efficient auction mechanisms for supply chain procurement. Manag. Sci. **51**(3), 467–482 (2005)
9. Clarke, E.H.: Multipart pricing of public goods. Public Choice **2**, 17–33 (1971)
10. Colini-Baldeschi, R., Goldberg, P., de Keijzer, B., Leonardi, S., Turchetta, S.: Fixed price approximability of the optimal gain from trade. In: Devanur, N.R., Lu, P. (eds.) WINE 2017. LNCS, vol. 10660, pp. 146–160. Springer, Cham (2017). https://doi.org/10.1007/978-3-319-71924-5_11
11. Colini-Baldeschi, R., Goldberg, P.W., de Keijzer, B., Leonardi, S., Roughgarden, T., Turchetta, S.: Approximately efficient two-sided combinatorial auctions. In: EC, pp. 591–608. ACM, New York (2017)
12. Colini-Baldeschi, R., de Keijzer, B., Leonardi, S., Turchetta, S.: Approximately efficient double auctions with strong budget balance. In: SODA, pp. 1424–1443. Society for Industrial and Applied Mathematics, Philadelphia (2016)
13. Facebook: CEO Mark Zuckerberg testifies before senate on user data (2018). https://m.youtube.com/watch?v=qAZiDRonYZI
14. Feldman, J., Mirrokni, V.S., Muthukrishnan, S., Pai, M.M.: Auctions with intermediaries: extended abstract. In: EC, pp. 23–32. ACM, New York (2010)
15. Feldman, M., Gonen, R.: Double-sided markets with strategic multi-dimensional players. CoRR abs/1603.08717 (2016). http://arxiv.org/abs/1603.08717
16. Gonen, M., Gonen, R., Pavlov, E.: Generalized trade reduction mechanisms. In: EC, pp. 20–29. ACM, New York (2007)
17. Gonen, R., Egri, O.: DYCOM: a dynamic truthful budget balanced double-sided combinatorial market. In: The 16th Conference on Autonomous Agents and Multi Agent Systems (AAMAS), May 2017, Brazil, pp. 1556–1558 (2017)
18. Groves, T.: Incentives in teams. Econometrica **41**, 617–631 (1973)
19. McAfee, R.P.: A dominant strategy double auction. J. Econ. Theory **56**, 434–450 (1992)
20. Myerson, R.B., Satterthwaite, M.A.: Efficient mechanisms for bilateral trading. J. Econ. Theory **29**, 265–281 (1983)
21. Segal-Halevi, E., Hassidim, A., Aumann, Y.: MUDA: a truthful multi-unit double-auction mechanism. In: Proceeding of AAAI (2018)

22. Stavrogiannis, L.C., Gerding, E.H., Polukarov, M.: Auction mechanisms for demand-side intermediaries in online advertising exchanges. In: AAMAS, pp. 5–9. International Foundation for Autonomous Agents and Multiagent Systems, Richland (2014)
23. Vickrey, W.: Counterspeculation, auctions and competitive sealed tenders. J. Financ. **16**, 8–37 (1961)

A Two-Stage Mechanism for Ordinal Peer Assessment

Zhize Li, Le Zhang, Zhixuan Fang, and Jian Li(✉)

Institute for Interdisciplinary Information Sciences,
Tsinghua University, Beijing, China
{zz-li14,le-zhang12,fzx13}@mails.tsinghua.edu.cn,
lijian83@mail.tsinghua.edu.cn

Abstract. Peer assessment is a major method for evaluating the performance of employee, accessing the contributions of individuals within a group, making social decisions and many other scenarios. The idea is to ask the individuals of the same group to assess the performance of the others. Scores or rankings are then determined based on these evaluations. However, peer assessment can be biased and manipulated, especially when there is a conflict of interests. In this paper, we consider the problem of eliciting the underlying ordering (i.e. ground truth) of n strategic agents with respect to their performances, e.g., quality of work, contributions, scores, etc. We first prove that there is no deterministic mechanism which obtains the underlying ordering in dominant-strategy implementation. Then, we propose a *Two-Stage Mechanism* in which truth-telling is the *unique* strict Nash equilibrium yielding the underlying ordering. Moreover, we prove that our two-stage mechanism is asymptotically optimal, since it only needs $n+1$ queries and we prove an $\Omega(n)$ lower bound on query complexity for any mechanism. Finally, we conduct experiments on several scenarios to demonstrate that the proposed two-stage mechanism is robust.

Keywords: Mechanism design · Peer assessment · Nash equilibrium

1 Introduction

Peer assessment is a commonly adopted solution for group evaluation without an independent arbiter, e.g., MOOC student assignments scoring [16], research proposal evaluation etc. Despite the pervasive success of peer assessment, there remain issues and controversies, especially on validity and reliability of peer review [21]. As the score of a participant is decided by the assessments given by others, one may manipulate the outcome by providing dishonest feedback. For example, students in a MOOC course usually conduct peer assessments by grading others' homeworks (e.g., percentile scores), and their scores are based on the average of all submitted assessments on their homeworks. In such cases, a student may be able to obtain a better ranking by giving worse evaluations of other students.

X. Deng (Ed.): SAGT 2018, LNCS 11059, pp. 176–188, 2018.
https://doi.org/10.1007/978-3-319-99660-8_16

Not surprisingly, professionals also suffer from unreliable, or "lottery-like" peer review results [30], e.g., irresponsible or derogatory comments appear in academic proposals, and even double-blind review cannot guarantee fairness [18]. In business or academic fields, it is almost inevitable that the reviews have an undisclosed conflict of interests. Under these circumstances, the fairness of peer assessments, even from experts, should be questioned [22,28].

The focus of this paper is to reveal the underlying ordering (ground truth) of the strategic agents. This work is mainly motivated by the applications in which the agents have a strong incentive to manipulate the system by not telling the truth. For example, a direct application is to rank the contributions in a relatively small team, where team members work collaboratively on a project and have a common opinion of the ranking of each member's contribution. Note that the bonus of each employee is indeed assessed by his/her team members in some companies. Thus the leader needs to know the ranking of all members. In this paper, we propose a two-stage mechanism to reveal the ground truth.

Previous work on the problem related to peer assessment, in particular to peer review, studied different setting in which their goal is to select a 'reasonable' aggregated ordering (or a subset) by partition-selection steps (see e.g., [2,10]). For example, although the mechanism given in [2,3] is strategyproof, it is not guaranteed to reveal the *true* underlying ordering even if all the agents share the same opinion of the ordering. Basically their algorithm divides the agents into disjoint clusters. Then the agents in one cluster give evaluations for the agents in other clusters. With these evaluations, the algorithm gets a kind of *value* for each cluster. Finally the top k agents are drawn from the clusters with proportion to their values. Consider a simple example, where $n = 4, k = 1$ and all agents hold the same ordering $\langle 1, 2, 3, 4 \rangle$. Their algorithm divides the agents into two clusters (e.g., $C_1 = \{1, 2\}$ and $C_2 = \{3, 4\}$), and then both clusters get the same value 0.5 according to the Borda score adopted in their algorithm. Finally, their allocation algorithm selects an agent with the highest score in C_1 with probability 0.5 and selects an agent with the highest score in C_2 with probability 0.5, i.e., there is a 0.5 probability to return the wrong agent 3.

1.1 Our Contributions

In this paper, we make the following technical contributions:

1. We prove that there is no deterministic mechanism which obtains the underlying ordering in a dominant-strategy implementation.
2. Although mechanisms can be designed in which Nash equilibria exist, they do not guarantee to obtain the underlying ordering. Under a reasonable assumption, we show that there is a mechanism in which truth-telling is the *unique* strict Nash equilibrium and it leads to the underlying ordering, except that there is an arbitrarily small probability of disorder between the last two agents. Such a disorder is proved to be inevitable if a mechanism has a strict Nash equilibrium leading to the underlying ordering.
3. We prove a lower bound of query complexity for any mechanism, which indicates that our two-stage mechanism is asymptotically optimal.

4. The experimental results on several scenarios demonstrate that our two-stage mechanism is very robust.

1.2 Related Work

Extracting accurate grading results from non-strategic participants has been studied in previous work, where grading errors are treated as noise or systematic bias. Wilson [31] eliminated rater bias and error by regression. Ross et al. [27] calibrated rating bias by solving quadratic programming, while Piech et al. [24] used Gibbs sampling and expectation-maximization to infer parameters of assumed probabilistic grading models. In addition, ordinal methods have also been considered to obtain more robust ranking results instead of cardinal evaluations [19]. Raman and Joachims [25,26] used a maximum likelihood estimator based on the classic Mallows model [20] and Bradley-Terry model [6]. Mi and Yeung [23] used the probabilistic graphical models to boost the grading performance.

Although many mechanisms have been proposed to improve ranking accuracy in peer assessment, there still remains a critical challenge when the agents are strategic. Similar strategic cases have been studied in a variety of forms. Alon et al. [1], Kurokawa et al. [17] and Aziz et al. [2] considered it as a social choice or voting and designed strategyproof mechanisms. Jurca and Faltings [13,14] used monetary incentives to guarantee that truthful reporting is a Nash equilibrium. Gao et al. [11] also used rewards to incentivize truth-telling at equilibrium in peer-prediction mechanisms. Carbonara et al. [8] used a Stackelberg audit game [4,5], associating security games with punishment to incentivize honest reporting. Kahng et al. [15] designed an impartial rank aggregation rule which has a small relative error with some other (nonimpartial) rank aggregation rules. Note that this peer assessment problem is also related to crowdsourcing [7,9,32].

2 Model and Results

2.1 Model

The problem of ordinal peer assessment is formally defined as follows. Let $A = \{1, 2, \ldots, n\}$ denote the set of strategic agents. Let r_i denote the ranking of agent i with respect to its performance (e.g. contribution, score, etc.). W.l.o.g. assume that $r_i = i$ for all $i \in A$, i.e., the underlying ordering is $\langle 1, 2, \ldots, n \rangle$. The underlying ordering (ground truth) is a private common information shared among these n agents. The problem is that a third party, who is not aware of the underlying ordering, wants to obtain it via a mechanism by adopting peer assessment. A mechanism will output an ordering (i.e., a permutation) of A by asking some queries to the agents. We emphasize that our goal is to reveal the *exact* underlying ordering, not a 'reasonable' aggregated ordering (or a subset) as the previous work studied. Thus the strategic agents should share the same underlying ordering, but they may report untruthful answers. Now, we formally define the query operation and the mechanism as follows.

Definition 1 (Query q). $q(i, A')$: *Ask agent i to report the best agent in A', where $A' \subseteq A$.*

Note that this defined query is sufficient to obtain the necessary information, e.g., the pairwise comparison query is such a query when $|A'| = 2$.

Definition 2 (Mechanism $\mathcal{M}$). $\mathcal{M} : (A, \mathcal{Q}) \mapsto \mathcal{P}$, *where A denotes the set of all strategic agents, $\mathcal{Q}$ is the set of any sequence of queries asked by $\mathcal{M}$ and $\mathcal{P}$ denotes the outcome of $\mathcal{M}$ which is the set of all permutations of A. $\mathcal{M}$ outputs an ordering $O \in \mathcal{P}$ according to the sequence of queries $Q \in \mathcal{Q}$ and the corresponding reported answers to Q.*

Definition 3 (Deterministic/Randomized Mechanism). *$\mathcal{M}$ is a deterministic mechanism if it always outputs a deterministic ordering for a given Q and its corresponding reported answers. $\mathcal{M}$ is a randomized mechanism if it may output a randomized ordering according to a distribution over $\mathcal{P}$.*

Naturally, mechanisms using fewer queries are more efficient. Now, we describe the actions of the agents. Every agent is self-interested, only caring about its own ranking in the output ordering of $\mathcal{M}$. We define the *payoff* of an agent to be its (expected) ranking in the outcome of $\mathcal{M}$ (if $\mathcal{M}$ is randomized). Their strategies are to report the answers for the queries asked by $\mathcal{M}$. Note that they may report untruthful answers.

Definition 4 (Strict NE). *A strategy profile is a strict Nash equilibrium if no agent can unilaterally switch to another strategy without reducing its payoff.*

2.2 Main Results

Our goal is to reveal the ground truth, i.e., obtain the underlying ordering. The first thing coming to mind is to design deterministic mechanisms in which all agents have dominant strategies. A *dominant strategy* means that it always achieves the best payoff no matter what the other agents do. Unfortunately, this is impossible, as shown in the following theorem.

Theorem 1 (Impossibility Theorem). *There is no deterministic mechanism which obtains the underlying ordering in a dominant-strategy implementation.*

Note that the Gibbard–Satterthwaite theorem [12,29] does not apply here since the condition is not satisfied, e.g., the preference relation induced by the utility function is *not* antisymmetric since two outcomes having ranked an agent at the same position have no difference for the agent. Also note that our theorem is different from [3] since they need randomization to guarantee the number of selected agents is *exact* k. The proof of our impossibility theorem can be found in the full version of this paper.

According to Theorem 1, the deterministic dominant-strategy implementation is too stringent to be achievable. Nonetheless, one can still obtain that the agents reporting the truth (truth-telling) is a Nash equilibrium, as in the following lemma.

Lemma 1. *There is a mechanism in which truth-telling is a Nash equilibrium.*

Proof. One such mechanism is the naive dictatorship, i.e., the mechanism returns a predefined ordering whatever the agents answer to the queries. Truth-telling obviously is a Nash equilibrium since no agent can affect its payoff by changing its strategy (answer). Note that this mechanism only outputs the true underlying ordering with a very small probability $1/n!$. □

According to the above lemma, although truth-telling and Nash equilibrium are easy to achieve, we still do not get the underlying order yet. Then, we provide a simple mechanism (randomly choose three agents from A, then let them report the best agent and take the majority answer (if it does not exist, we uniformly pick one from A) as the best agent a^*, finally let a^* report the remaining ordering of $A - \{a^*\}$ to obtain a complete ordering (output)) and it is not hard to verify that truth-telling is a Nash equilibrium and this Nash equilibrium leads to the underlying ordering (the details can be found in our full version). However, there are many Nash equilibria in this simple mechanism, e.g., three random agents all report the second-best agent or other agents. Thus this mechanism will output wrong orderings with high probability since most Nash equilibria are bad as they lead to wrong orderings.

Hence, we want to consider the *strict* Nash equilibrium (more extremely, *unique*) that yields the underlying ordering since the strict Nash equilibrium (see Definition 4) implies stability. However, the following lemma indicates a negative result.

Lemma 2. *The underlying ordering cannot be obtained with probability 1 by any strict Nash equilibrium of any mechanism.*

Proof. Assume that there is a strict Nash equilibrium which leads to the underlying ordering with probability 1. Consider the last agent n, the expected ranking of agent n is n in this strict Nash equilibrium. However, if it changes its strategy, it does not get a strictly worse ranking than n as the lowest ranking is n anyway. This contradicts the definition of strict Nash equilibrium. □

Fortunately, if a tiny disorder between agents n and $n-1$ is allowed, a strict Nash equilibrium is still possible. The point is that in the proof above, agent n always ranks the last. As agent n has no lower place to go down, the intuition is to let agent n have a tiny probability to go upward. Besides, we need the following reasonable assumption to get a simple enough mechanism.

Assumption 1. *If the ranking of an agent is fixed in the outcome, then the agent will report truthfully thereafter.*

We propose a two-stage mechanism (Algorithm 1) in Sect. 3, satisfying the following main theorem. We analyze this mechanism and provide the proof sketch in Sect. 3. The detailed proof can be found in our full version.

Theorem 2 (Main Theorem). *Under Assumption 1, there is a mechanism in which truth-telling is the unique strict Nash equilibrium and it leads to the underlying ordering, except for an arbitrarily small probability of disorder between the last two agents. Further, the number of queries asked by the mechanism is* $n+1$.

Moreover, the proposed two-stage mechanism is asymptotically optimal as indicated by the following lower bound theorem (the proof can be found in our full version).

Theorem 3 (Lower Bound). *Any mechanism capable of retrieving the underlying ordering requires* $\Omega(n)$ *queries in the worst case.*

3 The Two-Stage Mechanism

Note that the problem is trivial if $n = 1$. When $n = 2$, it is impossible to distinguish the two agents. For $n > 3$, we propose the *Two-Stage Mechanism* which is described in Algorithm 1. We defer $n = 3$ to the end of this section. In Stage 1 of the two-stage mechanism, we use *self-loop* to denote the reported answer when an agent reports itself, and let *irrational answer* denote the answer when an agent i reports another agent j while agent j does not report itself.

Algorithm 1 Two-Stage Mechanism

Input: $A = \{1, 2, \ldots, n\}$, $n > 3$;
Output: An ordering (permutation) of A;
1: **Stage** 1 :
2: Randomly choose three agents from A, denoted as i, j and k.
3: Ask $q_1(i, A)$, $q_2(j, A)$ and $q_3(k, A)$.
4: **if** At least two of them report themselves **then**
5: Uniformly pick an agent from $A - \{i, j, k\}$ as a^*.
6: **else**
7: Create a multi-set $C = \{i', j', k'\}$, where i', j' and k' denote the answers reported by i, j and k respectively.
8: Remove irrational answers from C (e.g. $i' = i$, $j' = i$, and $k' = j$, C changes from $\{i, i, j\}$ to $\{i, i\}$. $k' = j$ is an irrational answer since agent j reporting i does not report itself.).
9: If there exists o in C, where o denotes an agent except $\{i, j, k\}$, then remove self-loop answers from C (e.g. $i' = i$, $j' = o$, and $k' = j$, C changes from $\{i, o, j\}$ to $\{i, o\}$ in the previous step, and again to $\{o\}$ in this step.).
10: Randomly select a^* from C (candidate set) if $C \neq \emptyset$, otherwise uniformly pick from $A - \{i, j, k\}$.
11: **end if**
12: **Stage** 2 :
13: Let a^* be ranked the first position and then let it report the remaining ordering of $A - \{a^*\}$, i.e., asking $q_4(a^*, A - \{a^*\})$, $q_5(a^*, A - \{a^*, \tilde{q}_4\})$ etc., where $\tilde{q}_i$ denotes the answer of q_i. Now we obtain a complete ordering O. Denote the ranking of agent x as $\bar{r}_x$.
14: If the following answers were reported in Stage 1: $i' = i, j' = i, k' = j$: $i \quad j \quad k$, then swap k with $k+1$ if $k < n$ in O, i.e. punish k to one position later.
15: Denote the last two agents in O as x and y. Slightly perturb $\bar{r}_y$ to $n-1$ (and hence $\bar{r}_x$ to n) with probability ϵ to obtain the final ordering $\widetilde{O}$ if y but not x is in $\{i, j, k\}$ and $i' = j' = k'$, where $\epsilon \in (0, 1)$ is arbitrarily small.
16: **return** $\widetilde{O}$.

Intuition: When an agent i is asked to report the best agent (a^*) in Stage 1, the agent only has three possible answers described below, where the last two are wrong answers:

1. report a^* (true answer);
2. report itself (self-loop answer);
3. or, report someone else (mostly irrational answers).

To enforce truth-telling strategy of all agents, we will ignore the wrong answers (Case 2 and 3) and punish the misreporting agents in our mechanism. Concretely, for the irrational answers, we ignore the answers in Line 8 of Algorithm 1 and punish the agents in Line 14. For the self-loops, we ignore the answers in Line 9 and indirectly punish the agents in Line 5. Note that there is an exception for agent a^*, i.e., a^* reporting a^* (itself) also belongs to Case 2. But this answer should only be considered as the Case 1 true answer.

Analysis: Now we analyze the two-stage mechanism in more details before moving to the proofs. The best agent a^* is selected in Stage 1. This process can be distinguished as two modes:

1. Mode A: Uniformly pick
   (a) At least two self-loops (i.e. reporting themselves):
   Uniformly pick one from $A - \{i, j, k\}$. We denote this case as Case ⓵ⓐ (1a), i.e., sub-item (a) in item 1. Similar denotations are used for other cases.
   (b) Candidate set $C = \emptyset$:
   Uniformly pick one from $A - \{i, j, k\}$. This case happens *exactly* when $i', j', k' \in \{i, j, k\}$, and no self-loop exists, i.e. all three answers are irrational.
2. Mode B: Pick candidates (Randomly select from C.)
   (a) $i', j', k' \in \{i, j, k\}$ and exactly one self-loop exists:
   W.l.o.g., let $i' = i$. C must only contain i. Because agents j and k do not report themselves, those answers reporting j or k are irrational. Consequently an agent not reporting itself will not be contained in C.
   (b) There exists o in C. Recall that o denotes an agent except $\{i, j, k\}$:
   C contains these o-type answers, and if it happens that a self-loop exists with another agent also reporting the "self-loop" agent, C will contain the self-loop, e.g. $i' = o$, $j' = j$, $k' = j$, $C = \{o, j\}$.

The intuition is that Mode B usually allows us to select the correct agent a^*, while Mode A usually is a bad case but rarely happens. Besides, note that Line 14 and 15 of Algorithm 1 do not affect the ranking of a^*, i.e., a^* (selected in Stage 1) always ranks the first in the outcome $\widetilde{O}$. This ensures that true answers to the queries of Stage 2 can be obtained according to Assumption 1.

Now, we move to the proof sketch part. The proof details can be found in our full version. To show truth-telling is the unique strict Nash equilibrium, it should at least be a strict Nash equilibrium as stated by the following lemma.

Lemma 3. *When $n > 3$, in Stage 1, the strategy profile consisting of the chosen agents reporting* 1 *(truth-telling) is a strict Nash equilibrium.*

Then, to show the uniqueness in the following theorem, we only need to show there are no other strict Nash equilibria. We prove this by contradiction, i.e., we show that any agent reporting 1 (true answer) is not strictly worse than others, and thus there is no other strict Nash equilibrium except truth-telling, otherwise the agent can change its answer to 1 without reducing its payoff which contradicts the definition of strict Nash equilibrium (Definition 4). Note that the number of queries is $(n+1)$ since Stage 1 uses 3 queries and Stage 2 using $n-2$ queries is sufficient since only one agent is remained.

Theorem 4. *When $n > 3$, the two-stage mechanism uses $n+1$ queries and yields that truth-telling is the unique strict Nash equilibrium and it leads to the underlying ordering, except that there is an arbitrarily small probability of disorder between the last two agents.*

The following corollary easily follows from the proof of Theorem 4 which is provided in our full version.

Corollary 1. *The only remaining strategy in Stage 1 for each agent is truth-telling using the iterated elimination of dominated strategies (IEDS) process from agent 1 to agent n.*

3.1 Two-Stage Mechanism for $n = 3$

For the $n = 3$ case, this situation follows the same paradigm as Algorithm 1. But note that formerly *Mode A uniformly pick* works perfectly when $n > 3$, but for $n = 3$, there is no agent for $A - \{i, j, k\}$. Thus, a slight difference is that we might need to identify the second-best agent, rather than the best agent, i.e. select the second-best agent, and fix its ranking as 2 in the outcome. This update obeys the "spirit" of Assumption 1.

We describe the selection process in Stage 1 by cases. Let s denote the number of "self-loop" agents, i.e. the agents reporting themselves.

1. $s = 3$. Uniformly pick a^* from $\{1, 2, 3\}$. We denote this case as Case ①.
2. $s = 2$. Select the only non-self-loop as the second-best agent.
3. $s = 1$. Select the only self-loop as a^*. Only two special sub-cases need *extra* processing.
   (a) The same as Line 14 of Algorithm 1, i.e. swap k and $k+1$ if $k < 3$.
   (b) It is similar to Line 15 of Algorithm 1, while here $x \notin \{i, j, k\}$ is not required. The slight perturbation has be done if three agents report the same one.
4. $s = 0$. The same as Case ①.

The only slight modifications (differ from $n > 3$ case) are that the algorithm now uniformly picks from $\{i, j, k\}$ instead of from $A - \{i, j, k\}$, and the circumstance for *two self-loops* is tackled differently (i.e., identify the second-best agent now).

We have a similar result for this $n = 3$ case, as stated in the following lemma. The proof is not very hard and we provide it in our full version.

Lemma 4. *When $n = 3$, the two-stage mechanism with a slight difference uses $n + 1$ queries and yields that truth-telling is the unique strict Nash equilibrium and it leads to the underlying ordering, except that there is an arbitrarily small probability of disorder between the last two agents.*

Now, the main theorem, i.e. Theorem 2, easily follows from Theorem 4 and Lemma 4.

4 Experiments

In this section, we conduct experiments on several scenarios to show our two-stage mechanism is very robust. Recall truth-telling is the unique strict Nash equilibrium in our mechanism, and the *strict* Nash equilibrium implies stability. Intuitively, agents who adopt non-truth-telling strategy will eventually find it more beneficial to report the truth. The reason is that at least agent a^* (the best agent) would like to adopt truth-telling. Consequently, other agents are more or less forced to be honest according to the mechanism. Hence the strategy of non-truth-telling naturally converges to the unique truth-telling equilibrium.

Concretely, we simulate the situation where there are many rounds for the agents to switch their strategies. Initially, a portion of agents is set to hold untruthful answers (randomly chosen). Let the noise factor denote the initial fraction of misreporting agents, hence equivalently, the initial *truthful ratio* is $1 - \mathsf{noise}$. Experiments are conducted on the number of agents $n = 10, 40, 70, 100$ and $\mathsf{noise} = 0.1, 0.5, 1$. In each round, all agents sequentially switch their answers to their *best responses* with respect to the two-stage mechanism, under the condition of other agents keeping their answers unchanged. Thus the switching process in each round consists of n iterations, i.e., in each iteration, exactly one agent switches to its best response. To compute the best response for an agent, as the two-stage mechanism is randomized, this is approximated by enumerating all answers of that agent and computing the average payoff for every enumerated answer (by running the algorithm 10000 times) and then chose the highest average payoff one.

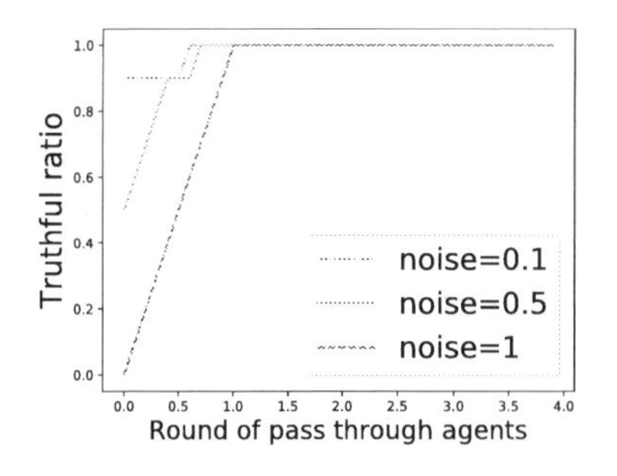

Fig. 1. $n = 10$

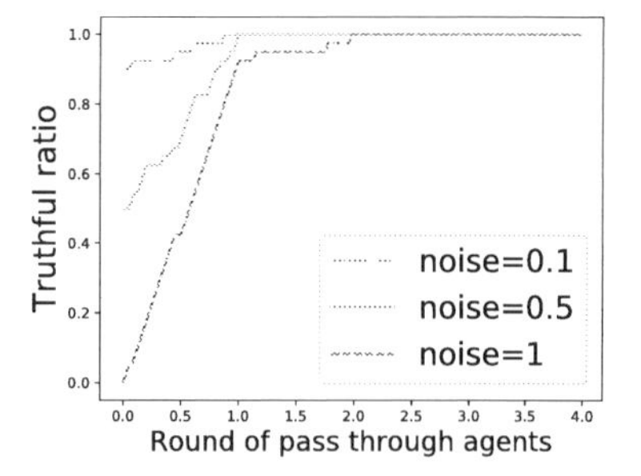

Fig. 2. $n = 40$

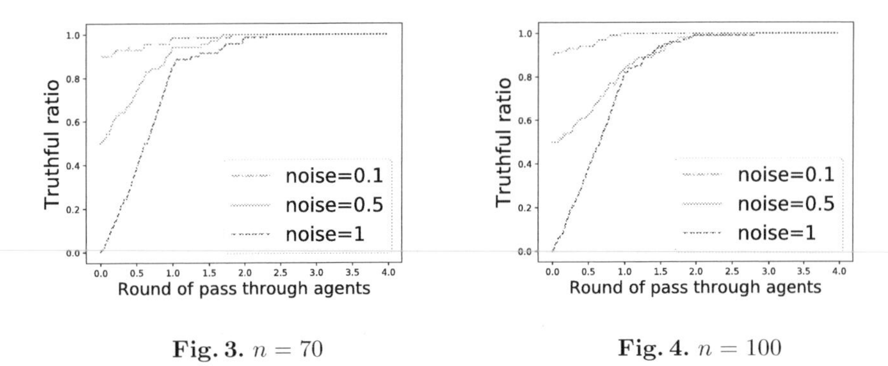

Fig. 3. $n = 70$

Fig. 4. $n = 100$

Figures 1, 2, 3 and 4 demonstrate the process of the strategies converging to the unique truth-telling equilibrium under different noise factors. Note that a truthful ratio equal to 1 means that all agents report true answers. The convergence processes in all figures are very similar, which means the number of agents does not affect the process much. Moreover, the total number of rounds in all figures are 4, which shows that this process converges very fast.

We use the following Figs. 5 and 6 to compare the convergence performance (speed) with respect to the number of agents. Particularly, if the initial fraction of misreporting agents (i.e., noise) is very large (see Fig. 6), the speed with which the strategies converge to the unique truth-telling equilibrium is almost independent of the number of agents (i.e., n). If the noise is not that large (see Fig. 5), the speed is inversely proportional to the number of agents. Nevertheless, they all converge quickly to the unique truth-telling equilibrium within four rounds.

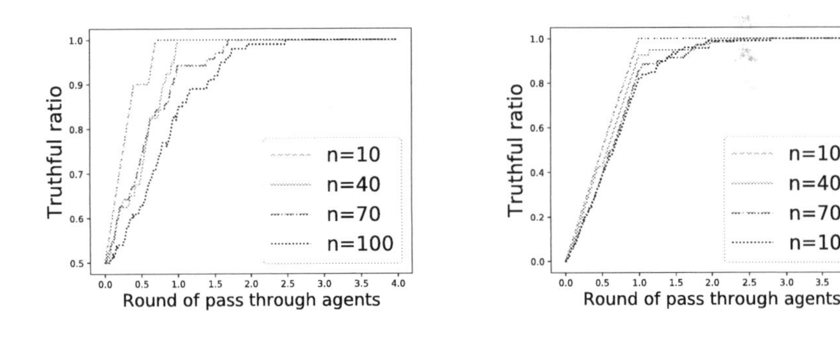

Fig. 5. noise $= 0.5$

Fig. 6. noise $= 1$

In conclusion, the number of agents (i.e., n) and the initial fraction of misreporting agents (i.e., noise) both do not affect the performance much, and the strategies of all agents converge quickly to the unique truth-telling equilibrium. The experimental results validate that our two-stage mechanism is very robust.

5 Conclusions

In this paper, we consider the problem of obtaining the underlying ordering (ground truth) among n strategic agents with respect to their performance by peer assessment. We first prove that there is no deterministic mechanism which obtains the underlying ordering in a dominant-strategy implementation. Then, we propose a two-stage mechanism in which truth-telling is the unique strict Nash equilibrium and it leads to the underlying ordering, except that there is an arbitrarily small probability of disorder between the last two agents. Note that such a disorder is proved to be inevitable if a mechanism has a strict Nash equilibrium leading to the underlying ordering. Moreover, our two-stage mechanism only needs $n + 1$ queries. We then prove an $\Omega(n)$ lower bound of query complexity for any mechanism, which indicates that our mechanism is asymptotically optimal. Finally, the experimental results demonstrate that the proposed query-optimal two-stage mechanism is also very robust.

Discussions: We discuss the applicability of our results in other query models. Theorem 1 (Impossibility Theorem) always holds for any query model since our proof is independent of the query model. For Lemma 2, it also always holds since the fact (the last-ranked agent can always cheat as we demonstrated in the proof) is irrelevant to the query model. As long as the number of possible answers to a single query is $O(n)$, Theorem 3 (Lower Bound) also holds. However, it might be violated if one allows a very powerful query model, e.g. requiring an agent to answer all the information it knows in just one query (the possible answers to a single query can be $n!$ in the worst case). For the extension, it is interesting to consider the situation of collusion.

Acknowledgments. This research is supported in part by the National Basic Research Program of China Grant 2015CB358700, the National Natural Science Foundation of China Grant 61772297, 61632016, 61761146003, and a grant from Microsoft Research Asia. The authors would like to thank Pingzhong Tang and the anonymous reviewers for their valuable comments.

References

1. Alon, N., Fischer, F., Procaccia, A., Tennenholtz, M.: Sum of us: strategyproof selection from the selectors. In: Proceedings of the 13th Conference on Theoretical Aspects of Rationality and Knowledge, pp. 101–110. ACM (2011)
2. Aziz, H., Lev, O., Mattei, N., Rosenschein, J.S., Walsh, T.: Strategyproof peer selection: mechanisms, analyses, and experiments. In: Proceedings of the 30th AAAI Conference on Artificial Intelligence (AAAI) (2016)
3. Aziz, H., Lev, O., Mattei, N., Rosenschein, J.S., Walsh, T.: Strategyproof peer selection using randomization, partitioning, and apportionment. arXiv preprint arXiv:1604.03632 (2016)
4. Blocki, J., Christin, N., Datta, A., Procaccia, A.D., Sinha, A.: Audit games. In: Proceedings of the Twenty-Third International Joint Conference on Artificial Intelligence (IJCAI), pp. 41–47 (2013)

5. Blocki, J., Christin, N., Datta, A., Procaccia, A.D., Sinha, A.: Audit games with multiple defender resources. In: Proceedings of AAAI, pp. 791–797 (2015)
6. Bradley, R.A., Terry, M.E.: Rank analysis of incomplete block designs: I. The method of paired comparisons. Biometrika **39**(3/4), 324–345 (1952)
7. Cao, W., Li, J., Tao, Y., Li, Z.: On top-k selection in multi-armed bandits and hidden bipartite graphs. In: Advances in Neural Information Processing Systems (NIPS), pp. 1036–1044 (2015)
8. Carbonara, A.U., Datta, A., Sinha, A., Zick, Y.: Incentivizing peer grading in MOOCs: an audit game approach. In: Proceedings of the Twenty-Fourth International Joint Conference on Artificial Intelligence (IJCAI) (2015)
9. Chai, C., Li, G., Li, J., Deng, D., Feng, J.: Cost-effective crowdsourced entity resolution: a partial-order approach. In: Proceedings of the 2016 International Conference on Management of Data, pp. 969–984. ACM (2016)
10. Fischer, F., Klimm, M.: Optimal impartial selection. In: Proceedings of the Fifteenth ACM Conference on Economics and Computation (EC), pp. 803–820. ACM (2014)
11. Gao, A., Wright, J.R., Leyton-Brown, K.: Incentivizing evaluation via limited access to ground truth: peer-prediction makes things worse. arXiv preprint arXiv:1606.07042 (2016)
12. Gibbard, A.: Manipulation of voting schemes: a general result. Econ. J. Econ. Soc. **41**, 587–601 (1973)
13. Jurca, R., Faltings, B.: Enforcing truthful strategies in incentive compatible reputation mechanisms. In: Deng, X., Ye, Y. (eds.) WINE 2005. LNCS, vol. 3828, pp. 268–277. Springer, Heidelberg (2005). https://doi.org/10.1007/11600930_26
14. Jurca, R., Faltings, B.: Mechanisms for making crowds truthful. J. Artif. Intell. Res. **34**, 209–253 (2009)
15. Kahng, A., Kotturi, Y., Kulkarni, C., Kurokawa, D., Procaccia, A.D.: Ranking wily people who rank each other. Technical report (2017)
16. Kulkarni, C., Wei, K.P., Le, H., Chia, D., Papadopoulos, K., Cheng, J., Koller, D., Klemmer, S.R.: Peer and self assessment in massive online classes. In: Plattner, H., Meinel, C., Leifer, L. (eds.) Design Thinking Research, pp. 131–168. Springer, Cham (2015). https://doi.org/10.1007/978-3-319-06823-7_9
17. Kurokawa, D., Lev, O., Morgenstern, J., Procaccia, A.D.: Impartial peer review. In: Proceedings of the Twenty-Fourth International Joint Conference on Artificial Intelligence (IJCAI) (2015)
18. Lee, C.J., Sugimoto, C.R., Zhang, G., Cronin, B.: Bias in peer review. J. Am. Soc. Inf. Sci. Technol. **64**(1), 2–17 (2013)
19. Liu, T.Y.: Learning to rank for information retrieval. Found. Trends Inf. Retr. **3**(3), 225–331 (2009)
20. Mallows, C.L.: Non-null ranking models. I. Biometrika **44**(1/2), 114–130 (1957)
21. Marsh, H.W., Jayasinghe, U.W., Bond, N.W.: Improving the peer-review process for grant applications: reliability, validity, bias, and generalizability. Am. Psychol. **63**(3), 160 (2008)
22. Merrifield, M.R., Saari, D.G.: Telescope time without tears: a distributed approach to peer review. Astron. Geophys. **50**(4), 4–16 (2009)
23. Mi, F., Yeung, D.Y.: Probabilistic graphical models for boosting cardinal and ordinal peer grading in moocs. In: Proceedings of AAAI, pp. 454–460 (2015)
24. Piech, C., Huang, J., Chen, Z., Do, C., Ng, A., Koller, D.: Tuned models of peer assessment in MOOCs. In: Educational Data Mining 2013 (2013)

25. Raman, K., Joachims, T.: Methods for ordinal peer grading. In: Proceedings of the 20th ACM SIGKDD International Conference on Knowledge Discovery and Data Mining, pp. 1037–1046. ACM (2014)
26. Raman, K., Joachims, T.: Bayesian ordinal peer grading. In: Proceedings of the Second (2015) ACM Conference on Learning @ Scale, pp. 149–156. ACM (2015)
27. Roos, M., Rothe, J., Scheuermann, B.: How to calibrate the scores of biased reviewers by quadratic programming. In: Proceedings of AAAI (2011)
28. Saidman, L.J.: Unresolved issues relating to peer review, industry support of research, and conflict of interest. Anesthesiology **80**(3), 491 (1994)
29. Satterthwaite, M.A.: Strategy-proofness and arrow's conditions: existence and correspondence theorems for voting procedures and social welfare functions. J. Econ. Theory **10**(2), 187–217 (1975)
30. Smith, R.: Peer review: a flawed process at the heart of science and journals. J. R. Soc. Med. **99**(4), 178–182 (2006)
31. Wilson, H.G.: Parameter estimation for peer grading under incomplete design. Educ. Psychol. Meas. **48**(1), 69–81 (1988)
32. Zhou, Y., Chen, X., Li, J.: Optimal PAC multiple arm identification with applications to crowdsourcing. In: International Conference on Machine Learning (ICML), pp. 217–225 (2014)

The Equilibrium Existence of a Robust Routing Game Under Interval Uncertainty

Xujin Chen[1,2], Xiaodong Hu[1,2], and Chenhao Wang[1,2,3](✉)

[1] Academy of Mathematics and Systems Science, Chinese Academy of Sciences, Beijing, China
[2] School of Mathematical Sciences, University of Chinese Academy of Sciences, Beijing, China
{xchen,xdhu}@amss.ac.cn
[3] Department of Computer Science, City University of Hong Kong, Kowloon, Hong Kong SAR, China
wangch@amss.ac.cn

Abstract. We study an atomic routing game in a network with interval uncertainty, where the cost of each edge is load-dependent, and can be any value in a given interval whose lower and upper limits are expressed as functions of the edge load. Each player would select a path from his source to his terminal in the network that minimizes his maximum regret, where given a path (strategy) profile of the game, the maximum regret of a player is the worst-case difference between his total cost (in the case of sum-type cost) or her bottleneck cost (in the case of bottleneck-type cost) and the optimum he could attain given the choices of other players in the profile and a priori knowledge about the actual realization of edge costs. A NE of this game, termed as *robust Nash equilibrium* (RNE), is a path profile under which no player can reduce his maximum regret by unilateral deviation. On the negative side, we show that the problem of deciding whether a given 3-player game with the sum-type costs has an RNE is NP-hard, even if the game is symmetric and all intervals have unit length. On the positive side, we characterize network topologies, for the game with either type of costs, that guarantee the RNE existence regardless of source-terminal locations and interval settings.

Keywords: Robust routing game · Atomic routing
Minimum maximum regret · Nash equilibrium · Interval uncertainty

1 Introduction

Atomic routing games have attracted a great deal of attention over recent years due to their various real-world applications, e.g., packets transmission in communication networks and vehicle routing in transport systems. In a typical setting

Research supported in part by NNSF of China under Grant No. 11531014.

X. Deng (Ed.): SAGT 2018, LNCS 11059, pp. 189–200, 2018.
https://doi.org/10.1007/978-3-319-99660-8_17

of the atomic routing game, each selfish player is associated to a pair of source and terminal vertices in the network; he selects a path from the source to the terminal in the network for his travel, aiming at a certain objective. The cost incurred to a player for traversing an edge in the network is typically load-dependent, e.g., the common delay suffered by all players on the edge increases as the edge becomes more congested. Two common player objectives extensively studied in the literature are the sum- and bottleneck-types [4,5]. The sum-type is minimizing the *total cost* a player spends in traveling along his chosen path, while the bottleneck-type concerns with minimizing the highest cost a player spends on an edge along his chosen path (called his *bottleneck cost*). When atomic routing games with either objective are played under complete information, pure-strategy Nash equilibria have been shown (under mild conditions on cost functions) to be natural outcomes of the interactions among selfish players [7,10].

The idealization of complete information is however impractical in a wide variety of real-world applications. Due to limited information, inherent fluctuations or unpredictable changes in the system (e.g., hardware failures, temporary construction projects, weather conditions, traffic accidents, etc.), players are often uncertain about some aspects of the game they are playing. Among many others, the interval uncertainty often arises when players only know an interval estimate on the cost of each edge before making decisions (and learn about the actual cost realization after their path selections). Under the interval uncertainty, a natural solution of the routing game might be each player selecting a risk-averse strategy which is "robust" against the worst-case realization of uncertain edge costs that could happen to him.

The Robust Routing Game Under Interval Uncertainty. We study in this paper the robust atomic routing game under interval uncertainty which adopts the popular robust criterion of *minimum maximum regret* [9,12,13]. In the game, each edge of the network is associated to an interval whose lower and upper limits are nondecreasing functions of the edge load. The cost of each edge can take any value from the associated interval, regardless of the values realized for other edges. The *maximum regret* of a player is the worst-case difference between his cost (total cost or bottleneck cost depending on the player objective of the game) and the optimum he could attain given a priori knowledge about the actual realization of edge costs. Each player aims at selecting a path from his source to his terminal that minimizes his maximum regret. We refer to such an atomic routing game as *robust routing game* Γ, and denote it by Γ^+ or Γ^b depending on whether the player objectives are the sum-type or bottleneck-type. If all players are able to choose their minimum-maximum-regret paths (strategies) at the same time, the resulting path (strategy) profile is called a *robust Nash equilibrium* (RNE) of the robust routing game. Given an RNE, no player can reduce his maximum regret by unilateral deviation, but for a fixed realization of edge costs, it might be possible for a player to reduce his cost by unilaterally deviating to another path.

Our Contributions. We provide both negative and positive results on the RNE existence of the robust routing game Γ. On the negative side, we present 2-player games which show that

- no matter whether the underlying network is directed (resp. planar) or not, the robust routing game Γ with either type objective does not necessarily admit any RNE, even if the game is symmetric.

Previously the RNE nonexistence was only known for the bottleneck objective and non-planar directed networks [14]. Building on the nonexistence we have discovered, we prove, with the help of an idea from [14], that

- the problem of deciding whether a given 3-player game Γ^+ on a directed network has an RNE is NP-hard, even if the game is symmetric and all intervals have unit length.

This stands in sharp contrast to the RNE existence of symmetric game Γ^b whose network is associated with intervals all of the same constant length [14].

On the positive side, we characterize network topologies that guarantee the RNE existence regardless of source-terminal locations and interval settings. A *cut-vertex* of a connected graph is a vertex whose removal will disconnect the graph. A *subdivision* of a graph G refers to a graph obtained from G by repeatedly subdividing edges (possibly none), where subdividing an edge (from vertex u to v) consists of deleting the edge, adding a new vertex w and two new edges from u to w and from w to v, respectively. For undirected graphs, we obtain the following complete characterization:

- Let G be a connected undirected graph. Then
  - every robust routing game Γ^+ on G admits an RNE if and only if every maximal connected subgraph of G without any cut-vertex is either an edge or a cycle;
  - every robust routing game Γ^b on G admits an RNE if and only if every maximal connected subgraph of G without any cut-vertex is either an edge or a cycle on 2 or 3 edges.

(Note that a pair of parallel undirected edges is considered as a cycle.) The result indicates that Γ^+ allows more chances for RNE existing than Γ^b does. The structures of directed graphs are relatively complicated. We establish the following necessary condition, which also implies a relatively narrow graphical class for ensuring the RNE existence.

- Let G be a connected directed graph. Then every robust routing game Γ^+ (resp. Γ^b) on G admits an RNE only if G has no subgraph isomorphic to a subdivision of G_1 or G_2 in Fig. 1.

These results partially answer a question of Werth *et al.* [14] concerning identifying graphs which guarantee the existences of RNE.

Related Work. The most related work to ours is the aforementioned paper [14] by Werth *et al.* which focused on the robust routing game Γ^b. The authors

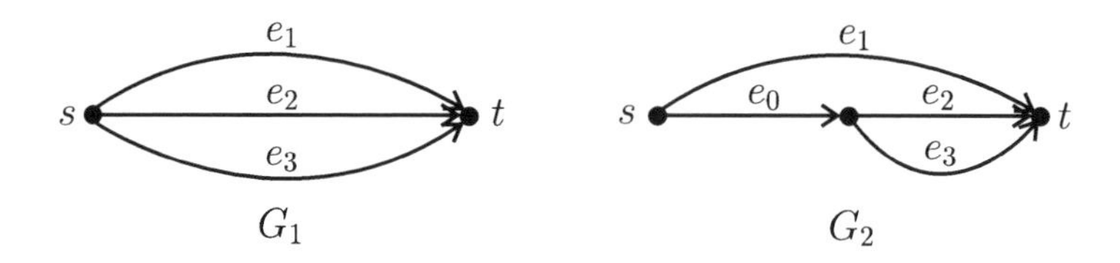

Fig. 1. Forbidden structures for the RNE existence in directed networks.

proved that it is NP-complete in general to decide whether a given 3-player game Γ^b has an RNE, even if the game is symmetric. The NP-membership of the decision problem is noteworthy; it follows from the polynomial-time solvability of computing a best response in Γ^b [2]. For the special case of symmetric Γ^b with uniform constant-length intervals, Werth *et al.* [14] proved that the Nash equilibria of a generalized bottleneck routing game are RNEs of the special game Γ^b. The so-called price of robustness was also investigated for Γ^b.

To the best of our knowledge, Aghassi and Bertsimas [1] were among the first who combined concepts from robust optimization with those from game theory to obtain distribution-free robust game model with incomplete information. In the game, given a set of possible values of the uncertain payoff functions, each player seeks to maximize his worst-case payoff. A mixed-strategy equilibrium is guaranteed to exist when the game is finite and has a bounded payoff uncertainty set. The vast majority of subsequent work on robust games adopted similar maximum-minimum-payoff (or minimum-maximum-cost) principles (e.g., [11]), with a few exceptions [12,14] concerning another widely accepted criterion of minimum maximum regret, which we consider in this paper for our robust routing game Γ.

The robust optimization counterpart to our robust routing game model, known as the *interval data minmax regret shortest* (resp. *bottleneck*) *path problem* [2,3], can be considered as a 1-player game Γ^+ (resp. Γ^b). It is to find a path from a given source to a given terminal that minimizes the maximum regret of the single player, where the intervals all have constant lower and upper limits. The shortest-path version of the robust optimization problem is strongly NP-hard [3,15], while the bottleneck-path version is polynomial-time solvable [2]. The complexity status instantly shows a difference between Γ^+ and Γ^b: the problem of deciding RNE existence for Γ^+ is not in NP unless P = NP, while the decision problem for Γ^b belongs to NP.

Paper Organization. Section 2 gives the formal mathematical model of our robust routing game. Section 3 presents several instances of game Γ without RNE, which particularly imply some necessary conditions for networks to ensure RNE existence. Section 4 proves the NP-hardness of determining the RNE-existence in game Γ^+, by a reduction from the directed 2-edge-disjoint-path problem. Section 5 characterizes all undirected graphs that guarantee the RNE existences. Section 6 provides concluding remarks on future research.

2 The Game Model

We study a game model for robust routing with interval uncertainty. By abuse of notation, we use Γ to denote both the game model and an instance of the game under consideration. Game Γ is played by a finite set U of players on an underlying (directed or undirected) graph $G = (V, E)$ with vertex set V and edge set E. All paths under consideration are simple, and they are directed if G is directed. Each player $i \in U$ is associated with a pair of source vertex s_i and terminal vertex t_i in G, and his strategy set $\mathscr{P}_i$ consists of all s_i-t_i paths (i.e., paths from s_i to t_i) in G. Let $\mathscr{P} = \times_{i\in U} \mathscr{P}_i$ denote the set of strategy profiles. We often identify a *path (strategy) profile* $\mathbf{p} = (P_i)_{i\in U} \in \mathscr{P}$, in which each player $i \in U$ chooses a path (strategy) $P_i \in \mathscr{P}_i$, with the *routing* in which each player $i \in U$ travels along P_i. Under the routing $\mathbf{p}$, the *load* on edge $e \in E$ is defined as the number $\mathbf{p}_e = |\{i \in U|e \in P_i\}|$ of players who go through e. For each $i \in U$, we use $\mathbf{p}_{-i} = (P_j)_{j\in U\setminus\{i\}}$ to denote the partial routing of players in $U \setminus \{i\}$. The game is called *symmetric* if all players have the same source and terminal.

In a typical setting of the game with complete information, where each edge $e \in E$ is equipped with a nondecreasing and nonnegative delay function $d_e : \{1, 2, \dots, |U|\} \to \mathbb{R}_{\geq 0}$, which maps the load of e to a nonnegative real number, called the *delay* of e. In contrast, game Γ is to be played under some scenario to be realized. This scenario is not known in advance, while some partial information is available. Given a routing of Γ, the delay on an edge depends on not only its load but also the scenario realized. Specifically, for each edge $e \in E$, the delay function $d_e : \{1, 2, \dots, |U|\} \times \Sigma \to \mathbb{R}_{\geq 0}$ is now a function of two variables, where Σ denotes the set of all realizable scenarios. Corresponding to each possible load $x \in \{1, 2, \dots, |U|\}$ on e, the delay $d_e(x, \cdot)$ on e, no matter which scenarios is realized, belongs to a given closed interval $[l_e(x), u_e(x)]$, where the lower limit $l_e(x) = \min\{d_e(x, \varsigma)|\varsigma \in \Sigma\}$ and upper limit $u_e(x) = \max\{d_e(x, \varsigma)|\varsigma \in \Sigma\}$ are both nondecreasing functions of x. It is assumed that the scenario realizations are "compact" over E, i.e., for any routing $\mathbf{p}$ and any values $m_e \in [l_e(\mathbf{p}_e), u_e(\mathbf{p}_e)]$, $e \in E$, there is a scenario $\varsigma \in \Sigma$ such that $d_e(\mathbf{p}_e, \varsigma) = m_e$ for all $e \in E$. The graph G along with the associated intervals $[l_e(x), u_e(x)]$, $e \in E$ is referred to as a *network*. The length $u_e(x) - l_e(x)$ of interval $[l_e(x), u_e(x)]$ is said to be *constant* (resp. *unit*) if there exists constant $\epsilon \geq 0$ (resp. $\epsilon = 1$) such that $u_e(x) - l_e(x) = \epsilon$ for each possible load x.

The game model Γ (or $\Gamma(G)$ to specify the underlying network G) consists of two sub-models, Γ^+ and Γ^b as specified below, for minimizing maximum regrets w.r.t. sum-type costs and bottleneck costs, respectively. For convenience, we often identify a subgraph (a path in particular) of G with its edge set.

The Sum-Type Costs. The most studied costs of individual players are their path delays, the sums of delays of edges on their paths. In our robust routing game, the delay depends not only the routing $\mathbf{p} = (P_i)_{i\in U}$ but on the scenario ς realized. Particularly, the *path delay* of player $i \in U$ is

$$d_i(\mathbf{p}, \varsigma) = \sum_{e\in P_i} d_e(\mathbf{p}_e, \varsigma).$$

The game Γ will be written as Γ^+ or $\Gamma^+(G)$ if players' costs are their path delays. Upon the reveal of the realized scenario ς, player i might regret his choice P_i in $\mathbf{p}$ because his unilateral change from P_i to another path $P_i' \in \mathscr{P}_i$ might incur to him a smaller cost, i.e., a shorter path delay $d_i((P_i', \mathbf{p}_{-i}), \varsigma)$, where $(P_i', \mathbf{p}_{-i})$ is the routing in which player i follows P_i', while others keep their paths in $\mathbf{p}_{-i}$ unchanged. The *regret* $r_i(\mathbf{p}, \varsigma)$ *of player* i *under routing* $\mathbf{p}$ *and scenario* ς is defined as the difference between his cost and the best he could attain assuming $\mathbf{p}_{-i}$ and ς. That is,

$$r_i(\mathbf{p}, \varsigma) = d_i(\mathbf{p}, \varsigma) - \min_{P_i' \in \mathscr{P}_i} d_i((P_i', \mathbf{p}_{-i}), \varsigma).$$

Removing the reference of scenario ς, the *maximum regret of player* i *under routing* $\mathbf{p}$ is defined to be his highest regret among all scenarios realizable, i.e.,

$$r_i(\mathbf{p}) = \sup_{\varsigma \in \Sigma} r_i(\mathbf{p}, \varsigma) = \sup_{\varsigma \in \Sigma} \left(d_i(\mathbf{p}, \varsigma) - \min_{P_i' \in \mathscr{P}_i} d_i((P_i', \mathbf{p}_{-i}), \varsigma) \right).$$

Each player aims at choosing his path to minimize his maximum regret given the choices of others.

The Bottleneck-Type Costs. Another type of individual costs often investigated in literature is a bottleneck one. Concerning this type of costs, we write Γ as Γ^b or $\Gamma^b(G)$, where given routing $\mathbf{p} = (P_i)_{i \in U} \in \mathscr{P}$ and scenario $\varsigma \in \Sigma$, the cost of player $i \in U$ is defined as

$$\text{his } \textit{bottleneck delay } b_i(\mathbf{p}, \varsigma) = \max_{e \in P_i} d_e(\mathbf{p}_e, \varsigma),$$

i.e., the maximum edge delay among his path P_i. Similar to Γ^+ introduced above, player $i \in U$ in Γ^b has a *regret*

$$r_i(\mathbf{p}, \varsigma) = b_i(\mathbf{p}, \varsigma) - \min_{P_i' \in \mathscr{P}_i} b_i((P_i', \mathbf{p}_{-i}), \varsigma)$$

under $\mathbf{p}$ and ς, and a *maximum regret* $r_i(\mathbf{p}) = \sup_{\varsigma \in \Sigma} r_i(\mathbf{p}, \varsigma)$ under $\mathbf{p}$. Again, each player wants to minimize his maximum regret.

Definition 1. *A routing* $\mathbf{p}$ *of* Γ *is called a* robust Nash equilibrium *(RNE) if no player* i *can decrease his maximum regret* $r_i(\mathbf{p})$ *by unilaterally changing his path.*

An RNE is actually a Nash equilibrium of game Γ, whose players pay for their maximum regrets. In the rest of this section, we present some preliminary facts that are useful for us to evaluate players' maximum regrets.

In order to calculate the maximum regrets $r_i(\mathbf{p}) = \sup_{\varsigma \in \Sigma} r_i(\mathbf{p}, \varsigma)$, $i \in U$, for game Γ with either type of cost, we identify some special scenarios from Σ. Recalling the aforementioned "compactness" of scenarios realizations, to each path P with at least one edge (particularly, P can be an edge), we may *associate* a *special* scenario $\varsigma^P \in \Sigma$, under which, given any routing $\mathbf{p}$, each edge $e \in P$

suffers from the highest possible delay $u_e(\mathbf{p}_e)$ and all other edges $e \notin P$ suffer from the lowest possible delays $l_e(\mathbf{p}_e)$. The following lemma confirms (for game Γ^+) the intuition that the maximum regret of a player can be observed in such a special scenario.

Lemma 2. *Given routing* $\mathbf{p} = (P_i)_{i \in U}$ *of game* Γ^+, *for every* $i \in U$ *we have*

$$r_i(\mathbf{p}) = r_i(\mathbf{p}, \varsigma^{P_i}) = d_i(\mathbf{p}, \varsigma^{P_i}) - \min_{P_i' \in \mathscr{P}_i} d_i((P_i', \mathbf{p}_{-i}), \varsigma^{P_i}).$$

Given a routing $\mathbf{p} = (P_i)_{i \in U}$ of game Γ^b and a scenario $\varsigma \in \Sigma$, an edge e on P_i is called i's *bottleneck* if e's delay $d_e(\mathbf{p}_e, \varsigma)$ equals i's bottleneck delay $b_i(\mathbf{p}, \varsigma)$.

Lemma 3 ([8,14]). *Given routing* $\mathbf{p} = (P_i)_{i \in U}$ *of game* Γ^b, *for every* $i \in U$, *there is an edge* $e_* \in P_i$ *such that* $r_i(\mathbf{p}) = r_i(\mathbf{p}, \varsigma^{e_*})$, *and* e_* *is* i*'s bottleneck under* $\mathbf{p}$ *and* ς^{e_*}.

3 Nonexistence of RNE

In this section, we present several 2-player instances of symmetric game Γ, none of which admits an RNE. Simple underlying topologies (parallel graphs, undirected cycles) or unit-length intervals are noteworthy features of the examples. Immediate corollaries of the nonexistence provide necessary conditions for graphical structures that could guarantee RNEs to exist.

3.1 A 2-player Game on a 3-edge Parallel Network

A parallel network G consists of a pair of vertices s and t connected by a set of parallel undirected edges or directed edges from s to t. On G, the games Γ^+ and Γ^b are identical. The games are symmetric in that all players have s and t as their common source and terminal, respectively.

Example 4. Consider the 2-player game Γ on the parallel network G_1 formed by three edges e_1, e_2, e_3 whose associated intervals are as shown in Table 1.

Lemma 5. *The symmetric 2-player game* Γ *on parallel network* G_1 *(directed or undirected) given in Example 4 does not admit any RNE.*

Table 1. The associated intervals $[l_{e_i}(x), u_{e_i}(x)]$ to parallel edges e_i, $i = 1, 2, 3$

	$x = 1$	$x \geq 2$
$i = 1$	$[0, 101]$	$[48, 200]$
$i = 2$	$[48, 52]$	$[51, 200]$
$i = 3$	$[51, 51]$	$[200, 200]$

This can be verified by using either Lemma 2 or Lemma 3 It has been known that game Γ^b in a directed network does not necessarily admit an RNE [14]. Example 4 extends the negative result to undirected networks. Moreover, it provides the following corollaries.

Corollary 6. *Let G_2 be the directed graph shown in Fig. 1. Then there is a 2-player symmetric game Γ on G_2 that does not admit any RNE.*

Corollary 7. *Let G be a connected directed graph. If every game Γ on G admits an RNE, then G has no subgraph isomorphic to a subdivision of G_1 or G_2 in Fig. 1.*

Corollary 8. *Let G be a connected undirected graph. If every game Γ on G admits an RNE, then every maximal connected subgraph of G that has no cut-vertex is either an edge or a cycle.*

3.2 A 2-player Game Γ^b on a 4-edge Undirected Cycle

Example 9. Consider game Γ^b is played by two players 1 and 2 in the undirected cycle formed by four edges $(s_1, s_2), (s_2, t_1), (t_1, t_2), (t_2, s_1)$, which are associated with intervals $[5, 5], [4x - 4, 6x - 5], [0, 12x - 12], [1, 7x - 6]$, respectively.

It can be shown that the above game Γ^b does not admit any RNE , which along with Corollary 8 implies the following.

Corollary 10. *Let G ba a connected undirected graph. If every game Γ^b on G admits at least one RNE, then every maximal connected subgraph of G that has no cut-vertex is either an edge or a cycle with at most three edges.*

3.3 2-player Games with Constant-Length Intervals

In this subsection, we first present a 2-player game Γ^+ (Example 11) that has no RNE . Then we modify it to be a 2-player symmetric game Γ^+ with unit-length intervals that does not admit any RNE (Corollary 13).

Example 11. Consider the game Γ^+ played by two players 1 and 2 on the acyclic directed network G depicted in Fig. 2, where the intervals for bounding edge costs are shown beside the corresponding edges. When the interval consists of a singleton, we simply write it as the singleton.

Example 12. We make modifications on the game $\Gamma^+(G)$ in Example 11 to construct a 2-player game Γ^+ on network $\tilde{G}$ which has *unit interval length.* Graph $\tilde{G} = (\tilde{V}, \tilde{E})$ is obtained from G by performing edge subdivisions as follows:

(i) For each $e \in E$ whose interval a positive integer length k (i.e., $u_e(x) = l_e(x) + k > l_e(x)$), we subdivide e into k edges $e_1, \ldots, e_k$ such they in series form a path from the tail of e to the head of e; we then associate e_1 with interval $[l_e(x), l_e(x) + 1]$ and e_i with interval $[0, 1]$ for every $i = 2, \ldots, k$.

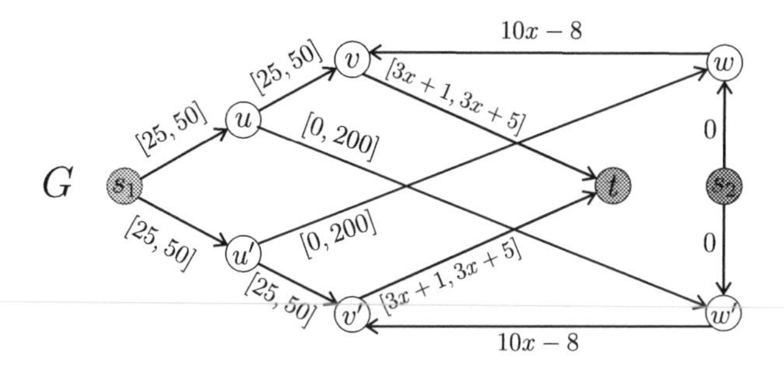

Fig. 2. A 2-player game Γ^+ without any RNE, where player $i \in \{1, 2\}$ travels from s_i to $t_i = t$.

(ii) We associate edges $(s_2, w), (s_2, w') \in \tilde{E}$ with interval $[0, 1]$.
(iii) We associate edges $(w, v), (w', v') \in \tilde{E}$ with interval $[\frac{11}{3}x^2 - \frac{8}{3}, \frac{11}{3}x^2 - \frac{5}{3}]$.

In $\Gamma^+(\tilde{G})$, two players i $(i = 1, 2)$ travel from s_i to $t_i = t$. It can be shown that this game admits no RNE. Furthermore, we can extend the nonexistence to a symmetric game, as the following corollary states.

Corollary 13. *There is a 2-player symmetric game Γ^+ with unit-length intervals that does not admit any RNE.*

Corollary 13 shows a sharp contrast to the RNE existence of the symmetric game Γ^b that is associated to intervals with uniform constant length [14].

4 Hardness of Determining RNE Existence

Building on the RNE nonexistence presented in Sect. 3, we are ready to prove the NP-hardness of determining RNE existences in game Γ^+ by reduction from the *directed* 2-*edge-disjoint-path* (2EDP) *problem*, which is NP-complete [6]. The same problem has been used in [14] to establish for similar hardness result for game Γ^b. Despite similarity in spirit of reduction, we need to come up with new ideas to get around difficulties unique for the sum-type costs. As have been seen in Sect. 3, game Γ^+ does have properties different from those of game Γ^b.

An instance of the directed 2EDP problem consists of a connected directed graph G' and four distinct vertices s', t', s'', t''. It is to determine whether G' contains a pair of edge-disjoint s'-t' path and s''-t'' path.

Theorem 14. *The problem of deciding whether a given* 3*-player symmetric game Γ^+ on a directed network has an RNE is NP-hard.*

Proof. We present a polynomial-time reduction from the directed 2EDP problem. Given an instance $(G' = (V', E'); (s', t'), (s'', t''))$ of the problem, we construct a 3-player game instance Γ^+ on network G as shown in Fig. 3.

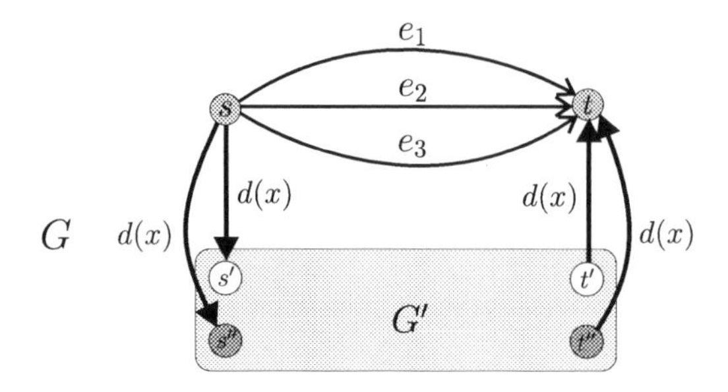

Fig. 3. The 3-player symmetric game $\Gamma^+(G)$, where three players travel from s to t.

Specifically, directed graph $G = (V, E)$ and the associated intervals are obtained as follows:

- Let parallel graph $G_1 = (\{s,t\}, \{e_1, e_2, e_3\})$ with associated intervals be as in Example 4 (see Table 1);
- Make a disjoint union of G_1 and G', and add edges $(s,s'), (s,s''), (t',t), (t'',t)$, so $V = \{s,t\} \cup V'$ and $E = \{e_1, e_2, e_3\} \cup E' \cup \{(s,s'), (s,s''), (t',t), (t'',t)\}$;
- Associate each edge $e \in E' \cup \{(s,s'), (s,s''), (t',t), (t'',t)\} = E \setminus \{e_1, e_2, e_3\}$ with the singleton interval $[l_e(x), u_e(x)] = \{d(x)\}$, where

$$d(x) = \begin{cases} \frac{1}{|E'|}, & \text{if } x = 1; \\ 600, & \text{otherwise.} \end{cases} \tag{1}$$

The player set of $\Gamma^+(G)$ consists of three players 1, 2, 3, who have common source s and common terminal t. We will show that $\Gamma^+(G)$ admits an RNE if and only if G' has edge-disjoint s'-t' path and s''-t'' path.

Let $\mathcal{P}$ denote the set of s-t paths in G that are edge-disjoint from $\{e_1, e_2, e_3\}$. Then $\mathcal{P} \cup \{e_1, e_2, e_3\} = \mathscr{P}_i$ for $i = 1, 2, 3$. Observe also that every path in $\mathcal{P}$ is either a concatenation of edge (s,s'), an s'-t' path in G', and edge (t',t), or a concatenation of edge (s,s''), an s''-t'' path in G', and edge (t'',t).

The high-level idea of our proof is simple: If $\Gamma^+(G)$ admits an RNE, then, as the 2-player game $\Gamma^+(G_1)$ in Example 4 admits no RNE (recall Lemma 5), two of the three players travel along paths in $\mathcal{P}$. In turn, the very high value 600 of delay $d(x)$ for all $x \geq 2$ (see (1)) enforces the two paths to be edge-disjoint. The s'-t' path and s''-t'' path contained in these two paths are as desired. Conversely, given a pair of edge-disjoint s'-t' path and s''-t'' path in G' whose total number of edges is minimized, we can show that the s-t paths in $\mathcal{P}$ containing them and edge e_3 form an RNE of $\Gamma^+(G)$. $\square$

Theorem 14 and its proof serve as a warm-up to help the reader understand the more technical proof (which is in the same spirit) of the following stronger hardness result.

Theorem 15. *The problem of deciding whether a given 3-player game Γ^+ has an RNE is NP-hard, even if the game is symmetric and all intervals have unit length.*

5 RNE-existence Characterization for Undirected Graphs

This section is devoted to establishing the following characterizations for undirected graphs to guarantee RNE existences for games Γ^+ and Γ^b.

Theorem 16. *Let G be a connected undirected graph. Then*

(i) every robust routing game Γ^+ on G admits an RNE if and only if every maximal connected subgraph of G that has no cut-vertex is either an edge or a cycle;
(ii) every robust routing game Γ^b on G admits an RNE if and only if every maximal connected subgraph of G that has no cut-vertex is either an edge or a cycle on 2 or 3 edges.

The "only if" part of (i) has been verified by Corollary 8. To see its"if" part, suppose G's graphical structure is as stated. If G is a tree, then nothing needs to be proved. So consider G being the edge-disjoint union of its cycles, denoted $C_1, \ldots, C_k$ (possibly none), and a forest which is spanned (induced) by the edges outside the cycles (if any). It is easy to see that the game $\Gamma^+(G)$ is decomposed into corresponding games Γ^+ on cycles $C_1, \ldots, C_k$ and the forest (if any). An RNE of $\Gamma^+(G)$ could be obtained if we could obtain an RNE for each of these smaller games and "weld" the RNEs into an RNE for the bigger game $\Gamma^+(G)$. This reduces to prove that game Γ^+ on an undirected cycle always possesses an RNE. We accomplish the task by showing that the game is a generalized ordinal potential game.

The "only if" part of (ii) has been proved by Corollary 10. By an argument similar to the above, the"if" part of (ii) follows from the RNE existence of games Γ^b on 2-edge undirected cycles (which are identical with Γ^+) and those on 3-edge undirected cycles.

6 Concluding Remarks

In this paper, we have studied, under the minimum-maximum-regret criterion, the equilibrium (referred to as RNE) existence of a robust routing game (model Γ) of unweighted atomic players in directed or undirected networks with interval uncertainty. We have obtained both negative and positive results for two sub-models Γ^+ and Γ^b, which concern with the sum- or bottle-type player costs, respectively. While the previous work [14] dealt with only Γ^b on directed networks, this paper provides a more complete picture about the general model Γ (in terms of RNE existence w.r.t. network topologies), exhibiting the similarities and differences between the two sub-models.

Our results show that, although Γ^+ turns out to be more tractable (in terms of RNE existence) than Γ^b, the class of network topologies that can guarantee RNE existence for either sub-model is not large. More conditions on interval limits, interval lengths, source-terminal locations are needed for deriving RNEs in more and larger classes of networks. The problem of deciding the RNE existence of Γ in a directed network has been shown to be NP-hard ([14] and this paper). It might be interesting to investigate the complexity of the problem when restricted to directed planar networks or undirected networks.

References

1. Aghassi, M., Bertsimas, D.: Robust game theory. Math. Program. **107**(1–2), 231–273 (2006)
2. Averbakh, I.: Minmax regret solutions for minimax optimization problems with uncertainty. Oper. Res. Lett. **27**(2), 57–65 (2000)
3. Averbakh, I., Lebedev, V.: Interval data minmax regret network optimization problems. Discret. Appl. Math. **138**(3), 289–301 (2004)
4. Banner, R., Orda, A.: Bottleneck routing games in communication networks. IEEE J. Sel. Areas Commun. **25**(6), 1173–1179 (2007)
5. Caragiannis, I., Galdi, C., Kaklamanis, C.: Network load games. In: International Symposium on Algorithms and Computation, pp. 809–818 (2005)
6. Fortune, S., Hopcroft, J.E., Wyllie, J.C.: The directed subgraph homeomorphism problem. Theor. Comput. Sci. **10**(2), 111–121 (1980)
7. Harks, T., Klimm, M., Möhring, R.H.: Strong equilibria in games with the lexicographical improvement property. Int. J. Game Theory **42**(2), 461–482 (2013)
8. Kasperski, A., Zielinski, P.: An approximation algorithm for interval data minmax regret combinatorial optimization problems. Inf. Process. Lett. **97**(5), 177–180 (2006)
9. Kouvelis, P., Gang, Y.: Robust Discrete Optimization and Its Applications. Kluwer Academic Publishers, Boston (1997)
10. Monderer, D., Shapley, L.S.: Potential games. Games Econ. Behav. **14**(1), 124–143 (1996)
11. Ordóñez, F., Stier-Moses, N.E.: Wardrop equilibria with risk-averse users. Transp. Sci. **44**(1), 63–86 (2010)
12. Piliouras, G., Nikolova, E., Shamma, J.S.: Risk sensitivity of price of anarchy under uncertainty. ACM Trans. Econ. Comput. **5**(1), 5:1–5:27 (2016). Article no 5
13. Savage, L.J.: The theory of statistical decision. Publ. Am. Stat. Assoc. **46**(253), 55–67 (1951)
14. Werth, T.L., Bttner, S., Krumke, S.O.: Robust bottleneck routing games. Networks **66**(1), 57–66 (2015)
15. Zielinski, P.: The computational complexity of the relative robust shortest path problem with interval data. Eur. J. Oper. Res. **158**(3), 570–576 (2004)

Online Trading as a Secretary Problem

Elias Koutsoupias and Philip Lazos(✉)

University of Oxford, Oxford, UK
{elias,filzos}@cs.ox.ac.uk

Abstract. We consider the online problem in which an intermediary trades identical items with a sequence of n buyers and n sellers, each of unit demand. We assume that the values of the traders are selected by an adversary and the sequence is randomly permuted. We give competitive algorithms for two objectives: welfare and gain-from-trade.

1 Introduction

We study the problem of facilitating trade between n buyers and n sellers that arrive online. We consider one of the simplest settings in which each trader, buyer or seller, is interested in trading a single item, and all items are identical. Each trader has a value for the item; a seller will sell to any price higher than its value and a buyer will buy for any price lower than its value. Upon encountering a trader, the online algorithm makes an irrevocable price offer, the trader reveals its value and, if the value is at the correct side of the offered price the item is traded. After buying an item from a seller, the online algorithm can store it indefinitely to sell it to later buyers. Of course, the online algorithm can only sell to a buyer if it has at least one item at the time of the encounter.

We consider online algorithms that offer prices based on the sequence of past values and we assume that the online algorithm knows only the number of buyers and sellers, but not their values. The values of the sellers and buyers are selected *adversarially and are randomly permuted.* In that respect, the problem is a generalization of the well-known secretary problem. The secretary problem corresponds to the special case in which there are only buyers, the algorithm starts with a single item, and the objective is to maximize the total welfare, which is to give the value to a buyer with as high value as possible.

Extending this to both sellers and buyers, creates a substantially richer setting. One of the most important differences between the two settings is that besides the objective of maximizing the total welfare, we now have the objective of maximizing the gain-from-trade. For both objectives, the algorithm must buy from sellers with low values and sell to buyers with high values. The objective is that at the end, the items end up at the hands of the traders, sellers or buyers, with the highest values. The welfare of a solution is defined as the value of the buyers and sellers that have an item. The gain-from-trade of a solution is the

Supported by the ERC Advanced Grant 321171 (ALGAME).

X. Deng (Ed.): SAGT 2018, LNCS 11059, pp. 201–212, 2018.
https://doi.org/10.1007/978-3-319-99660-8_18

difference between the welfare at the end of the process minus the welfare at the beginning. At optimality the two objectives are interchangeable: an algorithm achieves the maximum welfare if and only if it achieves the maximum gain-from-trade. But for approximate solutions, the two objectives are entirely different, with the gain-from-trade being the more demanding one.

The Bayesian version of the problem, in which the values of the buyers and sellers are drawn from known probability distributions has been extensively considered in the literature. Optimal mechanisms for bilateral trading, that is, the offline case of a single seller and a single buyer, were first analysed by Myerson and Satterthwaite in [18] and played a pivotal role in the development of the area (see the section Related Work). The online Bayesian case was considered in [10], where the values are drawn from a known distribution but the sequence is adversarially ordered.

A generalization of our model is when the items may not be identical and each buyer may have different value for each one of them, i.e., each seller has a value for its item and each buyer has a vector of values, one for every pair buyer-seller. This is also a generalization of the well-studied online maximum-matching problem [13,15]. One can cast the online maximum-matching problem as the version in which the sellers arrive first and have zero value for their item. The optimal online algorithm for this problem has competitive ratio $1/e$, when the objective is the welfare (which in the absence of seller values is identical to the gain-from-trade). Our model is incomparable to the online maximum-matching problem: it is simpler in the sense that the items are identical (a single value for each buyer instead of a vector of buyer-item values), and at the same time more complicated in that the items are not present throughout the process, but they are brought to the market by sellers that have their own utility. The fact that in our model the buyer-item values are related, allows for a much better competitive ratio regarding the welfare, (almost) 1 instead of $1/e$. More importantly, our algorithm is truthful, while in contrast, no good truthful algorithm is known for the online maximum-matching problem, which remains one of the main open problems of the area. On the other hand, the introduction of sellers poses new challenges, especially with respect to the objective of the gain-from-trade.

There are also similarities between our model and the extension of the classical secretary problem to k secretaries. From an influential result by Kleinberg [14] we know that this problem has competitive ratio $1 - 1/\sqrt{k}$ which is asymptotically tight, and can be transformed into a truthful algorithm. This result depends strongly on the knowledge of k. In our case the equivalent measure, the *number of trades* is not known from the beginning and has to be learned, with a degree of precision that is crucial, especially for the gain-from-trade objective. The fact that the gain-from-trade is not monotone as a function of time highlights the qualitative difference between the two models; the gain-from-trade temporarily *decreases* when the algorithm buys an item, with the risk of having a negative gain at the end. The mix of buyers and sellers harshly penalizes wrong decisions and the monotone structure of the problem is disrupted.

1.1 Our Results

We consider the case when both the number of buyers and the number of sellers is n. For welfare we show a competitive ratio of $1 - \tilde{O}(n^{-1/3})$, where $\tilde{O}$ hides logarithmic factors.

Our online algorithm achieves a competitive ratio of $1-\tilde{O}(n^{-1/3})$ against the optimal benchmark. To achieve this, it has a sampling phase of length $\tilde{O}(n^{2/3})$ to estimate the *median* of the values of all traders, and then uses it as a price for the remaining traders. But if the optimal number of trades is small, such a scheme will fail to achieve competitive ratio almost one, because with constant probability there will not have enough items to sell to buyers with high values. To deal with this risk, the algorithm not only samples values at the beginning but it additionally buys sufficiently many items, $\tilde{O}(n^{2/3})$, from the first sellers[1], to balance the potential loss of welfare that results from removing items from sellers to the expected loss from not having enough items for valuable buyers.

The term $O(n^{-1/3})$ in the competitive ratio seems to be optimal for a scheme that fixes the price after the sampling phase and relates to the number of items needed to approximate the median to a good degree. It may be possible to improve this term to $O(n^{-1/2})$ by a more adaptive scheme, as in the case of the k-secretary problem [14]. It may be possible to remove the logarithmic factors from the competitive ratio, but we have opted for simplicity and completeness.

For the objective of gain-from-trade, we give a truthful algorithm that has a constant competitive ratio, assuming that the algorithm starts with an item. The competitive ratio is high, approximately 10^3, but it drops to a small constant when the optimal number of trades is sufficiently high. The additional assumption of starting with an item is necessary, because without it, no online algorithm can achieve a bounded competitive ratio.

The main difficulty of designing an online algorithm for gain-from-trade is that even a single item that is left unsold at the end has dramatic effects on the gain-from-trade. The online algorithm must deal with the case of many traders, large welfare, but few optimal trades and small gain-from-trade.

To address this problem, our algorithm, unlike the case of welfare, has a large sampling phase. It uses this phase to estimate the number of optimal trades and two prices for trading with buyers and sellers. If the expected number of optimal trades is high, the algorithm uses the two prices for trading with the remaining traders. But if the number is small, it runs the secretary algorithm with the item that it starts with. Our algorithm is ordinal, in the sense that it uses only the order statistics of the values not the actual values themselves. This leaves little space for errors and it may be possible that cardinal algorithms that use the actual values can do substantially better.

All omitted proofs can be found in the full version of the paper.

[1] Buying from the first sellers cannot be done truthfully unless the algorithm knows an upper bound on their value. But this is not necessary since there is an alternative that has minor effects on the competitive ratio: the algorithm offers each seller the maximum value of the sellers so far. This is a truthful scheme that buys from all but a logarithmic number of sellers, in expectation.

2 Related Work

The bilateral trade literature was initiated by Myerson and Satterthwaite in their seminal paper [18]. They investigated the case of a single seller-buyer pair and proved their famous impossibility result: there exists no truthful, individually rational and budget balanced mechanism that also maximizes the welfare (and consequently, the gain from trade). Subsequent research studied how well these two objectives can be approximated by relaxing these conditions. Blumrosen and Mizrahi [2] devised a $1/e$-approximate, Bayesian incentive compatible mechanism for the gain from trade assuming the buyer's valuation is monotone hazard rate. Brustle et al. expanded in this direction in [3] for arbitrary valuations and downwards closed feasibility constraints over the allocations. In the case where there are multiple, unit demand, buyers and sellers, McAfee provided a weakly budget balanced, $1 - 1/k$ approximate mechanism for the gain from trade in [16], where k is the number of trades in the optimal allocation. This was later extended to be strongly budget balanced by Segal-Halevi et al. in [19]. McAfee also proved a simple 2-approximation to the gain from trade if the buyer's median valuation is above the seller's [17]. This was significantly improved by Colini-Baldeschi et al. in [5] to $1/r$ and $O(\log(1/r))$, where r is the probability that the buyer's valuation for the item is higher than the seller's. Recently, Giannakopoulos et al. [10] studied an online variant of this setting where buyers and sellers are revealed sequentially by an adversary and have known prior distributions on the value of the items.

The random order model we are using has its origins in the secretary problem, where n items arrive in online fashion and our goal is to maximize the probability of selecting the most valuable, without knowing their values in advance. The matroid secretary problem was introduced by Babaioff et al. [1], with many recent results presented by Dinitz in [7]. Of particular interest to our problem are secretary problems on bipartite graphs. Here, the left hand side vertices of the graph are fixed and the right hand side vertices (along with their incident edges) appear online. The selected edges must form a (incomplete) matching and the goal is to maximize the sum of their weights. Babaioff et al. in [1] provided a $4d$-competitive algorithm for the transversal matroid with bounded left degree d, which is a special case of the online bipartite matching where all edges connected to the same left hand side vertex have equal value. This was later improved to 16 by Dimitrov and Plaxton [6]. The case where all edges have unrelated weights was first considered by Korula and Pal in [15] who designed a 8-competitive algorithm, which was later improved to the optimal $1/e$ by Kesselheim et al. [13]. Another secretary variant which is close to our work is when the online selects k items instead of one, where Kleinberg [14] showed an asymptotically tight algorithm with competitive ratio $1 - O(\sqrt{1/k})$.

The wide range of applications of secretary models (and the related prophet inequalities) have led to the design of posted price mechanisms, that are simple, robust, truthful and achieve surprisingly good approximation ratios. Hajiaghayi et al. introduced prophet inequality techniques in online auctions in [12]. The k-choice secretary described above was then studied in [11] which combined with

[14] yielded an asymptotically optimal, truthful mechanism. For more general auction settings, posted-price mechanisms have been used by Chawla et al. in [4] for unit demand agents and expanded by Feldman et al. in [9] for combinatorial auctions and [8] for online budgeted settings.

3 Model and Notation

The setting of the *random intermediation* problem consists of two sets $B = \{b_1, \ldots, b_n\}$ and $S = \{s_1, \ldots, s_n\}$ containing the valuations of the buyers and sellers. For convenience, we assume that they are all distinct. The intermediary interacts with a uniformly random permutation σ of $S \cup B$ which is presented one agent at a time, over $2n$ steps. The intermediary has no knowledge of $\sigma(t)$ before step t.

We study *posted price* mechanisms that upon seeing the identity of agent t offer price p_t. This price cannot depend on the entire valuation function; only the values within $\sigma(1) \ldots \sigma(t-1)$ which are revealed at this point. We buy or sell one item from sellers or buyers who accept our price, respectively. Of course, we can only sell items if we have stock available.

The set of sellers from whom we bought items during the algorithm's execution is $T_S = \{s \in S \mid \exists t\ \sigma(t) = s \leq p_t\}$ and similarly the set of buyers we sold to is $T_B = \{b \in B \mid \exists t\ \sigma(t) = b \geq p_t \wedge \text{we have items available at time } t\}$. Notice that these sets are random variables, depending on σ.

The social *welfare* of online algorithm A is the sum of the valuations of all agents *with* items: $\mathcal{W}_A(S, B) = \mathbb{E}\left[\sum_{s \in S \setminus T_S} s + \sum_{b \in T_B} b\right]$. The gain from trade (or GFT) produced by algorithm A throughout the run is the difference between the final and starting welfare: $GFT_A(S, B) = \mathbb{E}\left[\sum_{b \in T_B} b - \sum_{s \in T_S} s\right]$.

We are interested in the *competitive ratio* of our online algorithm A compared to the offline algorithm OPT. In this setting there are two different offline algorithms to compare against: optimal offline and sequential offline. They both know S, B, but the first can always achieve the maximum welfare, whereas the second operates under the same constrains as we, namely he can only perform trades permitted by σ, which is unknown. We say that algorithm A is ρ-competitive for welfare (or gain from trade) if for any S, B we have:

$$\mathcal{W}_A(S, B) \geq \rho \cdot \mathcal{W}_{\mathrm{OPT}}(S, B) - \alpha, \tag{1}$$

for some fixed $\alpha \geq 0$.

Often we will refer to the *matching* between a set of buyers and a set of sellers. Let $M(S, B) = \{\{S_1\} \cup \{B_1\}\}$, where $S_1 \subseteq S, B_1 \subseteq B$ is the set of sellers and buyers with whom we trade (or are matched, in the sense that the items move from sellers to buyers) in a welfare maximizing allocation and $m(S, B)$ the optimal gain from trade. Note that this does *not* contain pairs: only the set of each side of the matching. Similarly, let $M(S, B, q, p)$ be the matching generated by only trading with sellers valued below q and buyers above p. In a slight abuse of notation, we will use $|M(S, B)| = |S_1|$ for the size of the matching and $M(S, B) \cap M(S', B') = \{\{S_1 \cap S'_1\} \cup \{B_1 \cap B'_1\}\}$.

4 Welfare

In order to approximate the welfare, the online algorithm uses a sampling phase to approximate the median price. The two main challenges are estimating the median with a small sample and missing trades due to the online nature of the input. We begin with two probability concentration results, similar to the Azuma-Hoeffding inequality for the case *without* replacement.

Lemma 1. *Let $\mathcal{X} = \{x_1, \cdots, x_N\}$ where $x_i \in \{0,1\}$ and $x_1 = x_2 = \ldots = x_m = 1$. Consider sampling n values of $\mathcal{X}$ uniformly at random* **without** *replacement and let X_i be the value of the $i-th$ draw. For $Y = \sum_{i=1}^{n} X_i$, we have that for any $\epsilon > 0$:*

$$\Pr[|Y - \mathbb{E}[Y]| \geq \epsilon \mathbb{E}[Y]] \leq e^{-2\epsilon^2 \max\{m,n\} \frac{mn}{N^2}} \tag{2}$$

Similarly, we often encounter a situation where we are interested in the number of trades between n sellers and n buyers, arriving in a uniformly random permutation. Assuming we buy from all sellers, occasionally we would encounter a buyer without having any items at hand. This results shows that even though this is the case, few trades are lost.

Lemma 2. *The number of trades $M(\sigma)$ in a uniformly random sequence containing n buyers and n sellers satisfies:*

$$\mathbb{E}[M(\sigma)] \geq \frac{n-1}{n}\left(n - \sqrt{2n \log n}\right), \tag{3}$$

assuming all sellers are valued below all buyers.

All the machinery is now in place to analyze sequential algorithms in this setting. We first show a key property of the offline algorithm.

Proposition 1. *The optimal offline algorithm sets a price p, equal to the median of all the agents' valuations and trades items from sellers valued below p to buyers valued above p.*

The optimal sequential offline algorithm would not just trade at this price. We can modify this approach with a bias towards buying more items than needed, in order to maximize the probability of finding high valued buyers.

Lemma 3. *The sequential optimal offline is $\left(1 - O\left(\frac{\log n}{n^{1/3}}\right)\right)$-competitive against the optimal offline algorithm for welfare.*

The next step is to design an online algorithm without knowing p or $|M(S,B)|$ beforehand. The algorithm is as follows:

1. Record the first $8n^{2/3} \log n$ agents and calculate their median p'. Buy from all sellers during this sampling phase.
2. After the sampling phase:
   (a) Buy from seller s if $s \leq p'$.
   (b) Sell to buyer b is an item is available and $b \geq p'$.

For the analysis of this algorithm, we first need a concentration result on the sample median p'.

Lemma 4. *Let $X = \{1, \ldots, 2n\}$ and select $8n^{2/3} \log n$ elements from X* without *replacement. Then, their sample median M satisfies:*

$$\Pr[|M - n| \geq n^{2/3}] \leq O\left(\frac{1}{n}\right). \tag{4}$$

This shows that our sample median p' might have at most $n^{2/3}$ agents more on one side compared to the true median p. However, this loss is negligible asymptotically. We now show that buying from sellers during the sampling phase, before considering any buyers, can only increase the number of trades.

Lemma 5. *Let σ be a sequence containing n buyers and n sellers and $s\sigma'$ the exact same sequence where a seller has been moved to the front. Then we have $M(s\sigma') \geq M(\sigma)$, where $M(\sigma)$ is the number of items sold.*

Theorem 1. *This algorithm is $\left(1 - \tilde{O}\left(\frac{1}{n^{1/3}}\right)\right)$-competitive for welfare against the optimal offline.*

Proof. As before, let $M = |M(S, B)|$ be the size of the optimal offline matching. The following analysis assumes that the event of Lemma 4 did not occur and p and p' split the agents in two sets, differing by at most $n^{2/3}$. Given this, we analyze the algorithm in three steps. First show that we never buy too many items from highly valued sellers, therefore we keep most of the sellers' contribution to the final welfare. Then we show that we always match a high proportion of the valuable buyers by considering two cases: if there are few such buyers then they are matched to the sellers we obtained during the sampling phase, otherwise we have enough sellers below p' to match them to.

We introduce some notation useful to the analysis: let W be the set containing the top $n - n^{2/3}$ highest valued agents. Then let S_W, B_W be the number of sellers and buyers respectively in W and S'_W, B'_W be how many of them appeared after the sampling phase.

To show the competitiveness of our algorithm, it suffices find the fraction of W that is achieved at the end of the sequence: being $(1 - \tilde{O}(1/n^{1/3}))$-competitive against W, the top $n - n^{2/3}$ agents, implies a ratio of

$$\left(1 - \tilde{O}(1/n^{1/3})\right) \cdot \frac{n - n^{2/3}}{n} = 1 - \tilde{O}(1/n^{1/3})$$

against all n agents above the median and therefore the optimal offline.

We first show that we never lose too much welfare by buying from sellers. Given p', the only occasion on which a seller in W is bought is if he is amongst the first $8n^{2/3} \log n$ sellers. This event is clearly independent from the condition on p', meaning in expectation we keep

$$\mathbb{E}\left[S'_W\right] = S_W \left(1 - \frac{8n^{2/3} \log n}{n}\right) = S_W \left(1 - \frac{8 \log n}{n^{1/3}}\right)$$

highly valued sellers. Therefore, enough of the sellers' original value is kept. For the number of items I_S bought during the sampling phase, the following holds by Lemma 1:

$$\Pr\left[I_S \leq (1 - \frac{1}{2})4n^{2/3}\log n\right] \leq e^{-2\frac{1}{4}8n^{2/3}\log n \frac{n^2}{4n^2}} \leq e^{-n^{2/3}}. \tag{5}$$

To analyze the number of buyers in W matched, we consider two cases.
$\boldsymbol{B_W \leq n^{2/3}\log n}$: We first need to find $\mathbb{E}\left[B'_W\right]$, which is slightly more complicated, since we have conditioned on p' approximating the median. Given p', at least $4n^{2/3}\log n$ agents were above the median value during the sampling phase. Note that any of the buyers in W is above the median. Therefore, any of the agents in the upper $4n^{2/3}\log n$ half of the sampling phase could be replaced by a buyer in W. Thus, in a random permutation, we have:

$$\mathbb{E}\left[B'_W\right] \geq B_W\left(1 - \frac{4n^{2/3}\log n}{n - n^{2/3}}\right).$$

We might also consider up to $n^{2/3}$ extra buyers, if p' underestimated p. However, given that $I_S \geq 2n^{2/3}\log n$ with high probability, every buyer in B'_W will be matched with an item, giving the claimed competitive ratio for this case.

$\boldsymbol{B_W > n^{2/3}\log n}$: The last case is similar but somewhat more complicated and can be found in the full version.

5 Gain from Trade

Compared to the welfare, the gain from trade is a more challenging objective. Even for large n, the actual trades that maximize the GFT can be very few and quite well hidden. Buying from a single seller and being unable to sell could completely shatter the GFT, while it could have very little effect on the welfare.

The setting has to be slightly changed. We give the online (and offline) algorithm one extra, free item at the beginning to ensure that at least one buyer can acquire an item. This modification is necessary.

Theorem 2. *Starting with no items, the competitive ratio for the GFT is arbitrarily high.*

The algorithm starts by estimating the total volume of trades in an optimal matching by observing the first segment of the sequence. Using this information, two prices $\hat{p}, \hat{q}$ are computed, to be offered to agents in the second part. Being an ordinal mechanism, the goal is to maximize the number of trades *and leave no item unsold.* The online algorithm $A(c, \epsilon, N)$ contains parameters whose values will be specified later.

With probability $\frac{1}{2}$ ignore sellers and sell the item as in the normal secretary, otherwise continue ;
Split the sequence into two segments such that $\sigma = \sigma_1\sigma_2$, with $|\sigma_1| = c \cdot 2n$;
Let S_1, B_1 denote the sets of sellers and buyers of σ_1;
Calculate the welfare maximizing matching $M(S_1, B_1)$;
if $|M(S_1, B_1)| \leq N$ **then**
    Sell the item to the highest buyer as in the normal secretary problem and stop;
end
Set $\hat{p}, \hat{q}$ which only keeps $(1-\epsilon) \cdot c \cdot |M(S_1, B_1)|$ many matched pairs;
$i \leftarrow c \cdot 2n$, $k \leftarrow \emptyset$, $M \leftarrow \emptyset$;
`/* For the first half of` σ_2`, buy and sell items, keeping at most one in stock */`
while $i \leq c \cdot 2n + (1-c) \cdot 2n/2$ **do**
    if *$\sigma(i)$ is a seller, $k = \emptyset$ and $\sigma(i) \leq \hat{q}$* **then**
        $k \leftarrow \sigma(i)$;
    end
    if *$\sigma(i)$ is a buyer, $k \neq \emptyset$ and $\sigma(i) \geq \hat{p}$* **then**
        Sell to $\sigma(i)$;
        $k \leftarrow \emptyset$;
    end
    $i \leftarrow i + 1$;
end
For the second half of σ_2, just try to sell the last remaining item, if any;

The idea is to use the first part of the sequence to estimate the matching $M(S, B)$. If a large (in terms of pairs) GFT maximizing matching is observed, it is likely that a proportionate fraction of it will be contained in the second half. In that case, sellers and buyers are matched in non overlapping pairs to avoid buying too many items. Before moving on to the analysis of the algorithm, we need a small lemma on the structure of welfare maximizing matchings, to explain the prices set.

Lemma 6. *For any S, B and $S_1 \subseteq S, B_1 \subseteq B$:*

1. *$m(S, B)$ can be obtained by setting two threshold prices p, q and trading with buyers above and sellers below them.*
2. *Choosing $\hat{p} > p$ and $\hat{q} < q$ such that $|M(S, B, \hat{q}, \hat{p})| \geq \alpha|M(S, B)|$ for $\alpha < 1$ yields $m(S, B, \hat{q}, \hat{p}) \geq \alpha m(S, B)$.*
3. *$|M(S, B)| \geq |M(S_1, B_1)|$ and $m(S, B) \geq m(S_1, B_1)$.*

Theorem 3. *$A(c = 0.3, \epsilon = 0.2758, N = 114)$ is $O(1)$-competitive for the gain from trade against the optimal offline.*

Proof. Let $z = |M(S, B)|$. We bound the gain from trade for the case where σ_1, σ_2 contain their analogous proportion of $M(S, B)$ and show that the losses are insignificant otherwise. Let $S_M = \{s \in M(S, B)\}$ and $B_M = \{b \in M(S, B)\}$, the sets of agents comprising the optimal matching. By Lemma 6, we know that any seller in S_M can be matched to any buyer in B_M. Since we only care about the *size* of the matching in σ_1 and σ_2, not its actual value, we define $f(c, \epsilon, z)$ as:

$$f(c, \epsilon, z) = \Pr\left[\frac{|S_M \cap S_1|}{|S_M|} \geq c(1-\epsilon) \wedge \frac{|B_M \cap B_1|}{|B_M|} \geq c(1-\epsilon) \wedge \right. \tag{6}$$

$$\left. \frac{|S_M \cap S_2|}{|S_M|} \geq (1-c)(1-\epsilon) \wedge \frac{|B_M \cap B_2|}{|B_M|} \geq (1-c)(1-\epsilon)\right]. \tag{7}$$

We call this the *well mixed* probability, where an ϵ-approximate chunk of the matching appears in both parts. The two events are not independent.

It is useful to think the input as being created in two steps: first the *volume* of agents in S_1, B_1, S_2, B_2 is chosen and *afterwards* their exact values are randomly assigned. As such, a lower bound on the size of the offline matching provides the same bound on its gain from trade. We begin by bounding $f(c, \epsilon, z)$.

Lemma 7. *The probability the matching is well-mixed is*

$$f(c, \epsilon, z) \geq 1 - 2(e^{-2\epsilon^2 zc^2} + e^{-2\epsilon^2 z(1-c)^2})$$

Let p and q be the prices achieving the matching $M(S, B)$, by Lemma 6. We need to show that the prices $\hat{p}, \hat{q}$ computed achieve a constant approximation of $m(S_2, B_2)$. Since $M(S, B)$ is well mixed and by using Lemma 6 we have that:

$$|M(S, B)| \geq |M(S_1, B_1)| \geq |M(S_1, B_1, q, p)| \geq (1-\epsilon) \cdot c \cdot |M(S, B)|, \tag{8}$$

where the second inequality holds since $M(S_1, B_1)$ is a gain from trade maximizing matching and the third because at least a $(1-\epsilon) \cdot c$ fraction of $M(S, B)$ appeared in σ_1. In particular, we have that $M(S_1, B_1, q, p) \subseteq M(S_1, B_1)$ is the highest value part of $M(S_1, B_1)$ and $M(S_1, B_1, \hat{q}, \hat{p}) \subseteq M(S_1, B_1, q, p)$, thus $\hat{q} \leq q$ and $\hat{p} \geq p$ leading to:

$$|M(S_1, B_1, \hat{q}, \hat{p})| \geq (1-\epsilon)^2 c^2 |M(S, B)| \tag{9}$$

by 8. We need to find how many of the trades in $M(S_2, B_2, \hat{p}, \hat{q})$ are achieved by our algorithm. Let $\hat{S}_2 = \{s \mid s \in S_2 \wedge s < \hat{q}\}$ and $\hat{B}_2 = \{b \mid b \in B_2 \wedge b > \hat{p}\}$.

Lemma 8. *Assuming the matching is well mixed:*

$$\Pr\left[|\hat{S}_2| \geq ((1-c)(1-\epsilon) - \frac{1}{2})|S_M|\right] \geq 1 - 2^{-c^2(1-\epsilon)^2|S_M|}.$$

Clearly, Lemma 8 holds for buyers as well. The proof is almost identical, keeping in mind that buyers are ordered the opposite way.

Since this is an ordinal mechanism, we want to maximize the number of trades *provided no item is left unsold* as even a single unsold item ruins our gain from trade guarantee.

Lemma 9. *Let $x = |M(S_2, B_2, \hat{q}, \hat{p})|$ and $y = |S_2|+|B_2|-x$. Then, the probability that no item is left unsold is at least $1 - 2^{-x}$. Moreover, the expected number of trades in this case is:*

$$\frac{\frac{x+y/2-1}{2x+y-1} \cdot \frac{x}{2} - \frac{x}{2^x}}{1 - 2^{-x}} \approx x/4 \tag{10}$$

Everything is now in place to provide a lower bound on the gain from trade of the matching calculated by the algorithm. Assuming $z = |M(S, B)|$, we combine Lemmas 7, 8 and 9 to show that with probability at least

$$J(c, \epsilon, z) = f(c, \epsilon, z) \cdot (1 - 2^{-c^2(1-\epsilon)^2 z}) \cdot (1 - 2^{-((1-c)(1-\epsilon)-\frac{1}{2})z}) \tag{11}$$

the matching has size at least $((1 - c)(1 - \epsilon) - \frac{1}{2})z/4$. The matching is not a uniformly random subset of $M(S, B)$, but it is skewed to contain higher value trades since $\hat{p} > p$ and $\hat{q} < q$. Taking into account that we run a simple secretary algorithm with probability 1/2 and assuming we lose the highest valued seller $s^\star$ in our matching when the agents are not well mixed, the GFT is:

$$\frac{1}{2e}b^\star + \frac{J(c, \epsilon, z)}{2} \cdot \frac{((1 - c)(1 - \epsilon) - \frac{1}{2})m(S, B)}{4} - \frac{1 - J(c, \epsilon, z)}{2}s^\star \tag{12}$$

whereas the optimal offline GFT is at most $m(S, B) + b^\star$, where $b^\star$ is the most valuable buyer.

Therefore, c, ϵ and N are selected to maximize the minimum amongst all cases of z, which is picked by the adversary. For the full proof please refer to the full version of the paper. Computationally, we find that setting $c = 0.3, \epsilon = 0.2758$ and $N = 114$ yields $\rho \geq 1/1434$.

If we are given that $|M(S, B)|$ will be large, then this algorithm can be adapted to have greatly improved competitive ratio. In particular, setting $c = \epsilon = 0.01$ achieves $\rho \geq 1/17$ as $|M(S, B)| \to \infty$.

References

1. Babaioff, M., Immorlica, N., Kleinberg, R.: Matroids, secretary problems, and online mechanisms. In: Proceedings of the Eighteenth Annual ACM-SIAM Symposium on Discrete Algorithms, SODA '07, pp. 434–443. Society for Industrial and Applied Mathematics, Philadelphia, PA, USA (2007)
2. Blumrosen, L., Mizrahi, Y.: Approximating gains-from-trade in bilateral trading. In: Cai, Y., Vetta, A. (eds.) WINE 2016. LNCS, vol. 10123, pp. 400–413. Springer, Heidelberg (2016). https://doi.org/10.1007/978-3-662-54110-4_28
3. Brustle, J., Cai, Y., Wu, F., Zhao, M.: Approximating gains from trade in two-sided markets via simple mechanisms. In: Proceedings of the 2017 ACM Conference on Economics and Computation, pp. 589–590. ACM (2017)
4. Chawla, S., Hartline, J.D., Malec, D.L., Sivan, B.: Multi-parameter mechanism design and sequential posted pricing. In: Proceedings of the Forty-Second ACM Symposium on Theory of Computing, pp. 311–320. ACM (2010)

5. Colini-Baldeschi, R., Goldberg, P., de Keijzer, B., Leonardi, S., Turchetta, S.: Fixed price approximability of the optimal gain from trade. In: Devanur, N.R., Lu, P. (eds.) WINE 2017. LNCS, vol. 10660, pp. 146–160. Springer, Cham (2017). https://doi.org/10.1007/978-3-319-71924-5_11
6. Dimitrov, N.B., Plaxton, C.G.: Competitive weighted matching in transversal matroids. Algorithmica **62**(1), 333–348 (2012)
7. Dinitz, M.: Recent advances on the matroid secretary problem. SIGACT News **44**(2), 126–142 (2013)
8. Eden, A., Feldman, M., Vardi, A.: Online random sampling for budgeted settings. In: Bilò, V., Flammini, M. (eds.) SAGT 2017. LNCS, vol. 10504, pp. 29–40. Springer, Cham (2017). https://doi.org/10.1007/978-3-319-66700-3_3
9. Feldman, M., Gravin, N., Lucier, B.: Combinatorial auctions via posted prices. In: Proceedings of the Twenty-sixth Annual ACM-SIAM Symposium on Discrete Algorithms, SODA '15, pp. 123–135. Society for Industrial and Applied Mathematics, Philadelphia, PA, USA (2015)
10. Giannakopoulos, Y., Koutsoupias, E., Lazos, P.: Online market intermediation. In: Proceedings of the 44th International Colloquium on Automata, Languages, and Programming, ICALP '17, (2017). Full version in CoRR: abs/1703.09279
11. Hajiaghayi, M.T., Kleinberg, R., Parkes, D.C.: Adaptive limited-supply online auctions. In: Proceedings of the 5th ACM conference on Electronic commerce, pp. 71–80. ACM (2004)
12. Hajiaghayi, M.T., Kleinberg, R., Sandholm, T.: Automated online mechanism design and prophet inequalities. In: AAAI, vol. 7, pp. 58–65 (2007)
13. Kesselheim, T., Radke, K., Tönnis, A., Vöcking, B.: An optimal online algorithm for weighted bipartite matching and extensions to combinatorial auctions. In: Bodlaender, H.L., Italiano, G.F. (eds.) ESA 2013. LNCS, vol. 8125, pp. 589–600. Springer, Heidelberg (2013). https://doi.org/10.1007/978-3-642-40450-4_50
14. Robert Kleinberg. A multiple-choice secretary algorithm with applications to online auctions. In: Proceedings of the Sixteenth Annual ACM-SIAM Symposium on Discrete Algorithms, SODA '05, pp. 630–631. Society for Industrial and Applied Mathematics, Philadelphia, PA, USA (2005)
15. Korula, N., Pál, M.: Algorithms for secretary problems on graphs and hypergraphs. In: Albers, S., Marchetti-Spaccamela, A., Matias, Y., Nikoletseas, S., Thomas, W. (eds.) ICALP 2009. LNCS, vol. 5556, pp. 508–520. Springer, Heidelberg (2009). https://doi.org/10.1007/978-3-642-02930-1_42
16. McAfee, P.R.: A dominant strategy double auction. J. Econ. Theory **56**(2), 434–450 (1992)
17. McAfee, P.R.: The gains from trade under fixed price mechanisms. Appl. Econ. Res. Bull. **1**(1), 1–10 (2008)
18. Myerson, R.B., Satterthwaite, M.A.: Efficient mechanisms for bilateral trading. J. Econ. Theory **29**(2), 265–281 (1983)
19. Segal-Halevi, E., Hassidim, A., Aumann, Y.: SBBA: a strongly-budget-balanced double-auction mechanism. In: Gairing, M., Savani, R. (eds.) SAGT 2016. LNCS, vol. 9928, pp. 260–272. Springer, Heidelberg (2016). https://doi.org/10.1007/978-3-662-53354-3_21

Constrained Swap Dynamics over a Social Network in Distributed Resource Reallocation

Abdallah Saffidine[1] and Anaëlle Wilczynski[2(✉)]

[1] College of Engineering & Computer Science, Australian National University, Canberra, Australia
abdallah.saffidine@gmail.com
[2] PSL, CNRS, LAMSADE, Université Paris-Dauphine, Paris, France
anaelle.wilczynski@dauphine.fr

Abstract. We examine a resource allocation problem where each agent is to be assigned exactly one object. Agents are initially endowed with a resource that they can swap with one another. However, not all exchanges are plausible: we represent required connections between agents with a social network. Agents may only perform pairwise exchanges with their neighbors and only if it brings them preferred objects. We analyze this distributed process through two dual questions. *Could an agent obtain a certain object if the swaps occurred favourably? Can an agent be guaranteed a certain level of satisfaction regardless of the actual exchanges?* These questions are investigated through parameterized complexity, focusing on budget constraints such as the number of exchanges an agent may be involved in or the total duration of the process.

Keywords: Resource allocation · Distributed process
Social network · Parameterized complexity

1 Introduction

Reallocating resources among agents is a central question widely studied both in computer science and economics [1,3,10,14]. This problem refers to a particular setting of resource allocation, where the agents are initially endowed with items [6]. Resource reallocation models many real-life situations, like reallocating tasks between employees or reassigning time slots in schedules. In such examples, agents are often assigned a single task. With exactly one item per agent, the problem is known as *housing market* [1,19]. In this context, a central authority may decide how to redistribute the objects [4,19]. Alternatively, the agents may direct the reallocation by trading and negotiating among them in a distributed process [7–9]. Although largely studied in general resource reallocation [11,12,14,18], this approach has only recently been introduced in housing markets [10].

X. Deng (Ed.): SAGT 2018, LNCS 11059, pp. 213–225, 2018.
https://doi.org/10.1007/978-3-319-99660-8_19

In a distributed process, natural obstacles may inhibit the agents in the trades. Lack of trust may lead agents to adopt a greedy behavior so as to be immediately better off in their new acquisition. Logistics difficulties, e.g., communication and geographical distance, may also prevent some trades to occur. This can be modeled by restricting trades to the links of a social network [13, 17], with exchanges limited to the cliques [9] or edges [16] of the graph.

We consider this latter format: a housing market with exchanges between neighbors in a social network [16]. One of the main questions is the Reachable Object (RO) problem: Given a target agent A and a target object x, is there a sequence of exchanges ensuring A is eventually allocated x? This problem is very appealing but is NP-complete even when the network is a tree [16]. We attempt to mitigate this negative result by looking at more realistic constrained settings.

We draw inspiration from the fact that an agent may not be willing to perform a large number of swaps or to wait a long time before getting the desired object. We introduce natural budget constraints, the number of exchanges agents may make and the total duration of the process, and we perform a refined complexity analysis. Moreover, we introduce Guaranteed Level of Satisfaction (GLS), a problem related to RO but more realistic. GLS asks whether an agent can be guaranteed to be eventually allocated an item at least as good as the input target item, regardless of the exchanges other agents perform, provided they are rational swaps. While RO takes an optimistic perspective, GLS adopts a more pessimistic point of view beyond "lucky" exchange sequences. These problems naturally arise when we analyze the distributed process of exchanges in reallocation but they can also model concrete issues. As an example, consider an online exchange platform where users input in the system which item they hold as well as their preference. A user may request a target object to the centralized system which would then suggest a series of intermediate swaps to bring it to her. Even in such a context, restricted rational exchanges are relevant: geographical constraints can still prevent two agents to trade and the guarantee of getting a better object is essential as otherwise an agent could be left worse off than she started, should an intermediate agent exit the system during the process.

When parameterizing the problems by the maximal number of swaps per agent, we show intractability even for highly structured graphs. However, when constraining the duration of the process, we obtain more promising results: RO and GLS are tractable in a very relevant class of networks, namely bounded degree graphs, and in general the problems are tractable when the duration does not depend on the input size. These results contrast strongly with both previous work [16] and our first parameterization, and they focus on realistic scenarios: actual social networks have indeed bounded degrees and the time that an agent is willing to wait for a target object is independent from the input size.

We start with the swap dynamics model, RO and GLS, and some complexity background. Section 3 relates RO and GLS. Our results for the max-swaps and for the total-duration parameters are in Sects. 4 and 5 respectively.

2 Formal Framework

2.1 Swap Dynamics Model

Let N be a set of n agents, and M a set of n resources (or objects). Each agent $i \in N$ has ordinal strict preferences $\succ_i$ over the objects, i.e., $a \succ_i b$ means that agent i prefers object a to object b. An allocation σ is a bijection $\sigma : N \rightarrow M$, assigning to each agent exactly one object. The object assigned to agent i in σ is denoted by σ_i. Each agent is initially endowed with an object, and we denote by σ^0 the initial allocation. The agents are embedded in a social network, represented by an undirected graph $G = (N, E)$. An instance of swap dynamics model is a tuple $\mathcal{I} = (N, M, \succ, \sigma^0, G)$.

Agents can exchange their objects so as to obtain better objects, but not all exchanges are plausible. The possibilities depend on the social network and on the preferences of the agents. We only admit *swaps*, rational trades between neighbors. Formally, a swap in an allocation σ is a trade between two adjacent agents $(i, j) \in E$ such that the exchange is rational, i.e., $\sigma_i \succ_j \sigma_j$ and $\sigma_j \succ_i \sigma_i$.

A sequence of swaps is a sequence of allocations $(\sigma^0, \ldots, \sigma^t)$ such that a swap is performed between two consecutive allocations σ^i and σ^{i+1}. An allocation σ is *reachable* if there is a sequence of swaps leading to it, i.e., there exists a sequence $(\sigma^0, \ldots, \sigma^t)$ such that $\sigma^t = \sigma$. An allocation σ is *stable* if no swap is possible from σ. An object x is reachable for agent i if there is a sequence of swaps $(\sigma^0, \ldots, \sigma^t)$ where $\sigma_i^t = x$. *Swap dynamics* refers to a distributed process where agents may rationally exchange their objects when they are neighbors in the network, until a stable allocation is reached.

Example 1. Consider an instance where $n = 4$ with the following social network and preferences. The framed objects represent the initial object of each agent.

$A : b \succ c \succ \boxed{a} \succ d$ $\quad$ $C : d \succ a \succ b \succ \boxed{c}$

$B : c \succ a \succ \boxed{b} \succ d$ $\quad$ $D : a \succ b \succ \boxed{d} \succ c$

Initially, only the swaps between agents A and B, and B and C are possible. The rational swap (A, C) is not possible because the agents are not adjacent. The swap (A, D) is not possible because it is not rational for A. The sequence of exchanges (A, B), (B, C), and (C, D) gives rise to a reachable allocation where every agent gets her best object. This is stable: no further swap can be performed.

2.2 Questions

We investigate swap dynamics by analyzing two natural decision problems.

Reachable Object (RO):
Instance: $\mathcal{I} = (N, M, \succ, \sigma^0, G)$, $A \in N$, $x \in M$.
Question: Is there a sequence of swaps $(\sigma^0, \ldots, \sigma^t)$ such that $\sigma_A^t = x$?

Guaranteed Level of Satisfaction (GLS):
Instance: $\mathcal{I} = (N, M, \succ, \sigma^0, G)$, $A \in N$, $y \in M$.
Question: *For all* sequences of swaps $(\sigma^0, \ldots, \sigma^t)$ where σ^t is stable, does it hold that either $\sigma^t_A = y$ or $\sigma^t_A \succ_A y$?

When asking whether an agent can obtain some object, a large number of swaps may not be realistic. We thus study three variants of RO and GLS, referred to as {RO/GLS}-{max/sum/makespan}, where the quantity of swaps in a solution sequence is limited. In each variant, this quantity is measured differently, leading to different complexity-theoretic characterizations of the problem.

- *max*: Every agent is involved in no more than k swaps.
- *sum*: The total length of the sequence is no more than k.
- *makespan*: The makespan of the sequence is no more than k.

The *makespan* of a sequence of swaps is the minimum time that elapses from the beginning to the end, when we allow parallel swaps. This notion can be formalized as follows. Let $s = (\sigma^0, \ldots, \sigma^t)$ be a sequence of swaps. A parallel decomposition of s is a tuple of integers $\ell = (\ell_0, \ell_1, \ldots, \ell_m)$ of length $|\ell| = m$, such that $0 = \ell_0 < \ell_1 < \cdots < \ell_m = t$, and for all $0 \leq i < m$ the swaps between allocation σ^{ℓ_i} and allocation $\sigma^{\ell_{i+1}}$ do not involve the same agents. In other words, the swaps between σ^{ℓ_i} and $\sigma^{\ell_{i+1}}$ can be performed simultaneously. The makespan is the length m of the shortest parallel decomposition. Observe that the makespan of a sequence can be computed in linear time, and that the *sum* parameter is a worst case bound in case no parallel swaps take place.

2.3 Parameterized Complexity

Parameterized complexity aims at solving hard problems in time $f(k)n^{O(1)}$, called FPT time (fixed-parameter tractable), where n is the size of the instance, f is a computable function, and k is a *parameter* of the problem. Assuming the problem we are trying to solve is NP-hard, function f has to be superpolynomial, unless P $=$ NP. However, if our parameter k is *small* compared to the size of the instance n, we achieve that the blow-up is limited to the small value k.

Some problems are highly suspected not to admit any algorithm in time $f(k)n^{\mathcal{O}(1)}$ for any computable function f, and thus not to be FPT. There are hierarchies of complexity classes beyond FPT : W[1] $\subseteq$ W[2] $\subseteq \cdots \subseteq$ W[SAT] $\subseteq \ldots$, and A[1] $\subseteq$ A[2] $\subseteq \cdots \subseteq$ AW[SAT] $\subseteq \ldots \subseteq$ XP, where W[1] $=$ A[1], W[t] $\subseteq$ A[t] for any t > 1, and W[SAT] $\subseteq$ AW[SAT]. For instance, W[1] is the class of parameterized decision problems that can be solved by a nondeterministic single-tape Turing machine within k steps, and XP is the class of decision problems solvable in time $\mathcal{O}(n^{f(k)})$ for some computable function f.

As a rudimentary informal intuition, FPT and W[1] can be thought of as corresponding to P and NP in the parameterized world. For instance, CLIQUE, the problem of finding a clique of size k in a graph, is W[1]-complete for FPT reductions, where an FPT reduction may blow-up the instance size n only polynomially but the new parameter can be any computable function of the old parameter.

2.4 First-Order Logic

A vocabulary τ is a finite set of relation symbols. A finite structure $\mathcal{A}$ over τ consists of a finite set A, called the universe, and for each R in τ a relation over A. We assume a countably infinite set of variables. Atoms over vocabulary τ are of the form $x_1 = x_2$ or $R(x_1, \ldots, x_k)$ where $R \in \tau$ and $x_1, \ldots, x_k$ are variables. First-order (FO) formulas over τ are built from atoms over τ using standard Boolean connectives $\neg, \wedge, \vee$ and from quantifiers $\exists, \forall$ followed by a variable. Let φ be an FO formula. The variables of φ that are not in scope of a quantifier are its *free variables*. Let $\varphi(\mathcal{A})$ be the set of all assignments of elements of A to φ's free variables, such that φ is satisfied. $\mathcal{A}$ is a *model* of φ if $\varphi(\mathcal{A})$ is not empty. The class Σ_1 (resp. Σ_2) contains all FO formulas of the form $\exists x_1, \ldots, \exists x_k \varphi$ (resp. $\exists x_1, \ldots, \exists x_k \forall y_1, \ldots, \forall y_k \varphi$) where φ is a quantifier free FO formula.

Let Φ be a class of formulas. The model checking problem inputs a finite structure $\mathcal{A}$ and a formula $\varphi \in \Phi$ and asks whether the formula satisfies the model, $\varphi(\mathcal{A}) \neq \emptyset$. A natural parameter is the size of (a reasonable encoding of) φ. We will use one result bridging model checking and parameterized complexity.

Theorem 1. [15] *Model checking the existential fragment of first-order logic,* MC(Σ_1)*, is* W[1]*-complete. Model checking the second level of the hierarchy,* MC(Σ_2)*, is* A[2]*-complete.*

3 Relation Between RO and GLS

Reachable Object (RO) asks whether an agent A can obtain an object x by a sequence of swaps. RO is known to be NP-complete even for trees [16]. Guaranteed Level of Satisfaction (GLS) asks whether agent A is guaranteed to obtain object y or an object preferred to y in any stable reachable allocation. GLS is even more natural than RO since it offers guarantees for the agent and does not only focus on lucky configurations. It is close to the complementary of RO, and thus the study of RO also contributes to the understanding of GLS.

Proposition 2. *co-*RO *is linearly reducible to* GLS.

Proof (sketch). We reduce from the co-RO problem asking whether object x is unreachable for agent A. Let $\mathcal{I} = \{(N, M, \succ, \sigma^0, G), A, x\}$ be an instance of co-RO. An instance $\mathcal{I}' = \{(N', M', \succ', \sigma^{0\prime}, G'), A, y\}$ of GLS is constructed by adding an agent Y and an object y. The initial allocation $\sigma^{0\prime}$ is the same as σ^0 for all agents in N and assigns y to Y. The network G' has the same structure as G, with one more edge (Y, A). Denote by a the object of agent A in σ^0. If A prefers x to a, then denote by P_a the set containing a and the objects that are preferred to a and less preferred than x in $\succ_A$. Otherwise, P_a contains a and the objects that A prefers to a. Denote by P_x the set of objects that A prefers to x. Ranking $\succ'_A$ is constructed from $\succ_A$, by moving the objects in P_x to the end of $\succ'_A$, and by putting y at the top of $\succ'_A$. The agents in $N \setminus \{A\}$ keep the same preferences as in $\succ$ but rank y last, and Y only prefers the objects of P_a to y.

We claim that x is not reachable for A in $\mathcal{I}$ iff A obtains y or an object preferred to y in any reachable stable allocation in $\mathcal{I}'$, i.e., iff there is no stable reachable allocation where A gets z such that $y \succ_A z$. This part is omitted. □

Since RO is NP-complete even on trees [16] and the previous reduction adds only one agent and possibly one swap to an instance of RO, GLS is co-NP-hard on trees. Observe that GLS is in co-NP: after guessing a reachable stable allocation, checking whether agent A owns y or an object preferred to y is direct.

Corollary 3. GLS *is* co-NP *-complete even for trees.*

We refine the complexity of the problems using natural parameters: the number of swaps per agent and the length of the sequence. Although RO and GLS are close to be dual problems, they are indeed not complementary. GLS focuses on stable allocations. A k-bound on the sequence of swaps introduces a dependency on k on the notion of stability: a stable allocation is either stable in the standard meaning, or is reached after k swaps. Stability is not necessary in RO because for an assignment solution σ where agent A gets object x, all the stable allocations reachable from σ assign to A an object preferred to x or x itself.

4 Maximum Number of Swaps per Agent

Consider that the agents are not willing to perform an important number of swaps in the whole swap process. Surprisingly in this context, our two problems, RO and GLS, remain difficult even for a very small maximum number of swaps.

Theorem 4. *For fixed $k \geq 2$,* RO*-max is* NP*-complete, even on degree 4 graphs.*

Proof. Membership in NP is straightforward, as it is a special case of the unconstrained RO problem, known to be in NP.

For hardness, we start with $k = 2$ and reduce from (3, B2)-SAT — the restriction of SAT to instances where each clause contains three literals and each variable occurs exactly twice as a positive literal and twice as a negative literal. This variant of the propositional satisfiability problem is NP-complete [5].

We are given an instance of (3, B2)-SAT with n variables $\{x_1, \ldots, x_n\}$ and m clauses $\{C_1, \ldots, C_m\}$. We create a literal-agent Y_j^ℓ (resp., $\overline{Y_j^\ell}$) for each ℓth ($\ell \in \{1, 2\}$) occurrence of literal x_j (resp., $\overline{x_j}$), and a variable-agent Y_j for each variable x_j. Two clause-agents K_i and K_i' are created for each $0 < i < m$. Three other agents Y_0, K_0' and K_m are added. Each agent initially owns an object denoted by the lower-case version of her name, e.g., agent K_i gets object k_i.

In the network, we have the paths $[Y_{j-1}, Y_j^1, Y_j^2, Y_j]$ and $[Y_{j-1}, \overline{Y_j^1}, \overline{Y_j^2}, Y_j]$ for each $1 \leq j \leq n$, and the edge (K_i, K_i') for each $1 \leq i < m$. If the ℓth literal x_j (resp., $\overline{x_j}$) belongs to clause C_i, then we have the path $[K_{i-1}', Y_j^\ell, K_i]$ (resp., $[K_{i-1}', \overline{Y_j^\ell}, K_i]$). We connect K_m and Y_n. See for an example Fig. 1.

The preferences of the agents are given below. Notation $\{\ell_i\}$ stands for the literal-objects of clause C_i ranked in arbitrary order and $k(x_i^j)$ (resp. $k(\overline{x}_i^j)$)

for the object related to the clause in which the jth occurrence of x_i (resp. $\overline{x_i}$) appears. The objects that are not mentioned in the preferences are ranked in arbitrary order after the initial endowment.

$K'_0: \{\ell_1\} \succ \boxed{b} \succ [\ldots]$	
$K_i: k'_i \succ b \succ \boxed{k_i} \succ [\ldots]$	$Y^1_j: k(x^1_j) \succ b \succ y^2_j \succ a \succ \boxed{y^1_j} \succ [\ldots]$
$K'_i: \{\ell_{i+1}\} \succ b \succ \boxed{k'_i} \succ [\ldots]$	$Y^2_j: k(x^2_j) \succ b \succ y_j \succ a \succ \boxed{y^2_j} \succ [\ldots]$
$K_m: a \succ b \succ \boxed{k_m} \succ [\ldots]$	$\overline{Y}^1_j: k(\overline{x}^1_j) \succ b \succ \overline{y}^2_j \succ a \succ \boxed{\overline{y}^1_j} \succ [\ldots]$
$Y_0: \overline{y}^1_1 \succ y^1_1 \succ \boxed{a} \succ [\ldots]$	$\overline{Y}^2_j: k(\overline{x}^2_j) \succ b \succ y_j \succ a \succ \boxed{\overline{y}^2_j} \succ [\ldots]$
$Y_n: b \succ a \succ \boxed{y_n} \succ [\ldots]$	$Y_j: \overline{y}^1_{j+1} \succ y^1_{j+1} \succ a \succ \boxed{y_j} \succ [\ldots]$

We claim that all clauses are satisfiable iff object b reaches agent Y_n. The only way for Y_n to get hold of b is by swapping a with K_m. Object b can only reach K_m via clause-agents and literal-agents, while a can only reach Y_n via variable-agents and literal-agents. Agents perform at most two swaps, so no literal-agents can be involved in the move of both a and b.

Suppose that truth assignment ϕ satisfies all clauses. Let T_i be a literal-agent of clause C_i related to a true literal in ϕ. Since all clauses are satisfiable, object b can reach K_m via the path $[K'_0, T_1, K_1, K'_1, T_2, \ldots, T_{m-1}, K_{m-1}, K'_{m-1}, T_m, K_m]$. For variable x_j, let Z^1_j and Z^2_j be the literal-agents related to the literal of x_j that is false in ϕ. Clearly, these agents are not an agent T_i. It suffices for a to reach Y_n via the path $[Y_0, Z^1_1, Z^2_1, Y_1, \ldots, Y_{n-1}, Z^1_n, Z^2_n, Y_n]$.

Suppose now that object b is reachable for agent Y_n. By construction, the path of b to K_m goes through exactly one literal-agent per clause, while the path of a to Y_n goes through exactly two literal-agents associated with the same literal for each variable. Thus, the truth assignment of variables that sets to true the literals related to literal-agents in the path of object b, satisfies all the clauses.

If $k > 2$, we adapt the reduction via a delay gadget added to each agent. □

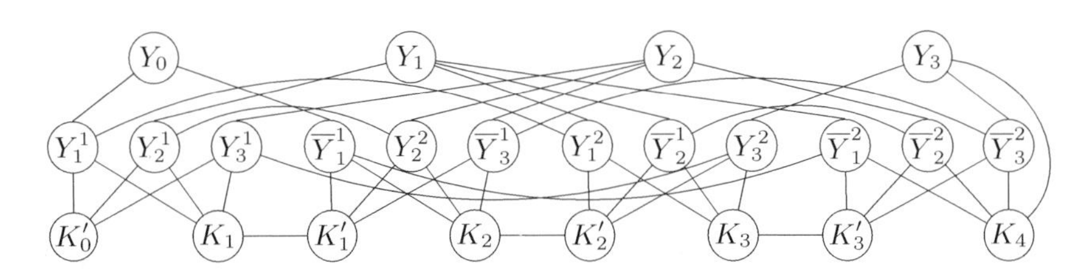

Fig. 1. Graph construction for an instance of (3, B2)-SAT with four clauses where $C_1 = (x_1 \vee x_2 \vee x_3)$, $C_2 = (\overline{x_1} \vee x_2 \vee \overline{x_3})$, $C_3 = (x_1 \vee \overline{x_2} \vee x_3)$, and $C_4 = (\overline{x_1} \vee \overline{x_2} \vee \overline{x_3})$.

From Proposition 2 and its proof, the same hardness exists for GLS, with an additional swap and one more neighbor for agent A who must obtain the object.

Corollary 5. *For $k \geq 3$, GLS-max is* co-NP*-complete, even on degree 5 graphs.*

One could think that the problem is easier when the structure of the network is restricted to trees. Yet, it is possible to prove that RO-max on trees is W[SAT]-hard. We leave out the lengthy formal proof but we state the main idea. We reduce from Monotone Weighted Satisfiability [2]. *Can an input propositional formula* φ *with no negations be satisfied with a truth assignment of weight* k? We build an instance of RO-max with a graph based on the syntax tree of φ where k chosen variable-objects must move to the occurrences of their corresponding variable as a prerequisite to given object x reaching given agent A. Since the variable-objects make up almost all the swaps, $O(k)$ swaps per agent suffice.

Globally, the problems remain difficult in very simple graphs even when the number of swaps per agent is limited. Fortunately, the parameters on the length of the sequence let us circumvent the general difficulty of the two problems.

5 Length of the Sequence of Swaps

Two parameters are used to bound the length of the sequence of swaps: the total number of exchanges and the makespan. Contrary to the previous parameter, they lead to circumscribe the problems into parameterized complexity classes that are not so high in the hierarchy, allowing tractability results when the parameters are bounded by a constant. Moreover, for bounded degree graphs, relevant in the context of social networks, we obtain fixed parameter tractability.

Theorem 6. RO-*sum and* RO-*makespan are* W[1]-*hard even for trees.*

Proof (sketch). We perform a reduction from CLIQUE, the problem of deciding whether there exists a clique of size k in a graph $G = (V, E)$ such that $V = \{1, ..., n\}$ and $|E| = m$. Assume that each edge in E is written (v, w) such that $v < w$, and consider the lexicographical order over E. Let us denote by e_i^1 and e_i^2 the first and second vertex of the ith edge. Let d_v be the degree of vertex v and $\delta_v(d)$, for $1 \leq d \leq d_v$, the dth edge incident to v. We construct an instance $\mathcal{I}'$ of RO (see Fig. 2 for an example) by creating:

- two connected agents X and Y, and two vertex-agents U_v^{vw} and U_w^{vw} for each edge $(v, w) \in E$, connected via a path $[Y, U_v^{vw}, U_w^{vw}]$.
- agents T and T^ℓ, for $1 \leq \ell \leq k$, representing the k vertices of the clique that we must choose. They are connected via a path to Y: $[Y, T^1, \ldots, T^k, T]$.
- agents A_v and A_v^ℓ, for $v \in V$ and $1 \leq \ell < k$, representing the choice of the $k-1$ edges of the clique that are incident to v if v belongs to the clique. They are connected via a path to Y for each v: $[Y, A_v^1, \ldots, A_v^{k-1}, A_v]$.
- agents $T^{\ell *}$ adjacent to T^ℓ, for $1 \leq \ell \leq k$, and agents $A_v^{\ell *}$ adjacent to A_v^ℓ, for $1 \leq \ell < k$ and $v \in V$. They are used to "validate" their associated agent by giving to her their initial object once they own an expected object.
- auxiliary agents used to facilitate the passage of some objects: if an agent B has a connected auxiliary agent $B^{[z]}$, then the swap with $B^{[z]}$ must precede a swap for getting an object associated with z. The auxiliary agents we use are agents $Y^{[vw]}$ corresponding to edge (v, w) and connected to Y, agents $Y^{[v]}$

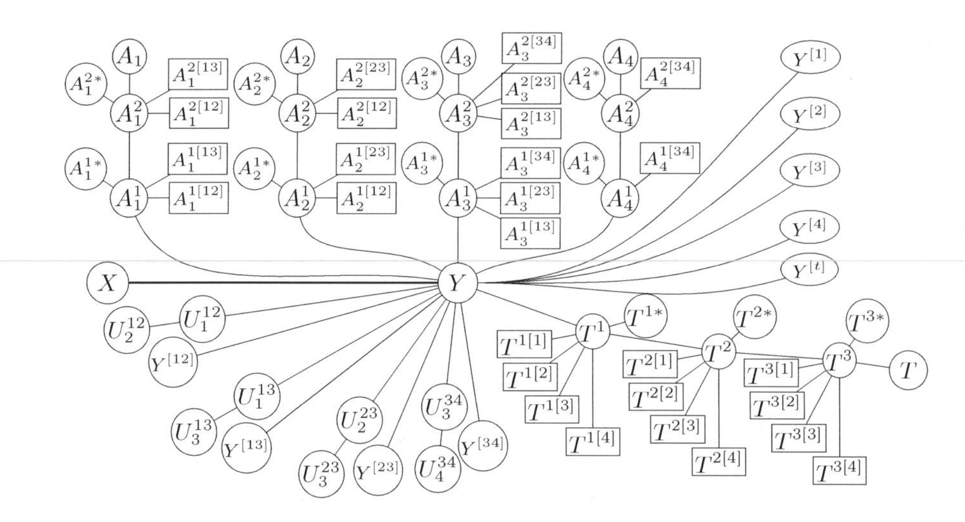

Fig. 2. Graph construction for an instance of CLIQUE with vertices $\{V_1, V_2, V_3, V_4\}$ and $k = 3$. The edges are: $\{V_1, V_2\}$, $\{V_1, V_3\}$, $\{V_2, V_3\}$ and $\{V_3, V_4\}$.

corresponding to vertex v and connected to Y, agent $Y^{[t]}$ corresponding to object t and connected to Y, agents $A_v^{\ell[\delta_v(d)]}$ corresponding to edge $\delta_v(d)$, for $1 \leq d \leq d_i$, and connected to agent A_v^ℓ, for $v \in V$ and $1 \leq \ell < k$, and agents $T^{\ell[v]}$ corresponding to vertex v and connected to agent T^ℓ for $1 \leq \ell \leq k$.

The initial object of an agent is denoted by the lower-case version of her name, e.g., agent $Y^{[v]}$ gets object $y^{[v]}$. The preferences of the agents are as follows (objects in brackets may not exist for all indices).

$$X: \quad t \succ \boxed{x} \succ [\dots] \qquad A_v : a_v^{k-1*} \succ \boxed{a_v} \succ [\dots]$$
$$U_v^{vw}: \quad a_v^{1[vw]} \succ u_w^{vw} \succ y^{[vw]} \succ \boxed{u_v^{vw}} \succ [\dots]$$
$$A_v^{\ell[\delta_v(d)]} : (a_v^{\ell+1[\delta_v(d_v)]}) \succ \cdots \succ (a_v^{\ell+1[\delta_v(1)]}) \succ a_v^\ell \succ \boxed{a_v^{\ell[\delta_v(d)]}} \succ [\dots]$$
$$T^{\ell[v]}: \quad (t^{\ell+1[v-1]}) \succ \cdots \succ (t^{\ell+1[1]}) \succ t^\ell \succ \boxed{t^{\ell[v]}} \succ [\dots]$$
$$Y^{[v]}: \quad (t^{1[v-1]}) \succ \cdots \succ t^{1[1]} \succ a_{e_m^2}^{1[e_m]} \succ \cdots \succ a_{e_1^2}^{1[e_1]} \succ \boxed{y^{[v]}} \succ [\dots]$$
$$Y^{[vw]}: \quad a_{e_m^2}^{1[e_m]} \succ \cdots \succ a_{e_1^2}^{1[e_1]} \succ y \succ \boxed{y^{[vw]}} \succ [\dots]$$
$$A_v^\ell: \quad y^{[v]} \succ (a_v^{\ell-1*}) \succ a_v \succ a_v^{\ell*} \succ u_v^{\delta_v(d_v)} \succ a_v^{\ell[\delta_v(d_v)]} \succ \cdots \succ$$
$$(a_v^{\ell+1[\delta_v(2)]}) \succ u_v^{\delta_v(2)} \succ a_v^{\ell[\delta_v(2)]} \succ (a_v^{\ell+1[\delta_v(1)]}) \succ u_v^{\delta_v(1)} \succ a_v^{\ell[\delta_v(1)]} \succ \boxed{a_v^\ell} \succ [\dots]$$
$$T^\ell: \quad y^{[t]} \succ (t^{\ell-1*}) \succ t \succ t^{\ell*} \succ a_n \succ t^{\ell[n]} \succ \cdots \succ$$
$$(t^{\ell+1[2]}) \succ a_2 \succ t^{\ell[2]} \succ (t^{\ell+1[1]}) \succ a_1 \succ t^{\ell[1]} \succ \boxed{t^\ell} \succ [\dots]$$
$$Y: \quad x \succ t \succ y^{[t]} \succ t^{1[n]} \succ a_n \succ y^{[n]} \succ \cdots \succ t^{1[1]} \succ a_1 \succ y^{[1]} \succ$$
$$a_{e_m^2}^{1[e_m]} \succ u_{e_m^2}^{e_m} \succ a_{e_m^1}^{1[e_m]} \succ u_{e_m^1}^{e_m} \succ y^{[e_m]} \succ \cdots \succ a_{e_1^2}^{1[e_1]} \succ u_{e_1^2}^{e_1} \succ a_{e_1^1}^{1[e_1]} \succ u_{e_1^1}^{e_1} \succ y^{[e_1]} \succ \boxed{y} \succ [\dots]$$

$$T: \quad t^{k*} \succ \boxed{t} \succ [\dots]$$
$$U_w^{vw}: y^{[vw]} \succ \boxed{u_w^{vw}} \succ [\dots]$$
$$A_v^{\ell*}: \quad u_v^{\delta_v(d_v)} \succ \cdots \succ u_v^{\delta_v(1)} \succ \boxed{a_v^{\ell*}} \succ [\dots]$$
$$T^{\ell*}: \quad a_1 \succ a_2 \succ \cdots \succ a_n \succ \boxed{t^{\ell*}} \succ [\dots]$$

We claim that there exists a clique of size k in graph G iff object x can reach agent Y within a total of $k^3 + 4k^2 + k + 2$ swaps or a makespan of $5k(k-1)/2 + 3k + 4$. An agent A_v^ℓ (or T^ℓ) is said to be "validated" if she obtains at a moment object $a_v^{\ell*}$ (or $t^{\ell*}$). We omit the details of the proof but the idea is that, let object x reach agent Y, all the k agents T^ℓ and all the $k-1$ agents

A_v^ℓ of k branches A_v need to be validated. The associated clique in graph G is given by the vertices v for which all the $k-1$ agents A_v^ℓ have been validated. All the agents A_v^ℓ, for $1 \leq \ell < k$, are validated if we can bring in the branch $k-1$ objects u_v^{vw} (or u_v^{wv}, following the order) representing an edge incident to v. Observe that the given budget allows bringing in the branches only $k(k-1)$ objects u_v^{vw} and the construction forces to choose u_w^{vw} if u_v^{vw} has been chosen. □

Combining the proofs of Theorem 6 and Proposition 2 leads to hardness for GLS.

Corollary 7. GLS-*sum and* GLS-*makespan are* co-W[1]-*hard even for trees.*

This W[1] -hardness for RO and GLS parameterized by the length of the sequence rules out the existence of FPT algorithms even in trees under standard complexity assumptions. However, the following results on the membership to respectively W[1] and co-A[2] show that the problems are not so hard. They are notably in XP for any graph, thus tractable when the parameter is a constant.

Theorem 8. RO-*sum is in* W[1].

Proof. An instance $\mathcal{I}$ of RO with a swap dynamics model $(N, M, \succ, \sigma^0, G)$, agent A and object x, and k as a total number of swaps, is transformed into an instance $\mathcal{I}' = (\mathcal{A}, \varphi)$ of MC(Σ_1), known to be W[1]-complete (Theorem 1). Structure $\mathcal{A}$ is an $(E, \succ, \sigma^0, A, X)$-structure with variables in $N \cup M$, where relations E, $\succ$, σ^0, A and X are defined as follows. The binary relation E over N^2 represents the edge set E. The ternary relation $\succ$ over $N \times M^2$ represents the preferences of the agents, i.e. $\succ(i, a, b)$ means that agent i prefers object a to b. For the sake of clarity, we write $a \succ_i b$ instead of $\succ(i, a, b)$. The binary relation σ^0 over $N \times M$ represents the initial allocation σ^0, i.e. $\sigma^0(i, z)$ means that i is initially endowed with z. Finally, the unary relations A and X respectively represent agent A and object x, i.e. $A(y)$ means that y is agent A and $X(y)$ means that y is object x.

The Σ_1-formula φ is defined as $\varphi = \exists x_0 \exists b_0 \exists x_1 \exists y_1 \exists a_1 \exists b_1 \ldots \exists x_k \exists y_k \exists a_k \exists b_k \left(\sigma^0(x_0, b_0) \wedge \bigvee_{0 \leq k' \leq k} \psi^{k'}\right)$ with

$$\psi^{k'} \equiv A(x_{k'}) \wedge X(b_{k'}) \wedge \bigwedge_{i=1}^{k'} \left(E(x_i, y_i) \wedge b_i \underset{x_i}{\succ} a_i \wedge\ a_i \underset{y_i}{\succ} b_i \wedge\ \mathrm{o}_i(x_i, a_i) \wedge \mathrm{o}_i(y_i, b_i)\right)$$

where for all i, $\mathrm{o}_i(q, r)$ stands for $\left(\sigma^0(q, r) \wedge \bigwedge_{j=1}^{i-1} x_j \neq q \wedge y_j \neq q\right) \vee \bigvee_{j=1}^{i-1} \left(\bigwedge_{p=j+1}^{i-1} x_p \neq q \wedge q_p \neq q\right) \wedge \left((x_j = q \wedge \mathrm{o}_j(y_j, r)) \vee \left(y_j = q \wedge \mathrm{o}_j(x_j, r)\right)\right)$.

One can prove by induction over i that formula $\mathrm{o}_i(q, r)$ is true iff object r is owned by agent q before i^{th} swap. Globally, formula $\psi^{k'}$ is true iff the sequence of exchanges between the agents (x_i, y_i) exchanging the objects (a_i, b_i), for $i \in \{1, \ldots, k'\}$, is a sequence of swaps leading to give object x to agent A. □

The same idea and a slightly different FO formula work for RO-makespan.

Proposition 9. RO-*makespan is in* W[1].

Proof (sketch). We reduce to MC(Σ_1) but face a new difficulty. We cannot quantify over all potential exchanges within makespan k: it would lead to a formula of size $\Omega(n)$. The crux of this proof is to observe that not all exchanges are relevant to decide the problem. Assume we process independent swaps in parallel for up to k time steps. Looking at it from the end, the only relevant swap in the last step k involves agent A, so we quantify over a single swap and ignore all concurrent ones. In the one-before-last, only swaps involving A or A's partner at step k may be relevant. So considering two swaps happening at step $k-1$ and ignoring all other concurrent ones suffice. All in all, we need to quantify over no more than 2^{k+1} exchanges. The rest is similar to that of Theorem 8. □

A similar reasoning is applied to GLS. We reduce GLS to model-checking FO formula using more sophisticated Σ_2 formulas.

Proposition 10. GLS-*sum/-makespan is in* co-A[2].

Proof (sketch). We reduce co-GLS to MC(Σ_2), known to be A[2]-complete (Theorem 1), following an approach similar to that of Theorem 8 and Proposition 9. □

The previous results show that RO and GLS are not "so hard" considering the length of the sequence as a parameter. Furthermore, for some natural classes of graphs, the problems are even tractable with respect to these parameters.

Proposition 11. RO/GLS-*sum/makespan are* FPT *on bounded degree graphs.*

Proof. The proof follows the idea developed for Proposition 9. Let Δ be the degree of G and consider the RO problem. At the kth step, the only relevant exchange involves agent A and a neighbor, so there are $\mathcal{O}(\Delta)$ possible swaps. The one-before-last step can only involve A or one her neighbor, therefore there are at most 2Δ possible swaps for RO-sum and at most $\Delta + \Delta^2$ for RO-makespan. This argument applies at any of the k steps, hence there are $\mathcal{O}(\Delta^k.k!)$ sequences of swaps for RO-sum and $\mathcal{O}(\Delta^{k^2})$ for RO-makespan, and it suffices to verify if one sequence assigns x to A. Concerning GLS, it suffices to test the reachability to A of any object x such that $y \succ_A x$, and so it just adds a factor of n. □

6 Conclusion and Perspectives

This article studies the distributed process of swap dynamics along a network for reallocating objects among agents. Two related problems are investigated: Reachable Object (RO), "can a given agent obtain a given object?", and Guaranteed Level of Satisfaction (GLS), "is a given agent guaranteed to get a given object or better?". Both problems are hard but the parameterized approach allows us to escape this difficulty for a relevant class of graphs.

We consider natural parameters constraining the number of swaps per agent or the duration of the sequence. Assuming that they remain small is reasonable in practice as the patience of the agents typically does not increase with the instance size. In the case of few swaps *per agent*, RO and GLS remain hard even on bounded degree graphs. So, this parameterization, although natural, does not help us to grasp the problems. However, considering the *total* length of the sequence, although both problems are intractable even for trees, this hardness is circumscribed to not "so hard" parameterized complexity classes, leading to the possibility of handling the problems when the parameters do not depend on the instance size, very natural assumption. Furthermore, unlike the first parameter, the length of the sequence permits to obtain fixed parameter tractability on bounded degree graphs, which typically model real social networks.

The parameterized approach allows progress in the understanding of the problems and leads to significant and realistic positive results. So far, we have considered restrictions on the network as well as on the solution size. A natural extension is to investigate the influence of a third dimension: constraints on the preference profile, e.g., single-peaked domains. Furthermore, assuming the full knowledge of the preferences and the network is not relevant in all the contexts. Relaxing this assumption could be a challenging future work.

References

1. Abdulkadiroğlu, A., Sönmez, T.: House allocation with existing tenants. J. Econ. Theory **88**(2), 233–260 (1999)
2. Abrahamson, K.A., Downey, R.G., Fellows, M.R.: Fixed-parameter tractability and completeness IV: on completeness for W[P] and PSPACE analogues. Ann. Pure Appl. Log. **73**(3), 235–276 (1995)
3. Aziz, H., Biró, P., Lang, J., Lesca, J., Monnot, J.: Optimal reallocation under additive and ordinal preferences. In: AAMAS. pp. 402–410 (2016)
4. Aziz, H., De Keijzer, B.: Housing markets with indifferences: a tale of two mechanisms. In: AAAI. pp. 1249–1255 (2012)
5. Berman, P., Karpinski, M., Scott, A.D.: Approximation hardness of short symmetric instances of max-3sat. Technical report (2004)
6. Chevaleyre, Y., et al.: Issues in multiagent resource allocation. Informatica **30**(1), 3–31 (2006)
7. Chevaleyre, Y., Endriss, U., Estivie, S., Maudet, N.: Multiagent resource allocation in k-additive domains: preference representation and complexity. Ann. Oper. Res. **163**(1), 49–62 (2008)
8. Chevaleyre, Y., Endriss, U., Lang, J., Maudet, N.: Negotiating over small bundles of resources. In: AAMAS. pp. 296–302 (2005)
9. Chevaleyre, Y., Endriss, U., Maudet, N.: Allocating goods on a graph to eliminate envy. In: AAAI. pp. 700–705 (2007)
10. Damamme, A., Beynier, A., Chevaleyre, Y., Maudet, N.: The power of swap deals in distributed resource allocation. In: AAMAS. pp. 625–633 (2015)
11. Dunne, P.E., Chevaleyre, Y.: The complexity of deciding reachability properties of distributed negotiation schemes. Theor. Comput. Sci. **396**(1–3), 113–144 (2008)
12. Dunne, P.E., Wooldridge, M., Laurence, M.: The complexity of contract negotiation. Artif. Intell. **164**(1–2), 23–46 (2005)
13. Easley, D., Kleinberg, J.: Networks, Crowds, and Markets: Reasoning about a Highly Connected World. Cambridge University Press, New York (2010)

14. Endriss, U., Maudet, N., Sadri, F., Toni, F.: Negotiating socially optimal allocations of resources. J. Artif. Intell. Res. **25**, 315–348 (2006)
15. Flum, J., Grohe, M.: Parameterized Complexity Theory. Springer, Berlin (2006)
16. Gourvès, L., Lesca, J., Wilczynski, A.: Object allocation via swaps along a social network. In: IJCAI. pp. 213–219 (2017)
17. Jackson, M.O.: Social and Economic Networks. Princeton University Press, Princeton (2008)
18. Sandholm, T.W.: Contract types for satisficing task allocation. In: AAAI Spring Symposium. pp. 23–25 (1998)
19. Shapley, L., Scarf, H.: On cores and indivisibility. J. Math. Econ. **1**(1), 23–37 (1974)

A Hashing Power Allocation Game in Cryptocurrencies

Yukun Cheng[1], Donglei Du[2], and Qiaoming Han[3](✉)

[1] School of Business, Suzhou University of Science and Technology, Suzhou, People's Republic of China
ykcheng@amss.ac.cn

[2] Faculty of Business Administration, University of New Brunswick, Fredericton, NB, Canada
ddu@unb.ca

[3] School of Data Science, Zhejiang University of Finance and Economics, Hangzhou, People's Republic of China
qmhan@zufe.edu.cn

Abstract. Various crypto-currencies backed by Blockchain technology are now springing up like mushrooms. Miners in these peer-to-peer networks compete to maintain the validity of the underlying ledgers to earn the bootstrapped crypto-currencies. With limited hashing power, each miner needs to decide how to allocate their resource to different crypto-currencies so as to achieve the best overall payoff. Together all the miners form a hashing power allocation game. We consider the setting of the game in which the miners are risk-neutral. We show that a unique pure Nash Equilibrium exists and can be computed efficiently in this settings.

1 Introduction

With the advancement of the blockchain technologies (a.k.a., distributed ledger technology), distributed applications (DApps) are burgeoning. Starting from the bitcoin, many altcoins have been proposed to achieve different goals. At the time of this writing, there are almost 1,600 cryptocurrencies with market capitalization totalling approximate $456 billion[1], and among them Bitcoin [5], Ethereum [9] and Ripple [26] are the top three market-caped crypto-currencies.

Miners in these peer-to-peer networks play the important role of maintaining the integrity of the underlying blockchains, incentivized to earn digital currencies and transaction fees. Mining involves executing a distributed consensus protocol on how to achieve agreement of the underlying ledger when there is no central authority in presence. Among them, proof-of-work (PoW) [20], proof-of-stake (PoS) [30] and proof-of-burn (PoB) [23] are the widely adopted consensus protocols by existing crypto-currencies.

For example, in the PoW framework, during a given average time period (e.g., every 10 min for Bitcoin), miners participate in a winner-take-all competition to

[1] According to https://coinmarketcap.com/.

X. Deng (Ed.): SAGT 2018, LNCS 11059, pp. 226–238, 2018.
https://doi.org/10.1007/978-3-319-99660-8_20

extend the next block on the longest block chain by solving some cryptographic hashing proof-of-work, a mathematical puzzle. As a concrete case, in Bitcoin network, the puzzle goes as follows [33]: given a difficulty $d > 0$, a challenge c and a nonce x (usually bit-strings), a function

$$F_d(c, x) \rightarrow \{\text{TRUE,FALSE}\}$$

is called a Proof-of-Work (PoW) function if it has the following two properties: (i) $F_d(c, x)$ is fast to compute, given d, c, and x; and (ii) for fixed parameters d and c, finding x such that $F_d(c; x) = \text{TRUE}$ is computationally difficult but feasible. The difficulty d is used to adjust the time to find such an x.

With miners equipped with certain computing power (a.k.a., hashing power in Bitcoin network) and a large number of different cryptocurrencies to mine, they are facing the challenge on how to allocate their computing power to compete in mining each cryptocurrency to maximize their expected payoffs. Due to the competitive nature of the mining protocol, all miners together form a non-cooperative allocation game. This work aims to answer the following questions associated with the aforementioned game: 1. Does Nash Equilibrium (NE) exist? 2. Is NE unique? 3. Can the NE be computed efficiently? We offer affirmative answers to all three questions for the risk-neutral miners.

For the game with risk-neutral miners, we show that the Nash Equilibrium allocation is unique and follows a proportional rule (Theorems 1 and 2) where each miner will allocate his total computing power to a given crypto-currency proportional to the percentage of the award among all currencies, while his expected revenue is proportional to the percentage of the hashing power possessed and the total award.

The equilibrium analysis of the allocation game is of both theoretical and practical relevance. On the theoretical side, we set up a succinct backbone model which admits a closed-form solution via non-trivial technical analysis. On the practical side, we provide insights which can help mining pool managers (such as BTC.com [7], AntPool [3], Slush Pool [25], ViaBTC [32] and BTC.TOP [8], etc.) or individual miners in making the most important operational/tactical decisions, namely how to allocate the hashing power when facing under reward and peer competition.

To filter out the most salient factors that are of managerial relevance, we made some simplifications in the modelling, such as the the assumption that the cost to purchase certain hashing power is independent from the price of the currencies. However, this type of deviations from the realism on one hand may be a good approximation to reality and on the other hand is to be expected in an early attempt to apprehend an otherwise complex problem. Also, this work focuses on static games, and leave the discussion of dynamic games to future research.

Several blockchain games (mainly non-cooperative in nature) are proposed in the recent literature to address and improve upon the limitations of existing distributed consensus mechanism in various crypt-currencies [4,11,14,17,21,30], while some other games (mainly non-cooperative in nature) focus on the application layer without invoking any protocol technicality, such as the mining pool games [10,13,14,24,29]. Our computing power allocation game is non-cooperative and focuses on the application layer; namely the allocation of mining resource. Furthermore, these games all deal with a single currency, which is a major difference from the game investigated in this work.

Our computing power allocation game is similar to the extensively-studied general blotto game in the game theory literature [1,2,15,16,27], but the two models have completely different utility functions to suit different applications in mind.

Our game can be considered as a special case of the games investigated in [12,31] in the context of P2P computing and post trading. However, our game possesses special structure that are lacking in the latter and hence admits stronger results. As a matter of fact, the games in [12,31] are so general that they only guarantees the existence of Nash equilibrium, while our game admits a unique pure Nash equilibrium with a closed-form solution.

The resource allocation nature is also relevant to the large literature on portfolio management [18], and the market equilibrium model, in particular the Fisher market [19,22]. However, the portfolio management literature usually assume that the supply of assets is independent from the allocation decision. And the Fisher market models focus on finding market-clearing prices and the allocation rule at market equilibrium.

The readers are referred to the survey by [6] for research perspectives and challenges for Bitcoin and cryptocurrencies.

2 The Computing Power Allocation Game

There are n miners $N = \{1, \ldots, n\}$ with computing powers $\mathbf{h} = (h_1, \ldots, h_n)$ (the cost to possess such a computing power, expressed in fiat currency such as US dollar). There are m cryptocurrencies $M = \{1, \ldots, m\}$ available for mining. Miner $i \in N$ allocates $x_{ij} \geq 0$ of his computing power to mine cryptocurrency j. Evidently $\sum_{j \in M} x_{ij} = h_i, i \in N$.

For each cryptocurrency j, the n miners play a winner-take-all game and the winner is rewarded with uncertain reward $\mathbf{R} = (R_1, \ldots, R_m)$ (expressed in fiat currency such as US dollar) with mean vector $\mathbb{E}[\mathbf{R}] = \mu^T = (\mu_1, \ldots, \mu_m)^T$. Miner $i \in N$ wins cryptocurrency $j \in M$ with probability proportional to its allocated computing power

$$p_{ij} = \frac{x_{ij}}{\sum_{\ell \in N} x_{\ell j}} \tag{1}$$

and his payoff for cryptocurrency $j \in M$ is given by

$$\pi_{ij}(x) = \begin{cases} R_j - x_{ij}, & \text{w.p. } p_{ij} \\ -x_{ij}, & \text{w.p. } 1 - p_{ij} \end{cases}$$

Therefore miner i's total payoff is given by

$$\pi_i(x) = \sum_{j \in M} \pi_{ij}(x) = \sum_{j \in M} R_j p_{ij} - \sum_{j \in M} x_{ij} = \sum_{j \in M} R_j \frac{x_{ij}}{\sum_{\ell \in N} x_{\ell j}} - h_i = R^T y_i(\mathbf{x}) - h_i$$

where $\mathbf{x} = (x_{ij})_{n\times m} \in \mathbb{R}_+^{n\times m}$ and

$$y_i(\mathbf{x}) = \begin{pmatrix} \frac{x_{i1}}{x_{11}+\ldots+x_{n1}} \\ \vdots \\ \frac{x_{im}}{x_{1m}+\ldots+x_{nm}} \end{pmatrix}, i \in N. \tag{2}$$

The mean of miner i's payoff is given as follows

$$\mathbb{E}_R[\pi_i(\mathbf{x})] = \mu^T y_i(\mathbf{x}) - h_i. \tag{3}$$

3 Main Result

The Nash Equilibrium for risk-neutral miners can be obtained by solving the following n optimization problems based on (3): for any given $i \in N$,

$$\begin{aligned} \max_{x_{i\cdot}\in\mathbb{R}_+^m} \quad & \mu^T y_i(\mathbf{x}) = \sum_{j\in M} \frac{x_{ij}}{\sum_{h=1}^n x_{hj}} \mu_j \\ \text{s.t.} \quad & \sum_{j\in M} x_{ij} = h_i \\ & x_{ij} \geq 0. \end{aligned} \tag{4}$$

Let $\mathbf{x}_i = (x_{ij})$ be the allocation of miner i and $\mathbf{x}_{-i}$ be the profile without miner i's allocation. For each risk-neutral miner, his object is to maximize the expected utility from all of cryptocurrencies. So given an allocation profile $\mathbf{x} = (x_{ij}) = (\mathbf{x}_1, \mathbf{x}_2, \cdots, \mathbf{x}_n)$, define $U_{ij}(\mathbf{x})$ to be the expected utility of miner i from cryptocurrency j, i.e. $U_{ij}(\mathbf{x}) = \frac{x_{ij}}{\sum_{h=1}^n x_{hj}} \mu_j$. Therefore the utility of miner i is $U_i(\mathbf{x}) = \sum_{j\in M} U_{ij}(\mathbf{x})$.

A power allocation profile $\mathbf{x}$ is a Nash equilibrium, if and only if no miner benefits by changing his strategy unilaterally. However, there is one difficulty, that is at the point where for some cryptocurrency $j \in M$, $x_{ij} = 0$ for each $i \in N$, U_i is discontinuous. So we cannot apply the standard method in [28] to study Nash equilibrium. We first propose the following lemma to show that Nash equilibrium could not be at the discontinuous point.

Lemma 1. *Given an allocation profile* $\mathbf{x}$. *If there is at least a cryptocurrency* $j \in M$ *with* $x_{ij} = 0$ *for each* $i \in N$, *then allocation* $\mathbf{x}$ *cannot be a Nash equilibrium.*

Proof. W.l.o.g. we assume that $x_{i1} = 0$ for each $i \in N$. Then there must exist another cryptocurrency, say $j = 2$, with $x_{i2} > 0$. Therefore, the allocation of miner i is $\mathbf{x}_i = (0, x_{i2}, x_{i3}, \cdots, x_{im})$. Let us consider another allocation $\mathbf{x}_i'$, with $x_{i1}' = \epsilon$, $x_{i2}' = x_{i2} - \epsilon$ and $x_{ij}' = x_{ij}$ for each $j = 3, \cdots, m$, in which

$$0 < \epsilon < \min\{x_{i2}, \frac{\mu_1 (\sum_{h\in N} x_{h2})^2}{\mu_1 \sum_{h\in N} x_{h2} + \mu_2 \sum_{h\neq i} x_{h2}}\}.$$

On one hand, allocation $\mathbf{x}_i'$ is feasible as $0 < \epsilon < x_{i2}$. On the other hand, if other miners remain their allocations unchanged and miner i reallocate his computing power as $\mathbf{x}_i'$ unilaterally, then miner i will obtain the whole reward from cryptocurrency 1 and his utility shall be

$$U_i' = \mu_1 + \frac{x_{i2} - \epsilon}{\sum_{h\in N} x_{h2} - \epsilon}\mu_2 + \sum_{j=3} \frac{x_{ij}}{\sum_{h\in N} x_{hj}}\mu_j.$$

The difference of utility is

$$\begin{aligned}
\Delta U_i = U_i' - U_i &= \mu_1 + \frac{x_{i2} - \epsilon}{\sum_{h\in N} x_{h2} - \epsilon}\mu_2 - \frac{x_{i2}}{\sum_{h\in N} x_{h2}}\mu_2 \\
&= \mu_1 - \frac{\sum_{h\in N, h\neq i} x_{h2}\mu_2\epsilon}{\sum_{h\in N} x_{h2}(\sum_{h\in N} x_{h2} - \epsilon)} \\
&= \frac{\mu_1(\sum_{h\in N} x_{h2})^2 - (\mu_1 \sum_{h\in N} x_{h2} + \mu_2 \sum_{h\neq i} x_{h2})\epsilon}{\sum_{h\in N} x_{h2}(\sum_{h\in N} x_{h2} - \epsilon)} \\
&> \frac{\mu_1(\sum_{h\in N} x_{h2})^2 - (\mu_1 \sum_{h\in N} x_{h2} + \mu_2 \sum_{h\neq i} x_{h2})\epsilon}{(\sum_{h\in N} x_{h2})^2} > 0.
\end{aligned}$$

The last inequality is from the definition of ϵ. Thus we can conclude that the allocation $\mathbf{x}$ in which for some $j \in M$, $x_{ij} = 0$ for each $i \in N$, is not a Nash equilibrium. □

Conveniently, we define the following condition of an allocation $\mathbf{x}$,

Condition 1 For each cryptocurrency $j \in M$, $\sum_{i\in N} x_{ij} > 0$.

Based on Lemma 1, it is sufficient for us to study the Nash equilibrium at such allocations satisfying Condition 1. From (4), we know the utility function of miner i is linear and the domain $\{(x_{i1}, x_{i2}, \cdots, x_{im}) | \sum_{j\in M} x_{ij} = h_i,\ x_{ij} \geq 0\}$ is convex. Then by the first-order optimality condition, there exists Lagrange multiplier $\alpha_i > 0$, such that

$$\frac{\partial U_i}{\partial x_{ij}} = \frac{\sum_{h\neq i} x_{hj}}{(\sum_{h=1}^{n} x_{hj})^2} \cdot \mu_j \begin{cases} = \alpha_i & \text{if } x_{ij} > 0, \\ \leq \alpha_i & \text{if } x_{ij} = 0. \end{cases}$$

From another perspective, at an equilibrium, if miner i allocates positive computing power for some cryptocurrencies, then he shall has the same marginal value on these cryptocurrencies. Otherwise, he may have lower marginal value. Therefore, we face another difficulty that how to characterize Nash equilibrium at $\mathbf{x}$ satisfying Condition 1, but $x_{ij} = 0$ for some $j \in M$ and $i \in N$. For this purpose, we consider a kind of restricted strategy at first, that is for miner i, he only changes his allocation between two cryptocurrencies j and k with

$$x_{ij}' = x_{ij} - \epsilon, \quad x_{ik}' = x_{ik} + \epsilon, \quad x_{i\ell}' = x_{i\ell},\ l \neq j, k, \tag{5}$$

where $x_{ij} > 0$ and $\epsilon > 0$. For convenience, we call such a kind of strategy as a *restricted strategy on cryptocurrencies j and k*.

Lemma 2. *In the hash power allocation game, if all miners are only permitted to play the restricted strategy, then an allocation $\mathbf{x}$ is a Nash equilibrium, if* *and only if for each miner $i \in N$, there is a constant α_i such that for each cryptocurrency j,*

$$\frac{\sum_{h \neq i} x_{hj}}{(\sum_{h=1}^{n} x_{hj})^2} \cdot \mu_j = \alpha_i. \tag{6}$$

Proof. Obviously, allocation $\mathbf{x}$ must satisfy Condition 1. W.l.o.g., assume miner i plays the restricted strategy on cryptocurrencies j and k and his new allocation $\mathbf{x}'$ is shown as (5). It is possible that $x_{ik} = 0$. Clearly,

$$U_{ij}(\mathbf{x}'_i, \mathbf{x}_{-i}) = \frac{x_{ij} - \epsilon}{\sum_{h=1}^{n} x_{hj} - \epsilon}\mu_j, \quad U_{ik}(\mathbf{x}'_i, \mathbf{x}_{-i}) = \frac{x_{ik} + \epsilon}{\sum_{h=1}^{n} x_{hk} + \epsilon}\mu_k.$$

Since others' allocations are unchanged and $x'_{i\ell} = x_{i\ell}$, $\ell \neq j, k$, we have $U_{i\ell}(\mathbf{x}'_i, \mathbf{x}_{-i}) = U_{i\ell}(\mathbf{x})$, $\ell \neq j, k$. Therefore

$$\begin{aligned}
\Delta U_i &= (U_{ij}(\mathbf{x}'_i, \mathbf{x}_{-i}) - U_{ij}(\mathbf{x})) + (U_{ik}(\mathbf{x}'_i, \mathbf{x}_{-i}) - U_{ik}(\mathbf{x})) \\
&= \left[\frac{x_{ij} - \epsilon}{\sum_{h=1}^{n} x_{hj} - \epsilon} - \frac{x_{ij}}{\sum_{h=1}^{n} x_{hj}}\right] \cdot \mu_j + \left[\frac{x_{ik} + \epsilon}{\sum_{h=1}^{n} x_{hk} + \epsilon} - \frac{x_{ik}}{\sum_{h=1}^{n} x_{hk}}\right] \cdot \mu_k \\
&= \frac{-\sum_{h \neq i} x_{hj}\mu_j\epsilon}{\sum_{h=1}^{n} x_{hj}(\sum_{h=1}^{n} x_{hj} - \epsilon)} + \frac{\sum_{h \neq i} x_{hk}\mu_k\epsilon}{\sum_{h=1}^{n} x_{hk}(\sum_{h=1}^{n} x_{hk} + \epsilon)}.
\end{aligned}$$

If the result of (6) holds, which means

$$\frac{\sum_{h \neq i} x_{hj}}{(\sum_{h=1}^{n} x_{hj})^2} \cdot \mu_j = \frac{\sum_{h \neq i} x_{hk}}{(\sum_{h=1}^{n} x_{hk})^2} \cdot \mu_k,$$

then it is easy to deduce that,

$$\frac{\sum_{h \neq i} x_{hj}\mu_j\epsilon}{\sum_{h=1}^{n} x_{hj}(\sum_{h=1}^{n} x_{hj} - \epsilon)} > \frac{\sum_{h \neq i} x_{hj}\mu_j\epsilon}{(\sum_{h=1}^{n} x_{hj})^2} = \frac{\sum_{h \neq i} x_{hk}\mu_k\epsilon}{(\sum_{h=1}^{n} x_{hk})^2} > \frac{\sum_{h \neq i} x_{hk}\mu_k\epsilon}{\sum_{h=1}^{n} x_{hk}(\sum_{h=1}^{n} x_{hk} + \epsilon)},$$

implying $\Delta U_i \leq 0$.

On the other hand, we shall prove (6) is the necessary condition for a Nash equilibrium allocation if each miner is only allowed to play the restricted strategy. For this purpose, we try to prove that once the result of (6) does not hold, miner i can get more utility by playing a restricted strategy.

W.l.o.g., suppose

$$\frac{\sum_{h \neq i} x_{hj}}{(\sum_{h=1}^{n} x_{hj})^2} \cdot \mu_j < \frac{\sum_{h \neq i} x_{hk}}{(\sum_{h=1}^{n} x_{hk})^2} \cdot \mu_k,$$

There must exist an arbitrarily small constant $\epsilon > 0$ such that

$$\frac{\sum_{h \neq i} x_{hj}\mu_j}{\sum_{h=1}^{n} x_{hj}(\sum_{h=1}^{n} x_{hj} - \epsilon)} < \frac{\sum_{h \neq i} x_{hk}\mu_k}{\sum_{h=1}^{n} x_{hk}(\sum_{h=1}^{n} x_{hk} + \epsilon)}.$$

So

$$\Delta U_i = \frac{-\sum_{h \neq i} x_{hj}\mu_j\epsilon}{\sum_{h=1}^{n} x_{hj}(\sum_{h=1}^{n} x_{hj} - \epsilon)} + \frac{\sum_{h \neq i} x_{hk}\mu_k\epsilon}{\sum_{h=1}^{n} x_{hk}(\sum_{h=1}^{n} x_{hk} + \epsilon)} > 0$$

It means that miner i can benefit by playing the restricted strategy on cryptocurrencies j and k and the current allocation $\mathbf{x}$ is not a Nash equilibrium.

In addition, because of the arbitrariness of j and k (even though $x_{ik} = 0$), we have the sufficient and necessary condition of any pure Nash equilibrium for the restricted strategy that there is a constant α_i for each miner $i \in N$ and

$$\frac{\sum_{h \neq i} x_{hj}}{(\sum_{h=1}^{n} x_{hj})^2} \cdot \mu_j = \alpha_i, \quad \forall j \in M.$$

is satisfied. □

Next let us turn to the characterization of any pure Nash equilibrium for more general strategy. W.l.o.g., suppose the new allocation $\mathbf{x}_i'$ after manipulation is

$$\begin{aligned} & x_{i1}' = x_{i1} + \epsilon_1, \; x_{i2}' = x_{i2} + \epsilon_2, \; \cdots, \; x_{ik}' = x_{ik} + \epsilon_k, \\ & x_{i(k+1)}' = x_{i(k+1)} - \epsilon_{k+1}, \; \cdots, \; x_{i(k+h)}' = x_{i(k+h)} - \epsilon_{k+h}, \\ & x_{i(k+h+1)}' = x_{i(k+h+1)}, \cdots, x_{im}' = x_{im}, \end{aligned} \tag{7}$$

where each $\epsilon_\ell > 0$, $x_{ij}' \geq 0$ for each $j \in M$, and $\sum_{\ell=1}^{k} \epsilon_\ell = \sum_{\ell=1}^{h} \epsilon_{k+\ell}$.

Algorithm A is proposed in Table 1, which constructs a series of intermediate allocations from $\mathbf{x}_i$ to $\mathbf{x}_i'$. For the sake of convenience, let $\mathbf{x}_i^0 = \mathbf{x}_i$, $\mathbf{x}_i^p = \mathbf{x}_i'$ and the intermediate allocations are denoted by $\mathbf{x}_i^1, \cdots, \mathbf{x}_i^{p-1}$.

Table 1. The Algorithm to Construct Intermediate Allocations

Algorithm A

Input: Allocations $\mathbf{x}_i^0 = \mathbf{x}_i$ and $\mathbf{x}_i^p = \mathbf{x}_i'$

Output: The intermediate allocations $\mathbf{x}_i^1, \mathbf{x}_i^2, \cdots, \mathbf{x}_i^{p-1}$.

1: **Set** $t := 1$, $r := 1$ and $q := 1$;

2: **While** $t \leq k$ and $r \leq h$;

3: **Set** $\eta_q = \min\{\epsilon_t, \epsilon_{k+r}\}$;

4: **Set** $x_{i\ell}^q = \begin{cases} x_{i\ell}^{q-1} + \eta_q, & \ell = t \\ x_{i\ell}^{q-1} - \eta_q, & \ell = k + r \\ x_{i\ell}^{q-1}, & \ell \neq t, k + r. \end{cases}$ and **Output** allocation $\mathbf{x}_i^q = (x_{i\ell}^q)$;

5: **If** $\eta_q = \epsilon_t = \epsilon_{k+r}$;
  Set $t := t + 1$, $r := r + 1$, $q := q + 1$ and go to line 2;

6: **Else**

7: **If** $\eta_q = \epsilon_t$
  Set $\epsilon_{k+r} := \epsilon_{k+r} - \eta_q$, $t := t + 1$, $q := q + 1$ and go to line 2;

8: **If** $\eta_q = \epsilon_{k+r}$
  Set $\epsilon_t := \epsilon_t - \eta_q$, $r := r + 1$, $q := q + 1$ and go to line 2.

Here we give an example to show how Algorithm A works.

Example 1. Suppose the strategic miner i changes his allocation to $\mathbf{x}'_i = (x_{i1} + 5, x_{i2} + 3, x_{i3} + 1, x_{i4} - 6, x_{i5} - 3, x_{i6})$. Then the intermediate allocations are

$$\begin{aligned}
\mathbf{x}_i^0 &= (x_{i1}, x_{i2}, x_{i3}, x_{i4}, x_{i5}, x_{i6});\\
\mathbf{x}_i^1 &= (x_{i1} + 5, x_{i2}, x_{i3}, x_{i4} - 5, x_{i5}, x_{i6});\\
\mathbf{x}_i^2 &= (x_{i1} + 5, x_{i2} + 1, x_{i3}, x_{i4} - 5 - 1, x_{i5}, x_{i6});\\
\mathbf{x}_i^3 &= (x_{i1} + 5, x_{i2} + 1 + 2, x_{i3}, x_{i4} - 5 - 1, x_{i5} - 2, x_{i6});\\
\mathbf{x}_i^4 &= (x_{i1} + 5, x_{i2} + 1 + 2, x_{i3} + 1, x_{i4} - 5 - 1, x_{i5} - 2 - 1, x_{i6}).
\end{aligned}$$

Obviously, at least one of indices t and r is increased by 1 at each step in Algorithm A. Because $\sum_{\ell=1}^{k} \epsilon_\ell = \sum_{\ell=1}^{h} \epsilon_{k+\ell}$, then Algorithm A must terminate at the case that $\eta_q = \epsilon_t = \epsilon_{k+r}$ and obtain the last allocation $\mathbf{x}'$. Thus Algorithm A can be finished in at most $k + h - 1$ steps, which implies the time complexity of Algorithm A is $O(n)$. Furthermore these intermediate allocations have several nice properties, which are necessary for us to obtain the result on Nash equilibrium.

Lemma 3. *Given the series of allocations $\mathbf{x}_i^0, \mathbf{x}_i^1, \cdots, \mathbf{x}_i^p$ from Algorithm A. For any two adjacent allocations $\mathbf{x}_i^{q-1}$ and $\mathbf{x}_i^q$*

1. $x_{it}^q = x_{it}^{q-1} + \eta_q$, $x_{i(j+r)}^q = x_{i(j+r)}^{q-1} - \eta_q$, *and* $x_{i\ell}^q = x_{i\ell}^{q-1}$, $\ell \neq t, k+r$;
2. *For any allocation $\mathbf{x}_i^q$, $q = 1, 2, \cdots, p$, there exist three cases:*
   - *Case 1.* $x_{it}^{q-1} = x_{it} + \omega$ *and* $x_{i(k+r)}^{q-1} = x_{i(k+r)}$, $\omega > 0$;
   - *Case 2.* $x_{it}^{q-1} = x_{it}$ *and* $x_{i(k+r)}^{q-1} = x_{i(k+r)} - \omega$, $\omega > 0$;
   - *Case 3.* $x_{it}^{q-1} = x_{it}$ *and* $x_{i(k+r)}^{q-1} = x_{i(k+r)}$.

Proof. The first claim is from line 4 in Algorithm A directly. The three cases in the second claim are right from line 5-8 in Algorithm A. □

Based on the previous analysis for the changing processes from $\mathbf{x}_i$ to $\mathbf{x}'_i$, the following theorem shows the sufficient and necessary condition for the existence of Nash equilibrium in Lemma 2 also holds, even though the miners are allowed to play more general strategy.

Theorem 1. *An allocation $\mathbf{x}$ in the hash power allocation game is a Nash equilibrium, if and only if for each miner $i \in N$ and any cryptocurrency j, there is a constant α_i satisfying*

$$\frac{\sum_{h \neq i} x_{hj}}{(\sum_{h=1}^{n} x_{hj})^2} \cdot \mu_j = \alpha_i. \tag{8}$$

Proof. Of course, $\mathbf{x}$ satisfies Condition 1. Let us suppose $\mathbf{x}$ to be a Nash equilibrium allocation, but Eq. (8) does not hold. By the proof in Lemma 2, miner i can change his allocation to a new one $\mathbf{x}'_i$ shown as (5), to improve his utility. It is a contradiction to the assumption that $\mathbf{x}$ is a Nash equilibrium.

On the other hand, we shall prove that once an allocation $\mathbf{x}$ satisfies (8), it must be a Nash equilibrium. W.l.o.g., suppose a strategic miner i changes his allocation to $\mathbf{x}'_i$ as (7). So we can obtain a series of allocations $\mathbf{x}_i^0, \mathbf{x}_i^1, \cdots, \mathbf{x}_i^p$ from Algorithm A. The first claim in Lemma 3,

$$x_{it}^q = x_{it}^{q-1} + \eta_q,\ x_{i(k+r)}^q = x_{i(k+r)}^{q-1} - \eta_q, \text{ and } x_{i\ell}^q = x_{i\ell}^{q-1},\ \ell \neq t, j+r,$$

shows that any two adjacent allocations $\mathbf{x}_i^q$ and $\mathbf{x}_i^{q-1}$ are the same, except for the t-th and $k+r$-th elements. It can be viewed as miner i plays a restricted strategy on two cryptocurrencies t and $k+r$ from $\mathbf{x}_i^{q-1}$ to $\mathbf{x}_i^q$. So we focus on the change between $\mathbf{x}_i^q$ and $\mathbf{x}_i^{q-1}$ by using the similar proof in Lemma 2. There are three cases for each $\mathbf{x}_i^q$, $q = 1, \cdots, p$, in the second claim of Lemma 3. Here we only concentrate on Case 1: $x_{it}^{q-1} = x_{it} + \omega$ and $x_{i(k+r)}^{q-1} = x_{i(k+r)}$, $\omega > 0$.

Suppose to the contrary that $\Delta U_i^q = U_i(\mathbf{x}_i^q, \mathbf{x}_{-i}) - U_i(\mathbf{x}_i^{q-1}, \mathbf{x}_{-i}) > 0$, then

$$0 < U_i(\mathbf{x}_i^q, \mathbf{x}_{-i}) - U_i(\mathbf{x}_i^{q-1}, \mathbf{x}_{-i}) =$$

$$\left(\frac{x_{i(k+r)} - \eta_q}{\sum_{g=1}^n x_{g(k+r)} - \eta_q} - \frac{x_{i(k+r)}}{\sum_{g=1}^n x_{g(k+r)}}\right)\mu_{k+r} + \left(\frac{x_{it} + \omega + \eta_q}{\sum_{g=1}^n x_{gt} + \omega + \eta_q} - \frac{x_{it} + \omega}{\sum_{g=1}^n x_{gt} + \omega}\right)\mu_t.$$

It is equivalent to

$$\frac{(\sum_{g\neq i} x_{gt})\mu_t}{(\sum_{g=1}^n x_{gt} + \omega)(\sum_{g=1}^n x_{gt} + \eta_q + \omega)} > \frac{(\sum_{g\neq i} x_{g(k+r)})\mu_{k+r}}{(\sum_{g=1}^n x_{g(k+r)})(\sum_{g=1}^n x_{g(k+r)} - \eta_q)}$$

$$\Leftrightarrow \eta_q < \frac{(\sum_{g\neq i} x_{gt})\mu_t(\sum_{g=1}^n x_{g(k+r)})^2 - (\sum_{g\neq i} x_{g(k+r)})\mu_{k+r}(\sum_{g=1}^n x_{gt} + \omega)^2}{(\sum_{g\neq i} x_{gt})\mu_t(\sum_{g=1}^n x_{g(k+r)}) + (\sum_{g\neq i} x_{g(k+r)})\mu_{k+r}(\sum_{g=1}^n x_{gt} + \omega)}. \quad (9)$$

Since

$$\frac{(\sum_{g\neq i} x_{gt}) \cdot \mu_t}{(\sum_{g=1}^n x_{gt} + \omega)^2} < \frac{\sum_{g\neq i} x_{gt} \cdot \mu_t}{(\sum_{g=1}^n x_{gt})^2} = \frac{\sum_{g\neq i} x_{g(k+r)} \cdot \mu_{k+r}}{(\sum_{g=1}^n x_{g(k+r)})^2},$$

where the equation is from condition (8), we can continue (9) to be

$$\eta_q < \frac{(\sum_{g\neq i} x_{gt})\mu_t(\sum_{g=1}^n x_{g(k+r)})^2 - (\sum_{g\neq i} x_{g(k+r)})\mu_{k+r}(\sum_{g=1}^n x_{gt} + \omega)^2}{(\sum_{g\neq i} x_{gt})\mu_t(\sum_{g=1}^n x_{g(k+r)}) + (\sum_{g\neq i} x_{g(k+r)})\mu_{k+r}(\sum_{g=1}^n x_{gt} + \omega)} < 0.$$

This contradicts to the condition that $\eta_q > 0$. Therefore $\Delta U_i^q \leq 0$ for Case 1.

By the similar analysis, we also can get $\Delta U_i^q \leq 0$ for other two cases. Thus each $\Delta U_i^q \leq 0$, $q = 1, \cdots, p$. It is not hard to see that the total difference ΔU_i can be partitioned as

$$\begin{aligned}\Delta U_i &= \left[U_i(\mathbf{x}_i^1, \mathbf{x}_{-i}) - U_i(\mathbf{x})\right] + \cdots + \left[U_i(\mathbf{x}'_i, \mathbf{x}_{-i}) - U_i(\mathbf{x}_i^{p-1}, \mathbf{x}_{-i})\right] \\ &= \Delta U_i^1 + \cdots + \Delta U_i^p. \end{aligned} \quad (10)$$

It implies $\Delta U_i \leq 0$, since each component $\Delta U_i^q \leq 0$ in (10). So miner i can not improve his utility by changing allocation from $\mathbf{x}_i$ to $\mathbf{x}'_i$. □

Based on the sufficient and necessary condition for a Nash equilibrium, we will propose the closed-form solution of a pure Nash equilibrium and prove the the uniqueness of such a pure Nash equilibrium in following theorem.

Theorem 2. *A hash power allocation profile* $\mathbf{x} = (x_{ij})$ *is a Nash equilibrium, if and only if it has the form as* $x_{ij} = \frac{\mu_j}{\sum_{\ell=1}^m \mu_\ell} \cdot h_i$, *for any* $i \in N$ *and* $j \in M$.

Proof. It is not hard to see that once each x_{ij} has the form as $x_{ij} = \frac{\mu_j}{\sum_{\ell=1}^m \mu_\ell} \cdot h_i$, then for any $j \in M$,

$$\frac{\sum_{h\neq i} x_{hj}}{(\sum_{h=1}^n x_{hj})^2} \cdot \mu_j = \frac{\sum_{h=1}^n h_h - h_i}{(\sum_{h=1}^n h_h)^2}(\sum_{\ell=1}^m \mu_\ell),$$

which is irrelevant to the cyprocurrency j and such a ratio can be defined as α_i. Then the allocation $\mathbf{x} = (x_{ij})$ with $x_{ij} = \frac{\mu_j}{\sum_{\ell=1}^m \mu_\ell} \cdot h_i$ is a Nash equilibrium by Theorem 1.

On the other hand, Theorem 1 tells us for any $j \in M$, $\frac{\sum_{h\neq i} x_{hj}}{(\sum_{h=1}^n x_{hj})^2} \cdot \mu_j = \alpha_i$. It implies

$$\sum_{i=1}^n \alpha_i = \sum_{i=1}^n \frac{\sum_{h\neq i} x_{hj}}{(\sum_{h=1}^n x_{hj})^2}\mu_j = \frac{(n-1)\sum_{h=1}^n x_{hj}}{(\sum_{h=1}^n x_{hj})^2}\mu_j = \frac{n-1}{\sum_{h=1}^n x_{hj}}\mu_j. \tag{11}$$

From Eq. (11), we continue to have

$$\sum_{i=1}^n \alpha_i = \frac{(n-1)\mu_1}{\sum_{h=1}^n x_{h1}} = \cdots = \frac{(n-1)\mu_m}{\sum_{h=1}^n x_{hm}}.$$

Then

$$\frac{\mu_1}{\sum_{h=1}^n x_{h1}} = \cdots = \frac{\mu_m}{\sum_{h=1}^n x_{hm}} = \frac{\sum_{j=1}^m \mu_j}{\sum_{j=1}^m \sum_{i=1}^n x_{ij}} = \frac{\sum_{j=1}^m \mu_j}{\sum_{i=1}^n h_i}.$$

So $\sum_{i=1}^n \alpha_i = (n-1) \cdot \frac{\sum_{j=1}^m \mu_j}{\sum_{i=1}^n h_i}$, which is a constant. On the other hand, from Eq. (11), we can get

$$\frac{\sum_{h=1}^n x_{hj}}{\mu_j} = \frac{n-1}{\sum_{i=1}^n \alpha_i} \quad \forall j \in M; \tag{12}$$

Then for any $j \in M$,

$$\alpha_i = \frac{\sum_{h\neq i} x_{hj}}{(\sum_{h=1}^n x_{hj})^2}\mu_j = \frac{\sum_{h\neq i} x_{hj}}{\mu_j}\left(\frac{\mu_j}{\sum_{h=1}^n x_{hj}}\right)^2 = \frac{\sum_{h\neq i} x_{hj}}{\mu_j}\left(\frac{\sum_{i=1}^n \alpha_i}{n-1}\right)^2, \tag{13}$$

where the last equality is from (12). Also Eq. (13) guarantees

$$\frac{\sum_{h\neq i} x_{hj}}{\mu_j} = \frac{(n-1)^2 \alpha_i}{(\sum_{i=1}^n \alpha_i)^2}. \tag{14}$$

Furthermore, the difference between (12) and (14) is

$$\frac{x_{ij}}{\mu_j} = \frac{n-1}{\sum_{i=1}^n \alpha_i} - \frac{(n-1)^2 \alpha_i}{(\sum_{i=1}^n \alpha_i)^2}. \tag{15}$$

The right side of (15) shows $\frac{x_{ij}}{\mu_j}$ is only related to index i which can be denoted by γ_i. So $x_{ij} = \gamma_i \mu_j$. In addition, by the condition of $\sum_{j=1}^{m} x_{ij} = h_i$, we have

$$\sum_{j=1}^{n} x_{ij} = \sum_{j=1}^{n} \gamma_i \mu_j = \gamma_i \sum_{j=1}^{n} \mu_j = h_i.$$

Therefore, $\gamma_i = \frac{h_i}{\sum_{j=1}^{n} \mu_j}$ and $x_{ij} = \frac{\mu_j}{\sum_{\ell=1}^{m} \mu_\ell} \cdot h_i$. It concludes this claim. □

Based on the result of Theorem 2 and the formation of each miner's expected payoff (3), we can easily get the following corollary.

Corollary 1 *Under the pure Nash equilibrium allocation* $x_{ij} = \frac{\mu_j}{\sum_{\ell=1}^{m} \mu_\ell} \cdot h_i$ *for any* $i \in N$ *and* $j \in M$*, each miner* i*'s expected payoff is* $\frac{h_i}{\sum_{\ell=1}^{n} h_\ell}(\sum_{\ell=1}^{m} \mu_\ell)$.

4 Conclusion

This paper discusses the issue of a hashing power allocation game in cryptocurrencies, in which there are n miners equipped with certain computing power and m different cryptocurrencies to be mined. Each miner shall allocate his computing power in mining the cryptocurrencies properly to compete with others to maximize his payoff. In this paper, we mainly consider the hashing power allocation game with the risk-neutral objective. We show that the Nash Equilibrium allocation of this game is unique and follows a proportional rule where each miner will allocate his total computing power to a given cryptocurrency proportional to the percentage of the award among all currencies, while his expected revenue is proportional to the percentage of the hashing power possessed and total award.

Besides, the risk-averse objective is also interesting for us to consider in the future. For each risk-averse miner i, he tries to minimize the uncertainty, that is to minimize the objective of $y_i(\mathbf{x})^T \Sigma y_i(\mathbf{x})$, subject to the constraint of $\sum_{j \in M} x_{ij} = h_i$. Here vector $y_i(\mathbf{x})$ is defined as (2) and $\Sigma \succ 0$ is the covariance matrix of uncertain reward $\mathbf{R}$, i.e. $\Sigma = \mathrm{Cov}[\mathbf{R}]$. For this kind of game, we are also concerned about the existence and uniqueness of a pure Nash equilibrium. In addition, how to compute a pure Nash equilibrium is our task too.

Acknowledgments. The first author's research is partially supported by the National Nature Science Foundation of China (No. 11301457). The second author's research is supported by the Natural Sciences and Engineering Research Council of Canada (NSERC) grant 06446, and NNSF of China 11771386 and 11728104. The third author's research is supported by NSFC (No. 10971187) and the NSF of 415 Zhejiang Province grant LQ12A01011.

References

1. Ahmadinejad, A.M., Dehghani, S., Hajiaghayi, M.T., Lucier, B., Mahini, H., Seddighin, S.: Computing equilibria of blotto and other games, from duels to battefields (2016)
2. Alpern, S., Howard, J.V.: Winner-take-all games: the strategic optimisation of rank. Oper. Res. **65**, 1165–1176 (2017)
3. AntPool. https://www.antpool.com/
4. Biais, B., Bisiere, C., Bouvard, M., Casamatta, C.: The Blockchain Folk Theorem. In: Social Science Electronic Publishing (2017)
5. Bitcoin. https://bitcoin.org/en/
6. Bonneau, J., Miller, A., Clark, J., Narayanan, A., Kroll, J.A., Felten, E.W.: Research perspectives and challenges for bitcoin and cryptocurrencies to appear. pp. 104–121 (2015)
7. BTC.com. https://btc.com/
8. BTC.TOP. http://btc.top/
9. Ethereum. https://www.ethereum.org/
10. Eyal, I.: The Miner's Dilemma, pp. 89–103. Computer Science (2015)
11. Eyal, I., Sirer, E.G.: Majority is not enough: bitcoin mining is vulnerable. Eprint Arxiv **8437**, 436–454 (2013)
12. Feldman, M., Lai, K., Zhang, L.: The proportional-share allocation market for computational resources. IEEE Trans. Parallel Distrib. Syst. **20**(8), 1075–1088 (2009)
13. Fisch, B., Pass, R., Shelat, A.: Socially optimal mining pools. In: International Conference on Web and Internet Economics, pp. 205–218 (2017)
14. Göbel, J., Keeler, H.P., Krzesinski, A.E., Taylor, P.G.: Bitcoin blockchain dynamics: the selfish-mine strategy in the presence of propagation delay. Perform. Eval. **104**, 23–41 (2016)
15. Goldberg, L.A., Goldberg, P.W., Krysta, P., Ventre, C.: Ranking games that have competitiveness-based strategies. In: ACM Conference on Electronic Commerce, pp. 335–344 (2010)
16. Hart, S.: Discrete colonel blotto and general lotto games. Int. J. Game Theory **36**(3–4), 441–460 (2008)
17. Kiayias, A., Koutsoupias, E., Kyropoulou, M., Tselekounis ,Y.: Blockchain mining games. In: ACM Conference on Economics and Computation, pp. 365–382 (2016)
18. Markowitz, H.: Portfolio selection. J. Financ. **7**(1), 77–91 (1952)
19. Mas-Colell, A., Whinston, M.D., Green, J.R.: Microeconomic Theory. Oxford University Press, Oxford (1995)
20. Nakamoto, S.: Bitcoin: a peer-to-peer electronic cash system. Consulted (2008)
21. Carlsten, M., Kalodner, H., Weinberg, S.M., Narayanan, A.: On the instability of bitcoin without the block reward. In: ACM Sigsac Conference on Computer and Communications Security, pp. 154–167 (2016)
22. Nisan, N., Roughgarden, T., Tardos, E., Vazirani, V.V.: Algorithmic Game Theory. Cambridge University Press, Cambridge (2007)
23. Proof of burn. https://en.bitcoin.it/wiki/proofofburn
24. Parham, R.: The predictable cost of bitcoin. Social Science Electronic Publishing (2017)
25. Slush Pool. https://slushpool.com/home/
26. Ripple. https://ripple.com/
27. Roberson, B.: The colonel blotto game. Econ. Theory **29**(1), 1–24 (2006)

28. Rosen, J.B.: Existence and uniqueness of equilibrium points for concave n-person games. Econometrica **33**(3), 520–534 (1965)
29. Rosenfeld, M.: Analysis of bitcoin pooled mining reward systems. Computer Science, December (2011)
30. Saleh, F.: Blockchain without waste: proof-of-stake (2017)
31. Shapley, L., Shubik, M.: Trade using one commodity as a means of payment. J. Polit. Econ. **85**(5), 937–968 (1977)
32. ViaBTC. https://viabtc.com/
33. Wattenhofer, R.: The Science of the Blockchain. CreateSpace Independent Publishing Platform, Charleston (2016)

Resource Based Cooperative Games: Optimization, Fairness and Stability

Ta Duy Nguyen[(✉)] and Yair Zick

School of Computing, National University of Singapore, Singapore, Singapore
nguyentaduy@u.nus.edu.sg, zick@comp.nus.edu.sg

Abstract. We study the class of *resource-based coalitional games*. We provide efficient algorithms to compute solution concepts for weighted voting games, threshold task games and r-weighted voting games; in particular, we compute approximately optimal coalition structures, and present non-trivial bounds on the cost of stability for these classes; in particular, we improve upon the bounds given in [2] for weighted voting games.

Keywords: Cooperative games · Cost of stability
Optimal coalition structure generation

1 Introduction

Several real-world scenarios require strategic agents to pool their resources in order to complete tasks. For example, in the EU council of members, a resolution requires support by at least 55% of member states, who must also represent at least 65% of the EU population; in computational domains, each agent has a certain amount of a computational resource (e.g. RAM, CPU cycles etc.) that are allocated to complete tasks. Such scenarios can be intuitively modeled as coalitional games based on the notion of *tasks* and *resources*, which we term *resource based cooperative games*. Resource-based games include the canonical class of *weighted voting games* (WVGs), as well as *threshold task games* (TTGs) [8], and *vector weighted voting games* [14]. Computing cooperative solution concepts in resource based cooperative games is computationally intractable, even for WVGs. However, there has been little effort towards computing approximate solution concepts for this class. This is where our work comes in.

1.1 Our Contribution

We propose efficient approximation algorithms for resource based cooperative games. We study three classes of games: weighted voting games (WVGs), threshold task games (TTGs) and r-weighted voting games (r-WVGs); for each class, we provide an efficient algorithm finding approximately optimal coalition structures, and bounds on the cost of stability. Additional results on r-TTGs appear in Appendix ??; proofs and additional details are in the appendix.

X. Deng (Ed.): SAGT 2018, LNCS 11059, pp. 239–244, 2018.
https://doi.org/10.1007/978-3-319-99660-8_21

1.2 Related Work

Weighted voting games are an extremely well-studied class of games: on the one hand, they are computationally succinct (requiring only n weights and a threshold to describe); on the other hand, computing solution concepts for WVGs is well-known to be computationally intractable [13–15,18]. The complexity of solution concepts for general cooperative games has been well studied in the literature, dating back to Deng and Papadimitriou [12]; more recent works include [6,10,16,17] (see [9,11] for an overview). Chalkiadakis et al. [8] introduce the class of threshold task games; however, they allow players to allocate partial resources to tasks. The only work we are aware of that studies a TTG model in the classic cooperative game setting is by Balcan et al. [5]. The optimal coalition structure generation problem is also well-studied (see Rahwan et al. [20] for a recent overview). Other related works include [3,4]. Anshelevich and Sekar [1] study stability under a similar model, where tasks are limited in supply. Bachrach et al. [2] introduce the cost of stability, and study the cost of stability for WVGs; however, they assume that coalition structures do not form, resulting in a higher cost of stability; other works studying the cost of stability include [7,19,21].

2 Preliminaries

A *cooperative game* $G = \langle N, v\rangle$ consists of a set of players $N = \{1, \dots, n\}$ and a *characteristic function* $v : 2^N \to \mathbb{R}$. Given a set of players $S \subseteq N$ (also known as a *coalition*), $v(S)$ is the value of S; we assume that $v(\emptyset) = 0$, and that v is monotone: if $S \subseteq T \subseteq N$ then $v(S) \le v(T)$. A *coalition structure* is a partition of N into disjoint coalitions. We say that a coalition structure CS^* is *optimal* if it maximizes social welfare; that is, CS^* maximizes $\sum_{S\in CS} v(S)$. Let $OPT(G)$ be the value of an optimal coalition structure over G. We refer to the problem of finding an optimal coalition structure (also known as the *coalition structure generation problem*) as OptCS. We say that a coalition structure CS^* is *β-optimal* for G if $v(CS^*) \ge \beta OPT(G)$.

Once players form coalitions, they need to find some reasonable way to divide the revenue they generate. Given a coalition structure CS, an *imputation* for CS is a vector $\boldsymbol{x} \in \mathbb{R}^n_+$ satisfying $\sum_{i\in S} x_i = v(S)$ for all $S \in CS$. The tuple $\langle CS, \boldsymbol{x}\rangle$ is called an *outcome* of G. Let $\mathcal{I}(G)$ be the set of all outcomes for G. The core is a subset of $\mathcal{I}(G)$ from which no coalition can deviate; that is,

$$Core(G) = \{\langle CS, \boldsymbol{x}\rangle \in \mathcal{I}(G) : x(S) \ge v(S), \forall S \subseteq N\}.$$

We observe that if $\langle CS, \boldsymbol{x}\rangle \in Core(G)$ then CS must be optimal (else at least one coalition S belonging to an optimal coalition structure CS^* can deviate). Unfortunately, the core of a game can be empty.

Example 2.1. Consider a 3 player game where $v(S) = 0$ if $|S| \le 1$ and is 1 otherwise. The optimal coalition structure has a value of 1. However, it is easy to check that if every coalition of size 2 receives a payoff of at least 1, the total payoff to all players must be at least $\frac{3}{2}$ (paying $\frac{1}{2}$ to every player).

As Example 2.1 implies, stabilizing a game may require an additional external subsidy. The minimal total payoff needed can be found by solving the following linear program

$$\min \sum_{i \in N} x_i, \text{ s.t. for all } S \subseteq N\colon x(S) \geq v(S). \tag{1}$$

Let V^* be the value of the optimal solution to (1); the *relative cost of stability* of a game G is $CoS(G) = \frac{V^*}{OPT(G)}$.[1]

Resource-based Cooperative Games:
Weighted voting games (WVGs) are the simplest class of resource-based cooperative games: a WVG is defined by a weight vector $\boldsymbol{w} \in Z_+^n$ and a threshold (or quota) $q \in \mathbb{Z}_+$, such that $v(S) = 1$ if $w(S) \geq q$, and $v(S) = 0$ otherwise.
Threshold task games (TTGs) are similar to WVGs. Just like in WVGs, each player has a weight $w_i \in \mathbb{Z}_+$; we now have a set of tasks $\mathcal{T} = \{t_1, \ldots, t_m\}$, where each task $t_j \in \mathcal{T}$ has a threshold $q_j \in \mathbb{Z}_+$ and a value $v_j \in \mathbb{Z}_+$. The characteristic function v is: $v(S) = \max_{j \in [m]}\{v_j : w(S) \geq q_j\}$.

We can further consider multiple resource *types*: players have r different resources, and each task requires a certain amount of each resource to be completed. More formally, v is an r-TTG if each player owns a resource vector $\boldsymbol{w}_i \in \mathbb{R}_+^r$; there is a set of tasks $\mathcal{T}$, where every task $t_j \in \mathcal{T}$ has a value v_j and a threshold vector $\boldsymbol{q}_j \in \mathbb{R}_+^r$. The value of a coalition $S \subseteq N$ is given by $v(S) = \max_{j \in [m]} \left\{v_j : \sum_{i \in S} \boldsymbol{w}_i \geq \boldsymbol{q}_j\right\}$. r-WVGs [14] (also known as vector WVGs) are a subclass of r-TTGs with a single task.

3 Warmup: Weighted Voting Games

We begin our exploration with WVGs, the most basic class of resource-based cooperative games. Computing an optimal coalition structure for WVGs is computationally intractable [13] (note that this immediately implies intractablity for TTGs, and in particular for r-TTGs). Thus, our first objective is to establish an approximation algorithm for the optimal coalition structure problem. Let G be a WVG with weights $\boldsymbol{w}$ and a threshold q. We begin by presenting a $\frac{1}{2}$-approximation algorithm for the optimal coalition structure problem (given in Appendix ??).

Theorem 3.1. *Given a WVG G, Algorithm ?? outputs a $\frac{1}{2}$-optimal coalition structure for G.*

Let us now turn to the problem of bounding the cost of stability in weighted voting games. The cost of stability in WVGs has been studied in [2]; however, their analysis does not assume the formation of optimal coalition structures. It turns out that doing so offers a significantly better approximation guarantee; in fact, the bound presented in Theorem 3.2 is tighter as $OPT(G)$ grows.

[1] In the original work defining the cost of stability [2], the cost of stability is defined as $V^* - OPT(G)$. Subsequent works (e.g. [7,19]) utilize the definition we use here.

Theorem 3.2. *Given a WVG G, $CoS(G) < \frac{3}{2} + \frac{1}{OPT(G)}$.*

The bound in Theorem 3.2 is not tight; consider a 3-player WVG where each player has a weight of 1, and the threshold is 2; this game is equivalent to the one in Example 2.1, and its cost of stability is $\frac{3}{2}$.

4 Threshold Task Games

We now turn our attention to threshold task games; recall that a TTG is given by a weight vector $\boldsymbol{w}$, and a set of tasks $\mathcal{T} = \{t_1, \ldots, t_m\}$, where each task t_j has a threshold q_j and a value v_j. The value of a coalition is the value of the best task that it can complete with its total weight. For general TTGs, we assume that the weight of each player is no more than the minimum threshold (i.e. that the value of single players is 0); this is a departure from the framework of Theorem 3.1 where single player coalitions were allowed.

Theorem 4.1. *Let G be a TTG, such that $v(\{i\}) = 0$ for all $i \in N$, and let v^* be the value of the most valuable task in $\mathcal{T}$; then there exists an efficient algorithm that outputs a coalition structure CS^* whose value is at least $\frac{1}{2}(OPT(G) - v^*)$.*

Leveraging the result in [8], Algorithm ?? (Appendix ??) outputs a $\frac{1}{2}$ approximation for the OPTCS problem in TTGs, and runs in pseudopolynomial time.

Theorem 4.2. *Given an n player TTG G, such that $v(\{i\}) = 0$ for all $i \in N$, Algorithm ?? outputs a $\frac{1}{2}$-optimal coalition structure for G, and runs in time polynomial in n and W, the sum of player weights.*

Whether there exists a truly polynomial time $\frac{1}{2}$-approximation algorithm for the optimal coalition structure problem for TTGs remains an open problem. Next, let us examine the cost of stability for TTGs.

Theorem 4.3. *For any TTG G, $CoS(G) \leq 2$; this bound is tight.*

5 r Weighted Voting Games

Let us now address the class of r-WVGs. Assuming multiple resource types significantly increases problem complexity; however, Algorithm ?? outputs an approximately optimal coalition structure of an r-WVG, with an approximation quality that exponentially decreases in r (see Appendix ??).

Theorem 5.1. *Given an r-WVG G, Algorithm ?? outputs a $\frac{1}{2\times 3^{r-1}}$-optimal coalition structure for G.*

Using Theorem 5.1 we can bound $CoS(G)$ when G is an r-WVG: i.e. an r-TTG with only one task. The full proof appears in Sect. ??.

Theorem 5.2. *If G is an r-WVG then $CoS(G) \leq 2 \times 3^{r-1}$.*

References

1. Anshelevich, E., Sekar, S.: Computing stable coalitions: approximation algorithms for reward sharing. In: Proceedings of the 11th Conference on Web and Internet Economics (WINE), pp. 31–45 (2015)
2. Bachrach, Y., Elkind, E., Meir, R., Pasechnik, D., Zuckerman, M., Rothe, J., Rosenschein, J.S.: The cost of stability in coalitional games. In: Mavronicolas, M., Papadopoulou, V.G. (eds.) SAGT 2009. LNCS, vol. 5814, pp. 122–134. Springer, Heidelberg (2009). https://doi.org/10.1007/978-3-642-04645-2_12
3. Bachrach, Y., Kohli, P., Kolmogorov, V., Zadimoghaddam, M.: Optimal coalition structure generation in cooperative graph games. In: Proceedings of the 27th AAAI Conference on Artificial Intelligence (AAAI), pp. 81–87 (2013)
4. Bachrach, Y., Meir, R., Jung, K., Kohli, P.: Coalitional structure generation in skill games. In: Proceedings of the 24th AAAI Conference on Artificial Intelligence (AAAI), vol. 10, pp. 703–708 (2010)
5. Balcan, M., Procaccia, A., Zick, Y.: Learning cooperative games. In: Proceedings of the 24th International Joint Conference on Artificial Intelligence (IJCAI), pp. 475–481 (2015)
6. Bistaffa, F., Farinelli, A., Chalkiadakis, G., Ramchurn, S.D.: A cooperative game-theoretic approach to the social ridesharing problem. Artif. Intell. **246**, 86–117 (2017)
7. Bousquet, N., Li, Z., Vetta, A.: Coalition games on interaction graphs: a horticultural perspective. In: Proceedings of the 16th ACM Conference on Economics and Computation (EC), pp. 95–112 (2015)
8. Chalkiadakis, G., Elkind, E., Markakis, E., Polukarov, M., Jennings, N.R.: Cooperative games with overlapping coalitions. J. Artif. Intell. Res. **39**(1), 179–216 (2010)
9. Chalkiadakis, G., Elkind, E., Wooldridge, M.: Computational Aspects of Cooperative Game Theory. Morgan and Claypool, San Rafael (2011)
10. Chalkiadakis, G., Greco, G., Markakis, E.: Characteristic function games with restricted agent interactions: core-stability and coalition structures. Artif. Intell. **232**, 76–113 (2016)
11. Chalkiadakis, G., Wooldridge, M.: Weighted voting games. In: Brandt, F., Conitzer, V., Endriss, U., Lang, J., Procaccia, A. (eds.) Handbook of Computational Social Choice, Chap. 16. Cambridge University Press (2016)
12. Deng, X., Papadimitriou, C.: On the complexity of cooperative solution concepts. Math. Oper. Res. **19**(2), 257–266 (1994)
13. Elkind, E., Chalkiadakis, G., Jennings, N.R.: Coalition structures in weighted voting games. In: Proceedings of the 18th European Conference on Artificial Intelligence (ECAI), pp. 393–397 (2008)
14. Elkind, E., Goldberg, L.A., Goldberg, P., Wooldridge, M.: On the dimensionality of voting games. In: Proceedings of the 23rd AAAI Conference on Artificial Intelligence (AAAI), pp. 69–74 (2008)
15. Elkind, E., Goldberg, L., Goldberg, P., Wooldridge, M.: On the computational complexity of weighted voting games. Ann. Math. Artif. Intell. **56**, 109–131 (2009)
16. Igarashi, A., Izsak, R., Elkind, E.: Cooperative games with bounded dependency degree. CoRR abs/1711.07310 (2017)
17. Lesca, J., Perny, P., Yokoo, M.: Coalition structure generation and CS-core: results on the tractability frontier for games represented by MC-nets. In: Proceedings of the 16th International Conference on Autonomous Agents and Multi-Agent Systems (AAMAS), pp. 308–316 (2017)

18. Matsui, T., Matsui, Y.: A survey of algorithms for calculating power indices of weighted majority games. J. Oper. Res. Soc. Jpn. **43**(1), 71–86 (2000)
19. Meir, R., Zick, Y., Elkind, E., Rosenschein, J.S.: Bounding the cost of stability in games over interaction networks. In: Proceedings of the 27th AAAI Conference on Artificial Intelligence (AAAI), pp. 690–696 (2013)
20. Rahwan, T., Michalak, T.P., Wooldridge, M., Jennings, N.R.: Coalition structure generation: a survey. Artif. Intell. **229**(Suppl. C), 139–174 (2015)
21. Resnick, E., Bachrach, Y., Meir, R., Rosenschein, J.S.: The cost of stability in network flow games. Math. Found. Comput. Sci. **2009**, 636–650 (2009)

Short Paper: Strategic Contention Resolution in Multiple Channels with Limited Feedback

George Christodoulou[1], Themistoklis Melissourgos[1(✉)], and Paul G. Spirakis[1,2]

[1] Department of Computer Science, University of Liverpool, Liverpool, UK
{G.Christodoulou,T.Melissourgos,P.Spirakis}@liverpool.ac.uk
[2] Computer Engineering and Informatics Department, University of Patras, Patras, Greece

Abstract. We consider a game-theoretic setting of contention in communication networks. In a *contention game* each of $n \geq 2$ identical players has a single information packet that she wants to transmit in a fast and selfish way through one of $k \geq 1$ multiple-access channels by choosing a protocol. Here, we extend the model and results of the single-channel case studied in [2] by providing equilibria characterizations for more than one channels, and giving specific anonymous, equilibrium protocols with finite and infinite expected latency. For our equilibrium protocols with infinite expected latency, all players, with high probability transmit successfully in optimal time, i.e. $\Theta(n/k)$.

Keywords: Contention resolution · Multiple channels
Acknowledgement-based protocol · Ternary feedback · Game theory

1 Introduction

The need for multiple channels in communications has become clear in today's technologies. Robustness and high throughput are two main goals that multiple-channels communication systems try to achieve, since dependence from a small group of nodes in a network as well as collision of packets that are transmitted on the same node are the issues from which single-channel broadcast communications suffer. Many works in the Electrical and Electronics Engineering community have so far considered *multi-channel* medium access control (MAC) protocols (e.g. [6]) which have been shown to achieve higher throughput and lower delay than the single-channel MAC protocols. The limited feedback in such systems is caused by the *multi-channel hidden terminal problem* [7]. To the authors' knowledge, *strategic* behaviour in such multi-channel systems is limited to the Aloha protocol [5], contrary to the case of single-channel systems (e.g. [1]).

P. G. Spirakis—The work of this author was partially supported by the ERC Project ALGAME.

X. Deng (Ed.): SAGT 2018, LNCS 11059, pp. 245–250, 2018.
https://doi.org/10.1007/978-3-319-99660-8_22

For equilibrium protocols, a desired property is *anonymity*, that is, protocols which do not use player IDs. If a players' protocol depended on her ID, then equilibria are simple, but can be unfair as well; scheduling each player's transmission through a priority queue according to her ID is an equilibrium. The only works on acknowledgement-based, equilibrium protocols, is by Christodoulou et al. [2,3] which consider only a single channel. Among other results, they give the unique, anonymous, equilibrium protocol with finite expected latency for 2 players, and an efficient protocol with infinite expected latency for at least 3 players. The existence of a symmetric equilibrium with finite expected latency remains an open problem, even for three players. However, for the settings with 2 and 3 transmission channels, we manage to present simple, anonymous protocols for up to 4 and 5 players respectively.

In this short paper, we examine the problem of *strategic contention resolution* in multi-channel systems, where obedience to a suggested protocol is not required. We provide two types of equilibrium protocols. The first type (Sect. 3) describes an anonymous, equilibrium protocol that yields finite expected time of successful transmission to a player. Similarly, the second type (Sect. 4) describes an anonymous, equilibrium protocol which yields infinite expected latency to a player but is also *efficient*, that is, all players transmit successfully within $\Theta(\frac{\#players}{\#channels})$ time with high probability. The latter result makes clear the advantage (with respect to time efficiency) that multiple channels bring to a system with strategic users, which is that the time until all players transmit successfully with high probability is inversely proportional to the number of available channels.

2 The Model and Definitions

Game Structure. We define a *contention game* as follows. Assume a set of players $[n] = \{1, 2, \ldots, n\}$ and a set of channels $K = \{1, 2, \ldots, k\}$. Each player has a single packet that needs to be sent through a channel in K, without caring about the identity of the channel. All players know n and K. Time is discrete, i.e. $t = 1, 2, \ldots$. The players that have not yet successfully transmitted their packet are called *pending* and initially all n players are pending. At any t, a pending player i has a set $A = \{0, 1, 2, \ldots, k\}$ of *pure strategies*: a pure strategy $a \in A$ is the action of choosing channel $a \in K$ to transmit her packet on, or no transmission ($a = 0$). At time t, a *(mixed) strategy* of a player i is a probability distribution over A that potentially depends on information that i has gained from the process based on previous transmission attempts. If exactly one player transmits on a channel in a given slot t, then her transmission is *successful*, she is no longer pending, and the game continues with the rest of the players. However, whenever two or more players try to transmit on the same channel at the same time slot, a collision occurs and they remain pending. The game continues until there are no pending players.

Transmission Protocols. Let $X_{i,t} \in A$ be the channel-indicator variable that keeps track of the identity of the channel where player i attempted transmission at time t; value 0 indicates no transmission attempt. An *acknowledgement-based* protocol uses very limited channel feedback. After each time step t, the information received by a player i who transmitted during t is whether her transmission was successful (in which case she gets an acknowledgement and exits the game) or whether there was a collision. Let $\vec{h}_{i,t}$ be the vector of the *personal transmission history* of player i up to time t, i.e. $\vec{h}_{i,t} = (X_{i,1}, X_{i,2}, \ldots, X_{i,t})$. A *decision rule* $f_{i,t}$ for a pending player i at time t, is a function that maps $\vec{h}_{i,t-1}$ to a strategy, with elements $\Pr(X_{i,t} = a|\vec{h}_{i,t-1})$ for all $a \in A$. For a player $i \in N$, a *(transmission) protocol* f_i is a sequence of decision rules $f_i = \{f_{i,t}\}_{t \geq 1} = f_{i,1}, f_{i,2}, \ldots$. When the context is clear enough we will drop some of the indices accordingly.

Individual Utility and Equilibria. For a *protocol profile* $\vec{f} = (f_1, f_2, \ldots, f_n)$, we denote the *expected latency* of player $i \in [n]$, given a history $\vec{h}_{i,t}$ by $C_i^{\vec{f}}(\vec{h}_{i,t})$. We say that $\vec{f}$ is an *equilibrium* if for any transmission history $\vec{h}_t$ the players cannot decrease their expected latency by unilaterally deviating after t.

3 Equilibria with Expected Latency $< \infty$

Nash Equilibria Characterization. Here we provide a characterization of general equilibria (both symmetric and asymmetric) for an arbitrary number of channels $k \geq 1$ and players $n \geq 2$.

Let $\vec{f} = (f_1, f_2, \ldots, f_n)$ be a tuple of acknowledgement-based protocols (not necessarily anonymous) for the n players. For a (finite) positive integer τ^*, and a given history $\vec{h}_{i,\tau^*} = (a_{i,1}, a_{i,2}, \ldots, a_{i,\tau^*})$, define for player i the protocol

$$g_i = g_i(\vec{h}_{i,\tau^*}) \triangleq \begin{cases} (\Pr\{X_{i,t} = a_{i,t}\} = 1), & \text{for } 1 \leq t \leq \tau^* \\ f_{i,t}, \quad \text{for } t > \tau^*. \end{cases} \tag{1}$$

We will call a personal history $\vec{h}_{i,\tau^*}$ *consistent with* the protocol profile $\vec{f}$ if there is a non-zero probability that $\vec{h}_{i,\tau^*}$ will occur for player i under $\vec{f}$. If $\vec{h}_{i,\tau^*}$ is consistent with $\vec{f}$ we call protocol $g_i(\vec{h}_{i,\tau^*})$ *consistent* with $\vec{f}$, and when clear from the context we write g_i instead. Also, we denote the set of all g_i's, that is, all $g_i(\vec{h}_{i,t})$'s for all $t \geq 1$, which are consistent with $\vec{f}$ by $\mathcal{G}_i^{\vec{f}}$.

In a protocol profile $\vec{f}$, a player is interested in her expected latency $C_i^{\vec{f}}(h_{i,0})$ at the start of the game, denoted by just $C_i^{\vec{f}}$.

Lemma 1 (Equilibrium characterization). *The following statements are equivalent:*

(i) $\vec{f}$ *is an equilibrium.*

(ii) $\forall$ *player* $i \in [n]$ $\begin{cases} (a) \quad C_i^{(\vec{f}_{-i}, g_i)} = C_i^{(\vec{f}_{-i}, r_i)} = C_i^{\vec{f}}, \quad \forall g_i, r_i \in \mathcal{G}_i^{\vec{f}}, \text{ and} \\ (b) \quad C_i^{(\vec{f}_{-i}, g_i)} \leq C_i^{(\vec{f}_{-i}, r_i)}, \quad \forall g_i \in \mathcal{G}_i^{\vec{f}}, r_i \notin \mathcal{G}_i^{\vec{f}}. \end{cases}$

Now we are ready to give anonymous, equilibrium protocols for 2 and 3 channels. Let us define the following memoryless protocol with parameter $k \in \{2, 3\}$ which corresponds to the number of channels.

Protocol $\mathbf{f^k}$: For any player i, every $t \geq 1$, and any transmission history,

$$f^k_{i,t} = \left(\Pr\{X_{i,t} = 0\} = 0, \quad \Pr\{X_{i,t} = a\} = \frac{1}{k}, \quad \forall a \in K \right). \quad (2)$$

n Players - 2 Transmission Channels. By employing our characterization, we show that for $k = 2$ channels, f^2 is an equilibrium protocol for $n \in \{2, 3, 4\}$ players (Theorem 1). The next two lemmata are easily proved by Markov chain analysis which is omitted due to lack of space.

Lemma 2. *When all $n \geq 2$ players use protocol f^2 the expected latency of any player is $2^n/n$.*

Lemma 3. *For $n \geq 5$ players, f^2 is not an equilibrium protocol. In fact, a better response for any player is to not transmit in $t = 1$ and then follow f^2.*

Theorem 1. *For $n \in \{2, 3, 4\}$ players and $k = 2$ channels, f^2 is an equilibrium protocol with expected latencies* 2, 8/3 *and* 4, *respectively.*

Proof Sketch. We show that the protocol profile where all n players use protocol f^2 is in equilibrium by showing that the condition (ii) of Lemma 1 holds. Starting with condition $(ii - a)$, assume a unilateral deviator i and an arbitrary protocol g_i consistent with $\vec{f}$. This protocol would dictate a history of transmissions $\vec{h}_{i,\tau^*}$ with only "1" and "2" in it for some arbitrary $\tau^* \geq 1$, and then continue following f^2. The process of any such protocol, from the perspective of i is modelled as a Partially Observable Markov Decision Process, which due to the anonymity and uniformity of f^2, reduces to a Markov chain that yields expected latency $2^n/n$.

For condition $(ii - b)$, suppose i chooses a protocol r_i that is not consistent with $\vec{f}$. This means that there must exist some time $t < \infty$ for which $\Pr\{X_{i,t} = 0\} > 0$. Let us focus on the smallest such t, namely $t_0 \triangleq \inf\{t : \Pr\{X_{i,t} = 0\} > 0\}$. Now if we consider some arbitrary history $\vec{h}_{i,t_0} = (a_{i,1}, a_{i,2}, \dots, a_{i,t_0})$ and its respective protocol $r_i = r_i(\vec{h}_{i,t_0})$ as in (1), one of two things can be true: either $a_{i,t_0} = 0$ or for $t > t_0$ protocol r_i is not identical to f^2. That is, we have the categories for r_i presented in Table 1. Note that the pairs of categories that r_i could be simultaneously are (1-I), (1-II), (2-I), and (2-II). By checking each of

Table 1. The categories of protocol $r_i(h_{i,t_0})$.

Category	Condition
Category 1	$a_{t_0} \neq 0$
Category 2	$a_{t_0} = 0$

Category	Condition
Category I	$\forall t > t_0$: $\Pr\{X_{i,t} = 0\} = 0$
Category II	$\exists t > t_0$: $\Pr\{X_{i,t} = 0\} > 0$

those cases and letting r_i be a best response, we show that no such protocol can yield expected latency to i lower than $2^n/n$. □

n Players - 3 Transmission Channels. Similarly, in the case with $k = 3$ channels, we employ our equilibria characterization and show that f^3, defined in (2), is an equilibrium protocol for $n \in \{2, 3, 4, 5\}$ players. However, now we do not have a closed-form expression for the expected latency of a player such as the one of Lemma 2, thus, in order to follow the same method as before, for each $n \in \{2, 3, 4, 5\}$ under examination we have to find its expected latency individually.

Theorem 2. *For $n \in \{2, 3, 4, 5\}$ players and $k = 3$ channels, f^3 is an equilibrium protocol with expected latencies $3/2$, $15/8$, $189/80$ and $597/200$, respectively.*

4 An Efficient Protocol with Expected Latency $= \infty$

In this section we give an anonymous, equilibrium protocol for the general case of $k \geq 1$ channels and any number of $n \geq 2k + 1$ players. For this, we employ the deadline idea introduced in [4] and consequently used in [2,3]. Our protocol has the property that the time until all players transmit successfully is $\Theta(n/k)$ with high probability, even though the expected latency is infinite.

Consider $k \geq 1$ transmission channels, $n \geq 2k + 1$ players, a fixed constant $\beta \in (0, 1)$ and a deadline t_0 to be determined consequently. The $t_0 - 1$ time steps are partitioned into $r + 1$ consecutive intervals $I_1, I_2, \ldots, I_{r+1}$ where r is the unique integer in $\left[-\log_\beta n/2 - 1, -\log_\beta n/2\right]$. For any $j \in \{1, 2, \ldots, r + 1\}$ define $n_j = \beta^j n/k$. For $j \in \{1, 2, \ldots, r\}$ the length of interval I_j is $l_j = \lfloor \frac{e}{\beta} n_j \rfloor$. Interval I_{r+1} is special and has length $l_{r+1} = n/k$. We define the following protocol.

> **Protocol g**: Every player among $1 \leq m \leq n$ pending players for $t \in I_j$ assigns transmission probability $1/\max\{n_j, k\}$ to each channel. Right before the deadline $t_0 = 1 + \sum_{j=1}^{r+1} l_j$ each pending player is assigned to a random channel equiprobably, and for $t \geq t_0$ always attempts transmission to that channel.

The proof of the following theorem is similar to that of Theorem 11 in [2] which considers the case with $k = 1$ channel, and is omitted due to lack of space.

Theorem 3. *Protocol g for $n \geq 2k + 1$ players and $k \geq 1$ channels, is an equilibrium protocol and it is also efficient.*

References

1. Eitan, A., Rachid, E.A., Tania, J.: Slotted aloha as a game with partial information. Comput. Netw. **45**, 701–713 (2004)
2. Christodoulou, G., Gairing, M., Nikoletseas, S.E., Raptopoulos, C., Spirakis, P.G.: Strategic contention resolution with limited feedback. In: ESA (2016)
3. Christodoulou, G., Gairing, M., Nikoletseas, S., Raptopoulos, C., Spirakis, P.: A 3-player protocol preventing persistence in strategic contention with limited feedback. In: Bilò, V., Flammini, M. (eds.) SAGT 2017. LNCS, vol. 10504, pp. 240–251. Springer, Cham (2017). https://doi.org/10.1007/978-3-319-66700-3_19
4. Fiat, A., Mansour, Y., Nadav, U.: Efficient contention resolution protocols for selfish agents. In: SODA (2007)
5. MacKenzie, A.B., Wicker, S.B.: Stability of multipacket slotted aloha with selfish users and perfect information. In: INFOCOM (2003)
6. Mo, J., So, H.W., Walrand, J.: Comparison of multichannel MAC protocols. IEEE Trans. Mob. Comput. **7**, 50–65 (2008)
7. So, J., Vaidya, N.H.: Multi-channel MAC for ad hoc networks: handling multi-channel hidden terminals using a single transceiver. In: Mobile Ad Hoc Networking and Computing. ACM (2004)

The Communication Burden of Single Transferable Vote, in Practice

Manel Ayadi[1,2(✉)], Nahla Ben Amor[1(✉)], and Jérôme Lang[2(✉)]

[1] LARODEC, Institut Supérieur de Gestion, Université de Tunis, Tunis, Tunisie
manel.ayadi@hotmail.com, nahla.benamor@gmx.fr
[2] CNRS, LAMSADE, Université Paris-Dauphine, Paris, France
lang@lamsade.dauphine.fr

Abstract. We study single-winner STV from the point of view of communication. First, we assume that voters give, in a single shot, their top-k alternatives; we define a version of STV that works for such votes, and we evaluate empirically the extent to which it approximates the standard STV rule. Second, we evaluate empirically the communication cost of the protocol for STV defined by Conitzer and Sandholm (2005) and some of its improvements.

1 Introduction

Single transferable vote (STV)[1] is an appealing voting rule: it is relatively easy to understand, it is not easy to manipulate, and it enjoys a very important normative property: clone-proofness. It is used in single-winner and multi-winner political elections in several countries. It fails to satisfy a number of other important properties,but in many contexts, being sensitive to cloning may be worse than the failure of these other properties. On the other hand, when compared to other rules that are widely used in practice (such as plurality, k-approval for small k, approval, or plurality with runoff), STV suffers from a significant drawback: its direct implementation requires an important amount of information to be communicated from the voters, because its input consists of a collection of complete rankings over candidates. Our aim is to get a more accurate idea of the precise amount of information that we need from the voters to compute or to approximate STV. We successively consider two contexts.

First, we assume that voters communicate, in a single shot, their top-k candidates, and we use an approximation of STV which needs only these top-k ballots as input.

Second, we consider interactive communication protocols, to be run between the central authority and the voters until the outcome of the vote is eventually determined. We study empirically the average communication complexity of the protocol defined by Conitzer and Sandholm [2], and of an improved variant of it.

This is a short version of our long paper submitted to SAGT 2018 (this submitted long version is available at https://goo.gl/Knd59d).

[1] For single-winner elections, STV is often called *instant runoff voting*.

X. Deng (Ed.): SAGT 2018, LNCS 11059, pp. 251–255, 2018.
https://doi.org/10.1007/978-3-319-99660-8_23

2 Approximating STV with Truncated Ballots

An election is a triple $E = (N, A, P)$ where $N = \{1, ..., n\}$ is the set of *voters*, A is the set of *candidates*, with $|A| = m$; and $P = (\succ_1, ..., \succ_n)$ is the *(preference) profile*, where for each i, $\succ_i$. A resolute voting rule maps any election to a single winner.

Given a prespecified linear order $\rhd$ over candidates, called *tie-breaking priority*, the $STV^{\rhd}$ rule proceeds in (up to $m-1$) rounds. (For brevity notation we will simply write STV, leaving $\rhd$ implicit.) In each round, the candidate with the smallest number of voters ranking them first is eliminated (using tie-breaking if necessary), and the votes who supported it now support their preferred candidate among those that remain.

Given $k \leq m$, a *top-k ballot* is a linear order of k among the m candidates in A. A *top-k profile* is a collection of n top-k ballots. Using truncated ballots as a way of reducing the amount of information in voting has been considered in a few recent works, especially [1,3,6–8].

For each $k \leq m$, STV_k is defined similarly as STV, but with top-k ballots as input. In each round, the candidate ranked first by the smallest number of voters is eliminated (using tie-breaking if needed). When all k candidates in a vote have been eliminated, the vote is ignored in later rounds (such a vote will be said to be *exhausted*). We repeat this process until there exists a candidate ranked first by the majority of non-exhausted truncated votes. STV_1 coincides with plurality, and STV_{m-1} (and STV_m) with STV.

In order to evaluate the quality of STV_k, we measure the frequency with which the approximation outputs the true winner using randomy generated date with the Mallows model, and then using real data.

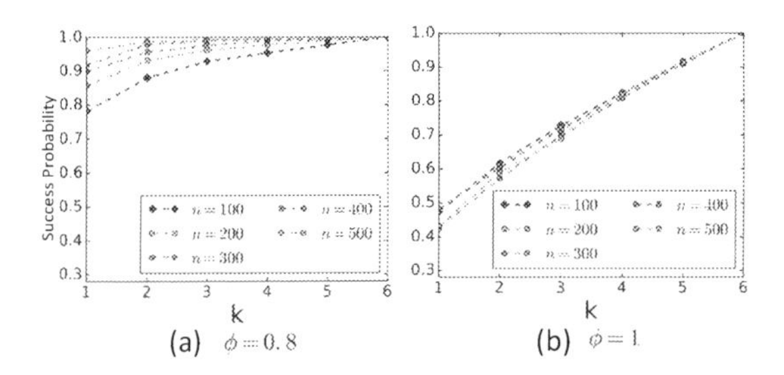

Fig. 1. Success probabilities of *top-k* voting for STV_k: $m = 7$ varying n, k and ϕ.

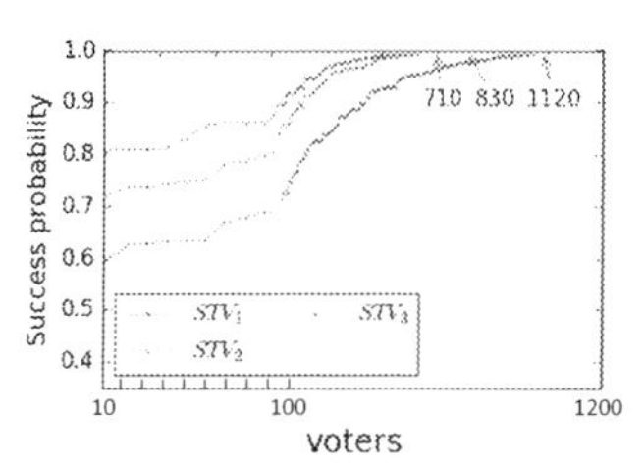

Fig. 2. Success probabilities of STV_k with Dublin data: varying $k \in \{1, 2, 3\}$ and n^* ($n^* < n$).

The Mallows ϕ model is described by two parameters: reference ranking σ and dispersion parameter $\phi \in [0, 1]$. The probability of a ranking r under this model is: $P(r; \sigma, \phi) = \frac{1}{Z}\phi^{d(r,\sigma)}$ where d is the Kendall tau distance and $Z = \sum_{r'} \phi^{d(r,\sigma)}$ is a normalization constant. We draw 1000 random profiles. then we simulate the elicitation of *top-k* ($k \in \{1 \ldots 6\}$) preferences $m = 7$ and let n and ϕ vary (Fig. 1).

Our results suggest that the winner is always predicted correctly when $\phi \leq 0.8$, $k = 2$ and with large n. When $\phi = 1$, the success rate is 82% with top-4 ballots of 500 voters. In all cases, *top*-2 ballots seem to be always sufficient to predict the correct STV winner with 100% accuracy with small values of ϕ and high number of voters.

Next, we use the Dublin data ($n = 3662$, $m = 12$) from the PrefLib library [5], with samples of n^* voters among n ($n^* < n$) where 1000 random profiles are constructed with n^* voters. Then, we consider the top-k ballots obtained from these profiles, where $k \in \{1, 2, 3\}$ over 12 candidates, and we compute the probability of selecting the correct winner (the winner of the complete profile of the n^* sampled votes) (Fig. 2). Our results suggest that predicting the correct winner with a small number of voters fails significantly often when k is too small ($k \leq \frac{1}{4}m$). Also, the performance increases with n. Indeed, $k = 1$ is sufficient to predict the correct winner when $n^* \geq 1120$.

Obviously, increasing the value of k leads to a decrease in the number of voters needed for correct winner selection for instance, when $k = \frac{1}{6}m$ (resp. $k = \frac{1}{4}m$) over 12 candidates, $n^* \geq 830$ (resp. $n^* \geq 710$) are needed to always output the correct result.

3 Communication Protocols for STV

Now, we allow for more sophisticated, *interactive* protocols where voters may report their preferences incrementally, when the central authority asks them to do so; on the other hand, we are not any longer interested in computing an approximation of STV, but in computing the real STV winner. With the aim of assessing the communication complexity of STV, Conitzer and Sandholm [2] a protocol for STV, which we call P_1:

1. each voter submits her most preferred candidate over the set of all available candidates to the central authority (C).
2. let $d \in A$ be the candidate ranked first by the fewest voters (using tie-breaking if necessary).
3. d is eliminated; all voters who had d as their current best candidate receive a message from C asking them to send their next preferred candidate among the remaining ones. For each of these voters, their vote is transferred to this next best remaining candidate.
4. this process is repeated until there exists a candidate x ranked first by more than 50% of the votes or only one candidate remains in the set of available candidates.

We say that $x \in A$ is an *immediate loser* if we know that x will be the next candidate eliminated after the currently eliminated one. Formally, let d be the candidate which is about to be eliminated, and U the set of remaining candidates (including d); candidate x is an *immediate loser* if for every $y \neq x, d$, either (1)

$S(y, P_U) > S(x, P_U) + S(d, P_U)$, or (2) $S(y, P_U) = S(x, P_U) + S(d, P_U)$ and $y \rhd x$.

Eliminating an immediate necessary loser during the execution of the protocol will never change the final outcome since we know exactly when it will be eliminated, then we can safely remove it.[2] This is the key property used in the next protocol, which we call P_2, which is an improvement over P_1: in P_2, the two first steps are similar as P_1. Then, if there is an immediate loser at this point, it is eliminated as well, together with d; from the set of available candidates. After d is eliminated, there may be *another* immediate loser; the process is repeated until there is no immediate loser. After removing all immediate losers in P_U, we select a voter whose top candidate is d or an immediate loser. We ask this voter to report her next preference among the available ones in U. Unlike P_1, P_2 queries one voter at a time since the new voter's preference may help to detect another immediate losers, thus reduce the set of available candidates. We repeat this process until we obtain a tops-only profile P with candidates among U for each voter. Finally, the process is repeated until there exists a candidate ranked first by more than 50% of the votes or only one candidate remains in U.

Now, we evaluate the average communication complexity of P_1 and P_2 using data generated from the Mallows ϕ model. Our objective is to determine the average communication complexity reported from voters in order to return the winner. We refer to P_{Worst} as the theoretical communication complexity.

For each experiment, we draw 1000 random profiles. We simulate the number of bits transferred between the central authority and the voters when with $m = 7$ and let n and ϕ vary (see Fig. 3). Results suggest that in practice, we can save a lot in communication costs compared to the theoretical complexity. Even with high ϕ, using P_2, we can save almost 50% of bits communicated. Also, our results suggest that when $\phi \leq 0.8$, P_2 is efficient to reduce the communication cost. When $\phi \geq 0.9$, from the results we can detect that P_1 and P_2 become closer in communication cost.

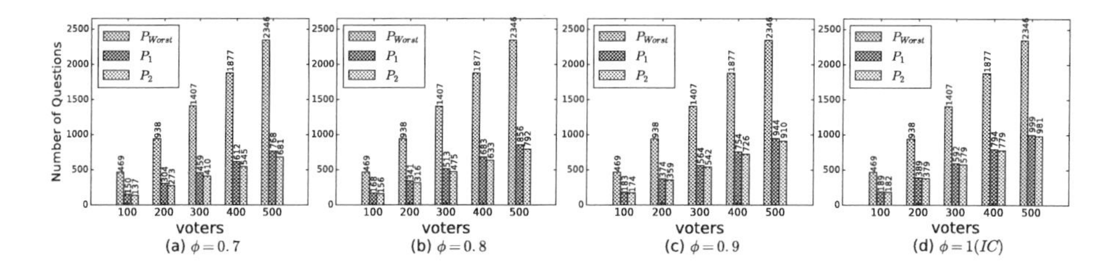

Fig. 3. Average communication cost with P_1, P_2 and P_{Worst}

[2] Jiang et al. [4] define a weaker version of necessary losers for STV in the context of a search algorithm for outputting all parallel universe STV winners.

References

1. Baumeister, D., Faliszewski, P., Lang, J., Rothe, J.: Campaigns for lazy voters: truncated ballots. In: Proceedings AAMAS, pp. 577–584 (2012)
2. Conitzer, V., Sandholm, T.: Communication complexity of common voting rules. In: Proceedings of the 6th ACM Conference on Electronic Commerce, pp. 78–87. ACM (2005)
3. Filmus, Y., Oren, J.: Efficient voting via the top-k elicitation scheme: a probabilistic approach. In: Proceedings of ACM Conference on Economics and Computation, pp. 295–312 (2014)
4. Jiang, C., Sikdar, S., Wang, J., Xia, L., Zhao, Z.: Practical algorithms for computing STV and other multi-round voting rules. In: EXPLORE-2017: The 4th Workshop on Exploring Beyond the Worst Case in Computational Social Choice (2017)
5. Mattei, N., Walsh, T.: PrefLib: a library for preferences http://www.preflib.org. In: Perny, P., Pirlot, M., Tsoukiàs, A. (eds.) ADT 2013. LNCS (LNAI), vol. 8176, pp. 259–270. Springer, Heidelberg (2013). https://doi.org/10.1007/978-3-642-41575-3_20
6. Naamani-Dery, L., Kalech, M., Rokach, L., Shapira, B.: Reducing preference elicitation in group decision making. Expert Syst. Appl. **61**, 246–261 (2016)
7. Oren, J., Filmus, Y., Boutilier, C.: Efficient vote elicitation under candidate uncertainty. In: Proceedings of IJCAI, pp. 309–316. AAAI Press (2013)
8. Skowron, P., Faliszewski, P., Slinko, A.: Achieving fully proportional representation: approximability results. Artif. Intell. **222**, 67–103 (2015)

Mechanism Design for Two-Opposite-Facility Location Games with Penalties on Distance

Xujin Chen[1,2], Xiaodong Hu[1,2], Xiaohua Jia[3], Minming Li[3], Zhongzheng Tang[1,2,3](✉), and Chenhao Wang[1,2,3]

[1] Academy of Mathematics and Systems Science, Chinese Academy of Sciences, Beijing, China
{xchen,xdhu,tangzhongzheng,wangch}@amss.ac.cn
[2] School of Mathematical Sciences, University of Chinese Academy of Sciences, Beijing, China
[3] Department of Computer Science, City University of Hong Kong, Kowloon Tong, Hong Kong SAR, China
{csjia,minming.li}@cityu.edu.hk

Abstract. This paper is devoted to the two-opposite-facility location games with a penalty whose amount depends on the distance between the two facilities to be opened by an authority. The two facilities are "opposite" in that one is popular and the other is obnoxious. Every selfish agent in the game wishes to stay close to the popular facility and stay away from the obnoxious one; its utility is measured by the difference between its distances to the obnoxious facility and the popular one. The authority determines the locations of the two facilities on a line segment where all agents are located. Each agent has its location information as private, and is required to report its location to the authority. Using the reported agent locations as input, an algorithmic mechanism run by the authority outputs the locations of the two facilities with an aim to maximize certain social welfare. The sum-type social welfare concerns with the penalized total utility of all agents, for which we design both randomized and deterministic group strategy-proof mechanisms with provable approximation ratios, and establish a lower bound on the approximation ratio of any deterministic strategy-proof mechanism. The bottleneck-type social welfare concerns with the penalized minimum utility among all agents, for which we propose a deterministic group strategy-proof mechanism that ensures optimality.

Keywords: Strategy-proof mechanism design · Facility location game

Research supported in part by NNSF of China under Grant No. 11531014, No. 11771365 and Shenzhen Research Institute, City University of Hong Kong.

X. Deng (Ed.): SAGT 2018, LNCS 11059, pp. 256–260, 2018.
https://doi.org/10.1007/978-3-319-99660-8_24

1 Introduction

The facility location game originally models the following scenario in practice: the central authority is going to build one or more facilities on a street (modeled as a line segment) where some selfish agents are located. The authority does not know the agents' exact locations, and thus conduct a survey for all agents. Each agent, who is required to report its own location, wishes to maximize its own utilities (e.g., minimizing its own distances to the facilities). The authority needs to design a mechanism, that maps the reported locations of agents to the locations where the facilities are to be opened. It is assumed that all agents know the mechanism that the authority adopts to aggregate agents' information to the final facility locations. Some agents might have incentive to misreport their locations. The goals of the authority are twofold: avoiding such misreports and maximizing some social welfare. The strategy-proofness of a mechanism guarantees that an agent cannot acquire more utility from misreporting, while the group strategy-proofness discourages simultaneous misreporting by any group of agents.

In this paper, we address (group) strategy-proof mechanism design for the *two-opposite-facility location game* proposed by [4], where two facilities to be opened have opposite characteristics for agents, that is, all agents want to stay as close as possible to one facility and stay as far away as possible from the other. Nevertheless, for some practical reasons, the two facilities should not be too far away to lose some connection. So, the distance between the two facilities cannot exceed a given constant C (referred to as *distance constant*). For instance, in order to save the cost of transportation and enhance garbage disposal efficiency, the government should set a limitation to the distance between the refuse collection point and the waste treatment plant. In our model, we relax the distance constraint to be a soft one by introducing to the central authority a penalty which equals a nonnegative coefficient λ times the amount of distance violation (w.r.t. to C). The more violation a location scheme incurs, the heavier penalty the authority carries. We evaluate mechanism efficiency in terms of optimizing certain social welfare – the sum-type (bottleneck-type) one of maximizing the penalized total (maximum) utility of all agents. We follow the convention to assume that the approximation ratio of a mechanism is always greater than or equal to one.

Related Work. The facility location games with one facility to be opened has been widely studied. Moulin [2] first characterized strategy-proof and Pareto efficient mechanisms in the line space. Procaccia and Tennenholtz [3] studied the facility location game on the line for both total utilities and minimum agent utilities, and derived several approximation bounds under the constraint of strategy-proofness. Study on the topic of two facilities can be found in [1,4] and references therein. In particular, for the aforementioned two-opposite-facility location game on a line segment with limited distance, Zou and Li [4] proposed a $(n/2)$-approximation and a n-approximation deterministic group strategy-proof mechanisms for an even number and an odd number of agents, respectively, where n is the number of agents. The approximation ratios for both mechanisms were proved to be the best that a deterministic strategy-proof mechanism can achieve.

Our Contributions. We investigate the two-opposite-facility location game model with a penalty whose amount depends on the distance between these two facilities. Henceforth, we abbreviate the game model and its restriction to a line segment as the 2OFLGP and 2OFLGP-L, respectively.

- For the sum-type social welfare, we design both randomized and deterministic group strategy-proof mechanisms for the 2OFLGP-L. The randomized one achieves an approximation ratio 2, by taking two possible optimal schemes respectively with probability $1/2$. The deterministic one achieves an approximation ratio $(k-1)R+1$ (when $n=2k$), or $2(k-1)R+1$ (when $n=2k-1$), where R is the ratio of the length of the line segment to the distance constant C. Furthermore, we prove that no deterministic strategy-proof mechanism can have an approximation ratio better than $k-1$, when $n=2k$ is even and penalty coefficient λ belongs to interval $(0,2)$.
- For the bottleneck-type social welfare, we propose a deterministic group strategy-proof mechanism for the 2OFLGP-L, that achieves the optimality.

2 Model

In the 2OFLGP, the decision maker wishes to build two facilities F_0 and F_1 with opposite preferences on some network, where all agents want to stay as close (far away) as possible to F_1 (from F_0). Let $N=\{1,2,\ldots,n\}$ denote the set of agents. Given the location x_i of agent $i \in N$ in the network, $\mathbf{x}=(x_1,\ldots,x_n)$ denotes the location profile of all agents. Given input $\mathbf{x}$, a *building scheme* $S=(y_0,y_1)$ is the output of a mechanism, where y_0 and y_1 are locations of F_0 and F_1, respectively. Denote by $d(x,y)$ the distance between the locations x and y in the network. For the scheme $S=(y_0,y_1)$, we define $|S|=d(y_0,y_1)$. Regarding the *distance constraint*: $|S| \le C$, where $C \ge 0$ is a given constant, we allow a moderate violation by paying a penalty. The penalty is measured by function $p(S) := \lambda(|S|-C)_+$ with a penalty coefficient $\lambda \ge 0$, where $(|S|-C)_+ = \max\{|S|-C, 0\}$. Given $\mathbf{x}$ and S, the utility of agent i is defined as the difference between its distances towards F_0 and F_1, i.e., $u(x_i,S)=d(x_i,y_0)-d(x_i,y_1)$. Each agent reports only its own location, and tends to maximize its utility by misreporting. The game model consists of two sub-models, for maximizing the sum-type social welfare – the penalized total utility of all agents and the bottleneck-type social welfare – the penalized minimum agent utility, respectively. Given a location profile $\mathbf{x}$ and building scheme $S=(y_0,y_1)$, the *sum-type social welfare* is defined as

$$su(S,\mathbf{x}) = \sum_{i=1}^{n}(d(x_i,y_0)-d(x_i,y_1)) - \lambda(|S|-C)_+, \tag{1}$$

and the *bottleneck-type social welfare* is defined as

$$mu(S,\mathbf{x}) = \min_{i\in N}(d(x_i,y_0)-d(x_i,y_1)) - \lambda(|S|-C)_+ \tag{2}$$

Given a mechanism that outputs a solution S for a location profile $\mathbf{x}$, we say the mechanism is *strategy-proof*, if for any agent $i \in N$ and its misreported location x'_i, we have $u(x_i, S) \geq u(x_i, S')$, where S' is the output of this mechanism with respect to input $\mathbf{x}' = (\mathbf{x}_{-i}, x'_i)$. In addition, it is *group strategy-proof* if for any group of agents $G \subseteq N$ and their misreported partial location profile $\mathbf{x}'_G$, $u(x_i, S) \geq u(x_i, S')$ holds for some $i \in G$, where S' is the output of the mechanism with respect to input $\mathbf{x}' = (\mathbf{x}_{-G}, \mathbf{x}'_G)$. For a randomized mechanism, it is *universally group strategy-proof* if it is a probability distribution over deterministic group strategy-proof mechanisms.

3 Mechanisms

In this section, we present mechanisms for the 2OFLGP-L on a line segment with length L. Without loss of generality assume the left end-point of the segment is 0 and the right end-point is L. The location of each agent or each facility is on this segment: $x_i, y_j \in [0, L]$ for $i \in N, j \in \{0, 1\}$, and the distance $d(x, y)$ between two locations x and y is their one-dimensional Euclidean distance $|x - y|$.

3.1 The Sum-Type Social Welfare

Given a location profile $\mathbf{x}$, define a function $g(y) = \sum_{i=1} g_i(y) = \sum_{i=1}^{n} d(x_i, y)$ over the domain $[0, L]$. Denote by $g'_-(y)$ and $g'_+(y)$ the left and right derivatives of $g(y)$, respectively. Given a scheme $S = (y_0, y_1)$, the social welfare $su(S, \mathbf{x}) = \sum_{i=1}^{n}(d(x_i, y_0) - d(x_i, y_1)) - \lambda(|S| - C)_+ = g(y_0) - g(y_1) - \lambda(|S| - C)_+$. Fixing $y_0 = 0$, we define $opt_l(\mathbf{x})$ by:

$$opt_l(\mathbf{x}) := \begin{cases} x_{m_1} & \text{if } x_{m_1} \leq C \\ C & \text{if } x_{m_1} > C \ \&\ \lambda \geq |g'_+(C)| \\ x_i & \text{if } x_{m_1} > C \ \&\ \lambda < |g'_+(C)| \end{cases}$$

where x_i is the unique solution of $|g'_-(x_i)| > \lambda$ and $|g'_+(x_i)| \leq \lambda$. Fixing $y_0^* = L$, we define $opt_r(\mathbf{x})$ by:

$$opt_r(\mathbf{x}) := \begin{cases} x_{m_2} & \text{if } L - x_{m_2} \leq C \\ L - C & \text{if } L - x_{m_2} > C \ \&\ \lambda \geq |g'_-(C)| \\ x_j & \text{if } L - x_{m_2} > C \ \&\ \lambda < |g'_-(C)| \end{cases}$$

where x_j is the unique solution of $|g'_+(x_j)| > \lambda$ and $|g'_-(x_j)| \leq \lambda$.

Mechanism 1. Given a location profile $\mathbf{x}$, output $S = (0, opt_l(\mathbf{x}))$ with probability α, and $S = (L, opt_r(\mathbf{x}))$ with probability $1 - \alpha$, where $\alpha \in [0, 1]$ is a constant.

Theorem 1. *Mechanism 1 is universally group strategy-proof, and achieves an approximation ratio* $\max\{\frac{1}{\alpha}, \frac{1}{1-\alpha}\}$ *for the 2OFLGP-L with the sum-type social welfare. (Taking* $\alpha = 1/2$*, Mechanism 1 achieves approximation ratio 2.)*

Mechanism 2. Given location profile $\mathbf{x}$, output $S = (0, opt_l(\mathbf{x}))$ if $opt_l \geq L - opt_r$; and output $S = (L, opt_r(\mathbf{x}))$ otherwise.

Theorem 2. *Let $R = L/C$. For the 2OFLGP-L with the sum-type social welfare, Mechanism 2 is group strategy-proof, and has an approximation ratio $(k-1)R+1$ when $n = 2k$, and an approximation ratio $2(k-1)R+1$ when $n = 2k-1$.*

The following complementary lower bound indicates the quality of the above approximation ratio in some special cases.

Theorem 3. *For $n = 2k$ and $\lambda \in (0, 2)$, any deterministic strategy-proof mechanism cannot have an approximation ratio smaller than $k-1$ for the 2OFLGP-L with the sum-type social welfare.*

3.2 The Bottleneck-Type Social Welfare

Denote by x_{e_1} and x_{e_2} the locations of leftmost and rightmost agents respectively. Define $v_l(\mathbf{x})$ to be x_{e_1} if $\lambda < 1$ and to be $\min\{C, x_{e_1}\}$ otherwise. Define $v_r(\mathbf{x})$ to be x_{e_2} if $\lambda < 1$, and to be $\max\{x_{e_2}, L - C\}$ otherwise.

Mechanism 3. If $v_l(\mathbf{x}) \geq L - v_r(\mathbf{x})$, output $(0, v_l(\mathbf{x}))$; otherwise, output $(L, v_r(\mathbf{x}))$.

Theorem 4. *Mechanism 3 is group strategy-proof and outputs an optimal building scheme for the 2OFLGP-L with the bottleneck-type social welfare.*

4 Conclusions

In this paper, we assume that the penalty is linear with the distance violations. It might be interesting to investigate the game with other kinds of penalties, e.g., those increasing exponentially with distance violations. We can also study the infinite line, $[0, +\infty]$ and $[-\infty, +\infty]$. For the line $[0, +\infty]$, under the sum-type objective, an optimal mechanism outputs $y_1 = \max_{1 \leq i \leq n} x_i$ and $y_0 = y_1 + C$ if $\lambda \geq n$; and outputs $S = (+\infty, \max_{1 \leq i \leq n} x_i)$ otherwise. This mechanism is group strategy-proof. Under the bottleneck-type objective, the results in Sect. 3.2 could be easily extended by analogous analyses.

References

1. Feigenbaum, I., Sethuraman, J.: Strategyproof mechanisms for one-dimensional hybrid and obnoxious facility location models. In: AAAI Workshop: Incentive and Trust in E-Communities (2015)
2. Moulin, H.: On strategy-proofness and single peakedness. Public Choice **35**(4), 437–455 (1980)
3. Procaccia, A.D., Tennenholtz, M.: Approximate mechanism design without money. In: ACM Conference on Electronic Commerce, pp. 177–186 (2009)
4. Zou, S., Li, M.: Facility location games with dual preference. In: International Conference on Autonomous Agents and Multiagent Systems, pp. 615–623 (2015)

An Optimal Strategy for Static Black-Peg Mastermind with Three Pegs

Gerold Jäger[1(✉)] and Frank Drewes[2]

[1] Department of Mathematics and Mathematical Statistics, University of Umeå, 901-87 Umeå, Sweden
gerold.jaeger@math.umu.se
[2] Department of Computing Science, University of Umeå, 901-87 Umeå, Sweden
drewes@cs.umu.se

Abstract. Mastermind is a famous game played by a *codebreaker* against a *codemaker*. We investigate its static (also called non-adaptive) black-peg variant. Given c colors and p pegs, the codemaker has to choose a secret, a p-tuple of c colors, and the codebreaker asks a set of questions all at once. Like the secret, a question is a p-tuple of c colors. The codemaker then tells the codebreaker how many pegs in each question are correct in position and color. Then the codebreaker has one final question to find the secret. His aim is to use as few of questions as possible. Our main result is an optimal strategy for the codebreaker for $p = 3$ pegs and an arbitrary number c of colors using $\lfloor 3c/2 \rfloor + 1$ questions.

A reformulation of our result is that the metric dimension of $\mathbb{Z}_n \times \mathbb{Z}_n \times \mathbb{Z}_n$ is equal to $\lfloor 3n/2 \rfloor$.

1 Introduction

Mastermind is a board game invented by Meirowitz in 1970 with applications in cryptography [3] and bioinformatics [4]. In the original version of the game, the so-called codemaker chooses a secret code consisting of 4 pegs and 6 possible colors for each peg. The so-called codebreaker must discover this code by making a sequence of guesses, called questions, until the secret has been found, using as few questions as possible. Each answer of the codemaker consists of black and white pegs, one black peg for each peg of the question which is correct in both position and color, and one white peg for each peg which is correct only in color. For Mastermind with p pegs and c colors the decision problem to decide whether a secret exists which satisfies a given set of questions and answers is $\mathcal{NP}$-complete [14]. An analysis of optimal strategies has been presented for original Mastermind [7,10], and for several of its variants, e.g., for the black-peg variant, where no white pegs are given in the answers [8,11], and the AB Game, where the secret and each question must not contain any color twice [12].

Static (or *Non-Adaptive) Mastermind* requires the codebreaker to ask all questions at the beginning of the game. The codebreaker then receives all answers and must be able to find the secret in a final question. Goddard [6] gave a

X. Deng (Ed.): SAGT 2018, LNCS 11059, pp. 261–266, 2018.
https://doi.org/10.1007/978-3-319-99660-8_25

$\lceil 2c/3+1 \rceil$-strategy for two pegs, a c-strategy for three pegs and a c-strategy for four pegs[1] all of which are optimal for sufficiently large c. Here, we consider Static Black-Peg Mastermind, i.e., the codebreaker only receives black pegs as answers, and thus only gets to know how many of the positions in each question are correctly colored. We present an optimal $\lfloor 3c/2 \rfloor + 1$-strategy for the case of $p = 3$ pegs and an arbitrary number c of colors. This continues the work in [9], where a $\lceil (4c-1)/3 \rceil$-strategy for $p = 2$ was presented, and in [5], where the static black-peg variant of the AB Game was studied and a $(\lceil 4c/3 \rceil - 1)$-strategy was given for the case $p = 2$[2] and additionally a $\mathcal{O}(n^{1.525})$-strategy for the case $p = c$, where the questions and secrets correspond to permutations.

Mastermind may be studied from the point of view of Game Theory, the theory that investigates the (effect of) strategic choices made by interacting opponents. While it is a one-player game, where only the codebreaker acts, it can be viewed as a two-player game by allowing the codemaker to change the secret before answering a question, in a way consistent with previous answers. To define a winning situation for each player, one may limit the number k of questions. The existence of winning strategies for either player would then depend on k. For Static Black-Peg Mastermind with 3 pegs and c colors our main result shows that the codebreaker has a winning strategy if and only if $k > \lfloor 3c/2 \rfloor$; otherwise, the codemaker has a winning strategy.

The metric dimension of an arbitrary (undirected and unweighted) graph is the minimal size of a set U such that every vertex v of the graph is uniquely determined by the vector of distances between v and the vertices in U. This concept occurred first in [1] and has since been studied in various papers. In [13] it was shown that the metric dimension of $(\mathbb{Z}_2)^n$ is asymptotically $\mathcal{O}(n/\log n)$, and in [2] that the metric dimension of $\mathbb{Z}_n \times \mathbb{Z}_n$ is $\lceil (4n-1)/3 - 1 \rceil$. It is easy to see that the minimal number of questions needed to win Static Black-Peg Mastermind on p pegs with c colors is equal to the metric dimension of $(\mathbb{Z}_c)^p$ plus 1. Thus, our optimal $\lfloor 3c/2 \rfloor + 1$-strategy gives that the metric dimension of $\mathbb{Z}_n \times \mathbb{Z}_n \times \mathbb{Z}_n$ is $\lfloor 3n/2 \rfloor$) for all n.

2 Preliminaries

We number the pegs by $1, 2 \ldots, p$, and the colors by $1, 2, \ldots, c$. For $r \in \mathbb{N}$, an r-*strategy* for Static Black-Peg Mastermind consists of $r-1$ questions $Q_1, Q_2, \ldots, Q_{r-1} \in \{1, 2, \ldots, c\}^p$. Such a strategy is *feasible* if every possible secret S is uniquely determined by the $r-1$ answers so that the codebreaker can ask the *final question* S to win the game. The strategy is called *optimal* if r is minimal. In the following consider only the case $p = 3$.

We implemented a computer program available at [15], to check for $p = 3$, small $c \in \mathbb{N}$ and $r \in \mathbb{N}$ all possible strategies, i.e., all combinations of questions, for being feasible. This program serves both as a corroboration of the feasibility of the main strategy for small c, and as a part of the optimality proof for all c.

[1] Note that in [6] the final question was not taken into account.

[2] The case $p = 1$ is trivial for both games: exactly c questions are needed.

3 The Main Result: A ($\lfloor 3c/2 \rfloor + 1$)-Strategy for $p = 3$

We introduce a ($\lfloor 3c/2 \rfloor + 1$)-strategy for each c, where we distinguish between the cases $c \equiv 0, 1, 2, 3 \bmod 4$. Note that for the number $k := \lfloor 3c/2 \rfloor$ of questions (without the final question) it holds that $k = 3 \cdot \frac{c}{2}$ if c is even, and that $k = 3 \cdot \frac{c-1}{2} + 1$ if c is odd. Table 1 shows examples of the four cases. Below, we use the superscript "$\times 2$" to denote a color that is repeated twice.

Strategy 1 (($\lfloor 3c/2 \rfloor + 1$)-strategy for p = 3 and c > 4, c ≡ 0 mod 4)[3]
Divide the k questions into three blocks B_1, B_2, B_3 of $k/3$ (= $c/2$) questions each.

1. *Peg 1 contains the colors* $1, 2, \ldots, k/3$ *in* B_1, *the colors* $(k/3+1)^{\times 2}, (k/3+2)^{\times 2}, \ldots, (k/2)^{\times 2}$ *in* B_2, *and the colors* $(k/2+1)^{\times 2}, (k/2+2)^{\times 2}, \ldots, c^{\times 2}$ *in* B_3.
2. *Peg 2 contains the colors* $(k/2+1)^{\times 2}, (k/2+2)^{\times 2}, \ldots, (2k/3)^{\times 2}$ *(i.e.,* B_3 *of peg 1) in* B_1, *the colors* $1, 2, \ldots, k/3$ *(i.e.,* B_1 *of peg 1) in* B_2, *and the colors* $k/2, (k/3+1)^{\times 2}, (k/3+2)^{\times 2}, \ldots, (k/2-1)^{\times 2}, k/2$ *(i.e., again* B_2 *of peg 1, but here shifted down by one question) in* B_3.
3. *Peg 3 contains the colors* $k/2, (k/3+1)^{\times 2}, (k/3+2)^{\times 2}, \ldots, (k/2-1)^{\times 2}, k/2$ *(i.e.,* B_2 *of peg 1, but shifted by one question) in* B_1, *the colors* $2k/3, (k/2+1)^{\times 2}, (k/2+2)^{\times 2}, \ldots, (2k/3-1)^{\times 2}, 2k/3$ *(i.e.,* B_3 *of peg 1, but shifted by one question) in* B_2, *and the colors* $1, 2, \ldots, k/3$ *(i.e.,* B_1 *of peg 1) in* B_3.

The shifting of questions is essential for this strategy and the following ones. E.g., if on peg 3 of Strategy 1 in Table 1a, the first eight colors were $9, 9, 10, 10, 11, 11, 12, 12$ instead of $12, 9, 9, 10, 10, 11, 11, 12$, the possible secrets $(1, 13, 10)$ and $(1, 14, 9)$ would get the same combination of answers: 2B, 1B, 1B, 1B, 14× 0B, i.e., the strategy would not be feasible. Without this shifting there would be colors which would occur in exactly the same set of questions, e.g., color 13 on peg 2 and color 9 on peg 3 would occur in the questions 1 and 2, and color 14 on peg 2 and color 10 on peg 3 would occur in the questions 3 and 4.

Strategy 2 (($\lfloor 3c/2 \rfloor + 1$)-strategy for p = 3 and c > 4, c ≡ 1 mod 4)[4]
Divide the k questions into three blocks B_1, B_2, B_3 of $(k+2)/3 (= (c+1)/2)$, $(k+2)/3$ and $(k+2)/3 - 2$ questions, respectively.

1. *Peg 1 contains the colors* $1, 2, \ldots, (k+2)/3$ *in* B_1, *and the colors* $((k+2)/3+1)^{\times 2}, ((k+2)/3+2)^{\times 2}, \ldots, c^{\times 2}$ *in* B_2 *and continuing throughout* B_3.
2. *Peg 2 contains the colors* $((k+1)/2+1)^{\times 2}, ((k+1)/2+2)^{\times 2}, \ldots, (2(k+2)/3-1)^{\times 2}, (k+1)/2$ *in* B_1, *the colors* $1, 2, \ldots, (k+2)/3$ *(i.e.,* B_1 *of peg 1) in* B_2, *and the colors* $((k+2)/3+1)^{\times 2}, ((k+2)/3+2)^{\times 2}, \ldots, ((k+1)/2-1)^{\times 2}, (k+1)/2$ *in* B_3.

[3] For $c = 4$, this strategy with 6 questions is not feasible, as the shifting step does not work. However, changing peg 3 of the third question from color 4 to color 3 leads to a feasible strategy with 6 questions.

[4] For $c = 1$, this strategy with 1 question is not defined, and the strategy with 0 questions (i.e., only the final question) is optimal.

3. *Peg 3 contains the colors* $(k+1)/2-2$, $((k+2)/3-1)^{\times 2}$, $((k+2)/3)^{\times 2}$, ..., $((k+1)/2-3)^{\times 2}$, $(k+1)/2-2$, $(k+1)/2-1$ *in* B_1, *the colors* $2(k+2)/3-1$, $((k+1)/2+1)^{\times 2}$, $((k+1)/2+2)^{\times 2}$, ..., $(2(k+2)/3-2)^{\times 2}$, $2(k+2)/3-1$, $(k+1)/2-1$ *in* B_2, *and the colors* 1, 2, ..., $(k+2)/3-2$ *in* B_3.

Strategy 3 $((\lfloor 3c/2 \rfloor)+1$**-strategy for** $p=3$ **and** $c \equiv 2 \bmod 4)$
Divide the k *questions into three blocks* B_1, B_2, B_3 *of* $k/3$ $(=c/2)$ *questions each.*

1. *Peg 1 contains the colors* 1, 2, ..., $k/3$ *in* B_1 *and the colors* $(k/3+1)^{\times 2}$, $(k/3+2)^{\times 2}$, ..., $c^{\times 2}$ *in* B_2 *and continuing throughout* B_3.
2. *Peg 2 contains the colors* 1, 2, ..., $k/3$ *(i.e.,* B_1 *of peg 1) in* B_2 *and the colors* $(k/3+1)^{\times 2}$, $(k/3+2)^{\times 2}$, ..., $(2k/3)^{\times 2}$ *in* B_3 *and continuing throughout* B_1.
3. *Peg 3 contains the colors* 1, 2, ..., $k/3$ *(i.e.,* B_1 *of peg 1) in* B_3 *and the colors* $(k/3+1)^{\times 2}$, $(k/3+2)^{\times 2}$, ..., $(2k/3)^{\times 2}$ *in* B_1 *and continuing throughout* B_2.

Strategy 4 $((\lfloor 3c/2 \rfloor)+1$**-strategy for** $p=3$ **and** $c \equiv 3 \bmod 4)$
Divide the k *questions into three blocks* B_1, B_2, B_3 *of* $(k+2)/3$ $(=(c+1)/2)$, $(k+2)/3$ *and* $(k+2)/3-2$ *questions, respectively.*

1. *Peg 1 contains the colors* 1, 2, ..., $(k+2)/3$ *in* B_1 *and the colors* $((k+2)/3+1)^{\times 2}$, $((k+2)/3+2)^{\times 2}$, ..., $c^{\times 2}$ *in* B_2 *and continuing throughout* B_3.
2. *Peg 2 contains the colors* $((k+2)/2)^{\times 2}$, $((k+2)/2+1)^{\times 2}$, ..., $(2(k+2)/3-1)^{\times 2}$ *in* B_1, *the colors* 1, 2, ..., $(k+2)/3$ *(i.e.,* B_1 *of peg 1) in* B_2, *and the colors* $(k+2)/2-1$, $((k+2)/3+1)^{\times 2}$, $((k+2)/3+2)^{\times 2}$, ..., $((k+2)/2-2)^{\times 2}$, $(k+2)/2-1$ *in* B_3.
3. *Peg 3 contains the colors* $(k+2)/2-2$, $((k+2)/3-1)^{\times 2}$, $((k+2)/3)^{\times 2}$, ..., $((k+2)/2-3)^{\times 2}$, $(k+2)/2-2$ *in* B_1, *the colors* $2(k+2)/3-1$, $((k+2)/2)^{\times 2}$, $((k+2)/2+1)^{\times 2}$, ..., $(2(k+2)/3-2)^{\times 2}$, $2(k+2)/3-1$ *in* B_2, *and the colors* 1, 2, ..., $(k+2)/3-2$ *in* B_3.

Theorem 1. *Strategies 1, 2, 3 and 4 are* feasible *and* optimal $(\lfloor 3c/2 \rfloor)+1$*-strategies for* $p=3$ *and for the corresponding* c *with* $c \neq 1, 4$.

Recall that the AB Game corresponds to Mastermind, but with the additional condition that both in the secret and questions the colors on different pegs must be pairwise distinct. By observing that the strategies in Theorem 1 only use questions containing three different colors each, we thus obtain the following.

Corollary 1. *For* $c=6$ *and for all* $c \geq 8$, *the Strategies 1, 2, 3 and 4 are also feasible strategies for the AB Game with* $p=3$ *and the considered* c.

4 Idea of Proof of Theorem 1

Feasiblity: For a given strategy, two questions are called *neighbors*, and *double neighbors*, if they have one color in common on exactly one peg and two pegs, respectively. A question Q is called a (a_1, a_2, a_3)*-question* for $a_1, a_2, a_3 \in \mathbb{N}$, if

Table 1. Examples for Strategies 1, 2, 3 and 4 with $p = 3$.

Peg	1	2	3
Q_1	1	13	12
Q_2	2	13	9
Q_3	3	14	9
Q_4	4	14	10
Q_5	5	15	10
Q_6	6	15	11
Q_7	7	16	11
Q_8	8	16	12
Q_9	9	1	16
Q_{10}	9	2	13
Q_{11}	10	3	13
Q_{12}	10	4	14
Q_{13}	11	5	14
Q_{14}	11	6	15
Q_{15}	12	7	15
Q_{16}	12	8	16
Q_{17}	13	12	1
Q_{18}	13	9	2
Q_{19}	14	9	3
Q_{20}	14	10	4
Q_{21}	15	10	5
Q_{22}	15	11	6
Q_{23}	16	11	7
Q_{24}	16	12	8

(a) $c = 16$ $k = 24$

Peg	1	2	3
Q_1	1	14	11
Q_2	2	14	8
Q_3	3	15	8
Q_4	4	15	9
Q_5	5	16	9
Q_6	6	16	10
Q_7	7	17	10
Q_8	8	17	11
Q_9	9	13	12
Q_{10}	10	1	17
Q_{11}	10	2	14
Q_{12}	11	3	14
Q_{13}	11	4	15
Q_{14}	12	5	15
Q_{15}	12	6	16
Q_{16}	13	7	16
Q_{17}	13	8	17
Q_{18}	14	9	12
Q_{19}	14	10	1
Q_{20}	15	10	2
Q_{21}	15	11	3
Q_{22}	16	11	4
Q_{23}	16	12	5
Q_{24}	17	12	6
Q_{25}	17	13	7

(b) $c = 17$, $k = 25$.

Peg	1	2	3
Q_1	1	14	10
Q_2	2	15	10
Q_3	3	15	11
Q_4	4	16	11
Q_5	5	16	12
Q_6	6	17	12
Q_7	7	17	13
Q_8	8	18	13
Q_9	9	18	14
Q_{10}	10	1	14
Q_{11}	10	2	15
Q_{12}	11	3	15
Q_{13}	11	4	16
Q_{14}	12	5	16
Q_{15}	12	6	17
Q_{16}	13	7	17
Q_{17}	13	8	18
Q_{18}	14	9	18
Q_{19}	14	10	1
Q_{20}	15	10	2
Q_{21}	15	11	3
Q_{22}	16	11	4
Q_{23}	16	12	5
Q_{24}	17	12	6
Q_{25}	17	13	7
Q_{26}	18	13	8
Q_{27}	18	14	9

(c) $c = 18$, $k = 27$.

Peg	1	2	3
Q_1	1	15	13
Q_2	2	15	9
Q_3	3	16	9
Q_4	4	16	10
Q_5	5	17	10
Q_6	6	17	11
Q_7	7	18	11
Q_8	8	18	12
Q_9	9	19	12
Q_{10}	10	19	13
Q_{11}	11	1	19
Q_{12}	11	2	15
Q_{13}	12	3	15
Q_{14}	12	4	16
Q_{15}	13	5	16
Q_{16}	13	6	17
Q_{17}	14	7	17
Q_{18}	14	8	18
Q_{19}	15	9	18
Q_{20}	15	10	19
Q_{21}	16	14	1
Q_{22}	16	11	2
Q_{23}	17	11	3
Q_{24}	17	12	4
Q_{25}	18	12	5
Q_{26}	18	13	6
Q_{27}	19	13	7
Q_{28}	19	14	8

(d) $c = 19$, $k = 28$.

for $i = 1, 2, 3$, the i-th color of Q occurs a_i times on the i-th peg (throughout the strategy). In Table 1a, $Q_3 = (3, 14, 9)$ and $Q_4 = (4, 14, 10)$ are neighbors (but not double neighbors), and both are $(1, 2, 2)$-questions,

As can be seen in Table 1, the four strategies consist only of $(1, 2, 2)$-questions, $(2, 1, 2)$-questions and $(2, 2, 1)$-questions, and no double neighbors exist.

In Table 1a, if the neighbors $Q_3 = (3, 14, 9)$ and $Q_4 = (4, 14, 10)$ both receive an answer $\geq 1B$, then the secret has the form $(?, 14, ?)$, unless one of the neighbors $Q_2 = (2, 13, 9)$ and $Q_5 = (5, 15, 10)$ also receives an answer $\geq 1B$, and in this case the secret has the form $(4, ?, 9)$ or $(3, ?, 10)$. The detailed proof needs an extensive case distinction and the introduction of the so-called *strategy graph*, where the questions correspond to vertices and two questions are connected by an edge, if they are neighbors.

Optimality: This proof relies on observations regarding the feasibility of the strategy, e.g., on each peg all but one color must occur, and there can be at most one $(1, 1, 1)$-question. We consider the cases of odd and even c and four subcases each, depending on how many of the three pegs contain c colors. Note that in the strategies of Table 1 all colors occur at least once on each peg, except for peg 3 of Strategies 2 and 4, where the colors $(k+1)/2$ and $k/2$, respectively, do not occur. This case must be excluded for Strategies 1 and 3 for even c as well, which makes the proof much more difficult for even c than for odd c. The detailed proof needs again several cases and the introduction of a new term, the so-called *proof questions*.

Due to the space limitation, we refer to the forthcoming full version of the paper for the feasibility and optimality proofs.

References

1. Blumenthal, L.M.: Theory and Applications of Distance Geometry. Clarendon Press, Oxford (1953)
2. Cáceres, J., Hernando, C., Mora, M., Pelayo, I.M., Puertas, M.L., Seara, C., Wood, D.R.: On the metric dimension of cartesian products of graphs. SIAM J. Discrete Math. **21**(2), 423–441 (2007)
3. Focardi, R., Luccio, F.L.: Guessing bank pins by winning a mastermind game. Theory Comput. Syst. **50**(1), 52–71 (2012)
4. Gagneur, J., Elze, M.C., Tresch, A.: Selective phenotyping, entropy reduction and the mastermind game. BMC Bioinform. (BMCBI) **12**, 406 (2011)
5. Glazik, C., Jäger, G., Schiemann, J., Srivastav, A.: Bounds for static black-peg AB mastermind. In: Gao, X., Du, H., Han, M. (eds.) COCOA 2017. LNCS, vol. 10628, pp. 409–424. Springer, Cham (2017). https://doi.org/10.1007/978-3-319-71147-8_28
6. Goddard, W.: Static mastermind. J. Comb. Math. Comb. Comput. **47**, 225–236 (2003)
7. Goddard, W.: Mastermind revisited. J. Comb. Math. Comb. Comput. **51**, 215–220 (2004)
8. Goodrich, M.T.: On the algorithmic complexity of the mastermind game with black-peg results. Inf. Process. Lett. **109**(13), 675–678 (2009)
9. Jäger, G.: An optimal strategy for static black-peg mastermind with two pegs. In: Chan, T.-H.H., Li, M., Wang, L. (eds.) COCOA 2016. LNCS, vol. 10043, pp. 670–682. Springer, Cham (2016). https://doi.org/10.1007/978-3-319-48749-6_48
10. Jäger, G., Peczarski, M.: The number of pessimistic guesses in Generalized Mastermind. Inf. Process. Lett. **109**(12), 635–641 (2009)
11. Jäger, G., Peczarski, M.: The number of pessimistic guesses in Generalized Black-Peg Mastermind. Inf. Process. Lett. **111**(19), 933–940 (2011)
12. Jäger, G., Peczarski, M.: The worst case number of questions in generalized AB game with and without white-peg answers. Discrete Appl. Math. **184**, 20–31 (2015)
13. Söderberg, S., Shapiro, H.S.: A combinatory detection problem. Am. Math. Mon. **70**(10), 1066–1070 (1963)
14. Stuckman, J., Zhang, G.Q.: Mastermind is NP-complete. INFOCOMP J. Comput. Sci. **5**, 25–28 (2006)
15. Source code of the computer program of this article. http://snovit.math.umu.se/~gerold/source_code_static_black_peg_mastermind_three_pegs.tar.gz

Tight Bounds on the Relative Performances of Pricing Mechanisms in Storable Good Markets

Gerardo Berbeglia[1], Shant Boodaghians[2], and Adrian Vetta[3](✉)

[1] Melbourne Business School, University of Melbourne, Melbourne, Australia
g.berbeglia@mbs.edu
[2] University of Illinois at Urbana-Champaign, Urbana, IL, USA
boodagh2@illinois.edu
[3] McGill University, Montreal, Canada
adrian.vetta@mcgill.ca

Abstract. In the storable good monopoly problem, a monopolist sells a storable good by announcing a price in each time period. Each consumer has a unitary demand per time period with an arbitrary valuation. In each period, consumers may buy none, one, or more than one good (in which case the extra goods are stored for future consumption incurring a linear storage cost). We compare the performance of two important pricing mechanisms on the profitability of the monopolist: pre-announced pricing mechanisms and price contingent mechanisms. In pre-announced pricing the prices in each time period are stated in advance; in a price contingent mechanism each price is stated at the start of the time period, and these prices are dependent upon past purchases. We prove that the monopolist can earn at most $O(\log T + \log N)$ times more profit by using a pre-announced pricing mechanism rather than a price contingent mechanism. Here T denotes the number of time periods and N denotes the number of consumers. This bound is tight; examples exist where the monopolist would earn a factor $\Omega(\log T + \log N)$ more by using a pre-announced pricing mechanism.

1 Introduction

The design and analysis of dynamic pricing mechanisms in a monopolistic environment is a fundamental topic in microeconomic theory, and one that has been studied in depth for at least thirty years; see, for example, [8]. In this setting, the two most studied pricing mechanisms are *pre-announced pricing* (also known as price-commitment) and *contingent pricing* (also known as no-commitment pricing, threat pricing, or subgame-perfect pricing). Under pre-announced pricing, the monopolist announces in advance the price at which goods can be purchased at every point in time along the time-horizon. If the monopolist uses a contingent pricing mechanism, the price in a given time period is only announced at the start of that specific period; moreover, the price may depend on the past history. One of the central questions in dynamic pricing is whether pre-announced pricing

X. Deng (Ed.): SAGT 2018, LNCS 11059, pp. 267–271, 2018.
https://doi.org/10.1007/978-3-319-99660-8_26

or contingent pricing provide higher profits for the monopolist. From a mathematical perspective, one difficulty in answering this question is that the tools required to study these mechanisms are quite different. Specifically, the study of pre-announced pricing relies on constrained optimization techniques whereas the study of contingent pricing relies heavily on game theoretic techniques.

In this paper we focus on a model for storable goods, i.e. goods that can be bought and stored for consumption in the future. A confounding feature in pricing storable goods is that a lower price may not only increase the current consumption (*the consumption effect*), but can also induce consumers to store additional goods for future consumption (*the stockpiling effect*).

[5] studied a storable good market where a monopolist sells a storable good to consumers with time-dependent demand over an arbitrary number of time periods. A key result they obtained is that consumer surplus and monopolist profits are higher under a pre-announced pricing mechanism than under a price-contingent mechanism. The model of [5] assumes that the good is infinitesimally **divisible**, by proposing that there is a continuum of non-atomic buyers or that there is a single consumer in the market who can always obtain some positive additional utility by consuming an additional fraction of the good. [2] studied the [5] model in the setting of **indivisible** goods. Specifically, they analyzed the cases where either there is a finite (possibly very large) number of buyers with a unitary demand per period, or, there is a single buyer with an arbitrary demand per period but who can only obtain value from an integral number of items. Surprisingly, for an indivisible good, in sharp contrast to the case of a divisible good, consumer surplus and monopolist profits may sometimes be lower under a pre-announced pricing mechanism than under a price-contingent mechanism. Indeed, the authors gave a simple two period and two consumer example where profits using a price contingent mechanism were more than 6% higher than could be achieved via a pre-announced pricing mechanism. More generally, [2] showed that the monopoly profits under a contingent pricing mechanism can be $\Omega(\log T + \log N)$ times more profitable than those obtained under pre-announced pricing mechanism, where T is the number of time periods and N is the number of consumers. Formally, let Π^{CP} and Π^{PA} denote the profits obtained by a contingent pricing mechanism and a pre-announced pricing mechanism, respectively. Then [2] proved:

Theorem 1. *For indivisible storable goods market with N consumers and T time periods, we have* $\Theta(\log T + \log N) \leq_{\exists} \frac{\Pi^{CP}}{\Pi^{PA}} \leq_{\forall} \Theta(\log T + \log N)$.

Here $\leq_{\exists}$ means that *there exists* a market instance, with T time periods and N consumers, such that the inequality is satisfied, and $\leq_{\forall}$ means that *for every* instance, with T time periods and N consumers, the inequality is satisfied.

This result suggests the use of a contingent pricing mechanism may be preferable for the monopolist. However, there are several reasons why the monopolist may not wish to use such a mechanism. For example, in practice the use of a threat-based pricing mechanism may not be popular with the firm's customers. In contrast, pre-announced pricing mechanisms are naturally transparent and fair. Furthermore, the optimal strategies in a contingent pricing mechanism are based upon subgame perfect equilibria; whether these actually arise in practice is debatable. Moreover, these equilibria may be very sensitive to initial conditions, so they may be less suited for use in incomplete information settings. Again, in contrast, the profitability of pre-announced pricing mechanisms do not change significantly given small changes in the instance, and computing (near)-optimal pre-announced pricing mechanisms can still be straight-forward in many incomplete information settings.

Given this, the aim of this work if to investigate in more detail the performance of pre-announced pricing mechanisms. As a first step, it turns out that contingent pricing mechanisms do **not** always outperform pre-announced pricing mechanisms. This we can see from the following example which will also serve to illustrate the storable good model (that we define formally in the full paper).

Consumer	Value at t=1	Value at t=2
I	1	0
II	0	2

Here we have a market with two consumers and two time periods. Consumer I only values the good in period 1 and consumer II only values the good in period 1. There is a storage cost of $\epsilon = 1$, so if the price in period 1 is low enough it may be attractive to the second consumer to purchase the good in period 1 and then store it for consumption in period 2.

In this example the monopolist will benefit from using a pre-announced pricing mechanism instead of a price-contingent mechanism. To see this, let's begin by analyzing the profit obtained under a contingent pricing mechanism. Without loss of generality, we may assume the monopolist has a cost 0 of producing the item; so the terms revenue and profits are interchangeable. First, suppose the monopolist wants to sell at least one unit in period 1. In that case, the price in period 1 needs to be at most $p_1 = 1$. In this case, consumer I would buy the item and a second item will be also be bought by consumer II. By storing the good for one period, the total cost to consumer II is 2. Consumer II would not be better off by waiting until the second period to buy. Because the monopolist does not have to commit to future prices under this mechanism, the optimal price the monopolist would then charge in period 2 is $p_2 = 2$.[1] Thus, the profit of the monopolist is 2. Second, if the monopolist decides not to sell any units in period 1 then consumer I won't buy at all. Further, the period 2 price would be $p_2 = 2$ which would provide the monopolist the same profits. Thus, under a contingent pricing mechanism the profit is $\Pi^{CP} = 2$.

We now show that there exists a pre-announced pricing strategy for the monopolist that can guarantee a profit strictly higher than 2. Suppose the monopolist pre-announces the prices $p_1 = 1$ and $p_2 = 1.99$. Naturally, consumer I would buy at period 1. However, note that consumer II would prefer to

[1] Ties can be broken in either direction by slightly modifying the price announced.

buy in period 2 as buying in period 1 and incurring the storage cost of $\epsilon = 1$ is worse than buying at 1.99 in period 2. Thus, the monopolist profit is $2.99 > 2$. So using a pre-announced pricing mechanism provides the monopolist a profit of almost 1.5 times higher than by using a price contingent policy.

This example illustrates that, as well as the practical advantages discussed above, there may be financial incentives for the monopolist to use a pre-announced pricing mechanism. Our main contribution is to quantify exactly how large this financial incentive may be. Specifically, in the full paper we prove:

Theorem 2. *For indivisible storable goods market with N consumers and T time periods, we have* $\Theta(\log T + \log N) \leq_{\exists} \frac{\Pi^{PA}}{\Pi^{CP}} \leq_{\forall} \Theta(\log T + \log N)$.

Hence, a pre-announced pricing mechanism can be $\Theta(\log T + \log N)$ times more profitable than a price-contingent mechanism and this bound is tight! Thus, there is a remarkable symmetry with Theorem 1. Together, Theorems 1 and 2 show that, in the storable good monopoly model, neither of the two pricing mechanisms can consistently provide a higher profit for the monopolist. Moreover, whilst the ratios in either direction can be arbitrarily large the performances of the two mechanism are equivalent to within a logarithmic factor.

1.1 Related Literature

The literature on monopoly pricing is extensive. [4] studied pre-announced pricing strategies for a monopoly model in which consumers are atomic and the number of items is limited (i.e. the number of items is less than the number of consumers). They proved that pre-announced pricing is beneficial for the monopolist, but this benefit decreases as the market size increases. [9] considered a monopoly model in a finite time horizon model where buyers arrive continuously and are heterogeneous in patience and valuation. The authors characterized the structure of the optimal contingent pricing policy. [1] studied a two-period monopoly model in which consumers arrive under a Poison process. The authors showed that the monopoly profits can increase by switching from a contingent pricing mechanism to a pre-announced pricing policy. [7] considered a two-period model of a durable good and analyzed the potential benefit of strategic capacity rationing under pre-announced pricing. Their main result is that rationing can be beneficial for the monopolist but only when consumers are risk-adverse. [3] studied a new pricing mechanism in which the seller commits to a price menu which states the future price as a function of the available inventory. They proved that there exists a unique equilibrium under this policy when there is a single unit of inventory, but multiple equilibria may exist otherwise. [10] studied a pricing mechanism in a problem where the monopolist sells a limited number of items over a finite time horizon. The authors found market conditions in which their proposed pricing strategy outperforms a pre-announced pricing. Recently, [6], studied pre-announced pricing mechanisms for a modified version of the storable goods model of [2] where goods can be stored for a limited time (seasonal goods).

References

1. Aviv, Y., Pazgal, A.: Optimal pricing of seasonal products in the presence of forward-looking consumers. MSOM **10**(3), 339–359 (2008)
2. Berbeglia, G., Rayaprolu, G., Vetta, A.: The storable good monopoly problem with indivisible demand. In: WINE, pp. 431–431 (2015)
3. Correa, J., Montoya, R., Thraves, C.: Contingent preannounced pricing policies with strategic consumers. Oper. Res. **64**, 251–272 (2016)
4. Dasu, S., Tong, C.: Dynamic pricing when consumers are strategic: analysis of posted and contingent pricing schemes. Eur. J. Oper. Res. **204**(3), 662–671 (2010)
5. Dudine, P., Hendel, I., Lizzeri, A.: Storable good monopoly: the role of commitment. Am. Econ. Rev. **96**(5), 1706–1719 (2006)
6. Ghomi, A.A., Borodin, A., Lev, O.: Seasonal goods and spoiled milk: Pricing for a limited shelf-life. In: AAMAS (2018)
7. Liu, Q., van Ryzin, G.: Strategic capacity rationing to induce early purchases. Manage. Sci. **54**(6), 1115–1131 (2008)
8. Roth, A.E.: Game-Theoretic Models Of Bargaining. Cambridge University Press, Cambridge (1985)
9. Su, X.: Intertemporal pricing with strategic customer behavior. Manage. Sci. **53**(5), 726–741 (2007)
10. Surasvadi, N., Tang, C., Vulcano, G.: Using contingent markdown with reservation to profit from strategic consumer behavior. In: POMS (2017)

What is the Optimal Deferral Number in Waitlist Mechanism

Zhou Chen[1], Qi Qi[1(✉)], Changjun Wang[2], and Wenwei Wang[1]

[1] Hong Kong University of Science and Technology, Kowloon, Hong Kong
{zchenaq,kaylaqi,wwangaw}@ust.hk, wcj@bjut.edu.cn
[2] Beijing Institute for Scientific and Engineering Computing,
Beijing University of Technology, Beijing, China

Waitlist is the most commonly used mechanism for public resource allocation, like public housing and organs for transplants. Currently, Hong Kong government adopts a waitlist mechanism for public rental housing. Eligible applicants are entitled to 3 housing offers (one at a time). Applicants who refuse all the 3 housing offers have to rejoin the waitlist system again.

Motivated by this allocation system, we study the waitlist mechanism with any number k of deferrals, which means each agent either takes the flat in k chances or rejoin the waitlist. Under a Markov Decision Process model, we investigate the optimal strategy for each agent in the system and characterize the equilibrium state of the system. The optimal strategy is determined by the agent's outside option that represents the urgence for public housing. Then we focus on measuring the performance of the waitlist mechanisms with varying number of deferral numbers, concerning four evaluating metrics: (i) idle waiting time, the number of periods taken between the time of registration for (i.e., joining) the waitlist and the time of receiving the first offer; (ii) match value, which is defined as an agent's expected value for a matched house conditioned on matching and shows whether the matched agents receive their desirable houses; (iii) match distribution, which evaluates how the low income families are satisfied by the public housing; (iv) social welfare, the total benefit for the whole society. The results depend on the distribution of all agents' outside options. When the probability density function (p.d.f.) of the outside options among the agents is concave, the idle waiting time is decreasing in k, the match value and social welfare is increasing in k; when p.d.f. is convex, the idle waiting time and match distribution are increasing in k, the match value is decreasing in k. This will help the government to decide a proper mechanism to improve the current system.

A full version of this paper is available at https://papers.ssrn.com/sol3/papers.cfm?abstract_id=3203280.

Work is supported in part by the Research Grant Council of Hong Kong (GRF Project No. 16213115 and GRF Project No. 16243516), and the National Natural Science Foundation of China (NSFC Grant No. 11601022).

X. Deng (Ed.): SAGT 2018, LNCS 11059, p. 273, 2018.
https://doi.org/10.1007/978-3-319-99660-8

Author Index

Amanatidis, Georgios 87
Amor, Nahla Ben 251
Antoniadis, Antonios 31
Ayadi, Manel 251

Berbeglia, Gerardo 267
Boodaghians, Shant 267

Chauhan, Ankit 137
Chen, Jing 43, 150
Chen, Wei 56
Chen, Xujin 189, 256
Chen, Zhou 273
Cheng, Yukun 226
Christodoulou, George 87, 245
Cristi, Andrés 31
Cseh, Ágnes 19

Drewes, Frank 261
Du, Donglei 226
Dziubiński, Marcin 82

Elkind, Edith 12

Fang, Zhixuan 176
Fearnley, John 87
Feldman, Moran 163
Fleiner, Tamás 19

Garg, Jugal 100
Gonen, Rica 163
Greenwald, Amy 113

Han, Qiaoming 226
Hof, Frits 69
Hu, Xiaodong 189, 256

Jäger, Gerold 261
Jia, Xiaohua 256

Kaklamanis, Christos 125
Kanellopoulos, Panagiotis 125
Kern, Walter 69
Koutsoupias, Elias 201
Kurz, Sascha 69

Lang, Jérôme 251
Lazos, Philip 201
Lenzner, Pascal 137
Li, Bo 43
Li, Jian 176
Li, Minming 256
Li, Yingkai 43
Li, Zhize 176

Markakis, Evangelos 87
McCauley, Samuel 150
McGlaughlin, Peter 100
Melissourgos, Themistoklis 245
Molitor, Louise 137

Nguyen, Ta Duy 239

Oyakawa, Takehiro 113

Patouchas, Dimitris 125
Paulusma, Daniël 69
Psomas, Christos-Alexandros 87

Qi, Qi 273

Roy, Jaideep 82

Saffidine, Abdallah 213
Shan, Xiaohan 56
Singh, Shikha 150
Spirakis, Paul G. 245
Sun, Xiaoming 56
Syrgkanis, Vasilis 113

Tang, Zhongzheng 256

Vakaliou, Eftychia 87
Vetta, Adrian 267

Wang, Changjun 273
Wang, Chenhao 189, 256
Wang, Wenwei 273
Wilczynski, Anaëlle 213

Yao, Andrew Chi-chih 1

Zhang, Jialin 56
Zhang, Le 176
Zick, Yair 239

Printed in the United States
By Bookmasters